GEOLOGICAL MAP
of the
UNITED STATES
CANADA &c.
COMPILED FROM THE STATE SURVEYS
OF THE U.S. AND OTHER SOURCES.
By
C. LYELL Esqr.

For authorities see description of Map.

Explanation of the Colouring

1 — Alluvium & Post pliocene
2 — Miocene
3 — Eocene
4 — Cretaceous
5 — Coal (oolite?) Virginia
6 — New Red Sandstone & Trap
7 — Coal Measures
8 — Carboniferous Limestone & Gypsum of Nova Scotia
9 — Old Red Sandstone or Devonian
10 — Hamilton group
11 — Helderberg series   } Upper
12 — Onondaga Salt group   } Silurian
13 — Niagara & Clinton Groups
14 — Hudson River Utica &c.   } Lower
15 — Limestone of Trenton &c. } Silurian
16 — Potsdam Sandstone

z — Sandstone of Lake Superior (age undetermined)

a — Hypogene (Granite Gneiss &c.)
b — Trap Rocks
c — Metamorphic Limestone

ELEMENTS OF GEOLOGY

Sir Charles Lyell, F.R.S., one of the founders of the geological sciences. (From a portrait by T. H. Maguire)
Throughout a long and laborious life he sought the means of deciphering the fragmentary records of the earth's history in the patient investigation of the present order of nature, enlarging the boundaries of knowledge and leaving on scientific thought an enduring influence.
O Lord how great are thy works and thy thoughts are very deep. Psalms XCII.5
(Inscription on Lyell's tomb in Westminster Abbey)

Elements of Geology

JAMES H. ZUMBERGE University of Nebraska
CLEMENS A. NELSON University of California, Los Angeles

third edition

John Wiley & Sons, Inc., New York • London • Sydney • Toronto

Plate II Map of the floor of part of the Atlantic Ocean. The depths below sea level and elevations above sea level are given in feet. The figures in parentheses are heights above the 16,000-foot average depth of abyssal plains. Exaggerated vertical scale, Mercator projection. To be used with reference to Chapters 7 and 14. (Reproduced through the courtesy of the National Geographic Society, Washington, D.C. © *The National Geographic Magazine,* 1968, all rights reserved.)

Plate III Map of the floor of part of the Pacific Ocean. The depths below sea level and elevations above sea level are given in feet. The figures in parentheses are heights above the 16,000-foot average depth of the abyssal plains. Exaggerated vertical profile, Mercator projection. To be used with reference to Chapters 7 and 14. (Reproduced through the courtesy of the National Geographic Society, Washington, D.C. © *The National Geographic Magazine,* 1969, all rights reserved.)

Library of Congress Catalogue Card Number: 79-180247

ISBN 0-471-98673-9

Printed in the United States of America.

10 9 8 7 6 5 4 3 2

To W. Charles Bell
Frank F. Grout
John W. Gruner
George M. Schwartz
Robert P. Sharp
George A. Thiel
Herbert E. Wright, Jr.
Professors of geology who provided pro-
fessional instruction, personal inspiration,
and scholarly guidance to us when we
were graduate students at the University of
Minnesota in the years immediately
following World War II.

Preface   This book is an outgrowth of *Elements of Geology,* Second Edition, by James H. Zumberge. The present edition, like the previous one, is designed for a one-semester course in physical and historical geology for the nonmajor. In this edition, Part One, "The Dynamic Earth," assumes a greater proportion of space (about two thirds of the book) than in the previous edition, in relation to Part Two, "The Geological Story."

In recent years, great strides have been made in many aspects of earth science. As a result, this volume includes more information than the previous edition. Each chapter has been revised; most chapters have been expanded; and some subjects have been shifted for pedagogical reasons. Major changes include the separation of an earlier chapter on earthquakes, volcanoes, and the earth's interior into two expanded chapters. Also we grouped the traditional ideas on geosynclines, mountain building, continental drift, and the new sea-floor spreading hypothesis into a single chapter, "Global Tectonics." The chapter "Climate, Weathering, and Soils" has been expanded. A new chapter, "Gravitational Movement of

Earth Materials," has been added. A brief treatment of desert formation and morphology has been included in the chapter on wind action and deserts. Pleistocene glacial history has been shifted from the chapter entitled "The Cenozoic Era" to an expanded treatment of glaciers and glaciation. A discussion of the oceans has been expanded and incorporated into the chapter, "Oceans and Shorelines."

In Part Two, "The Geological Story," the principal expansion has been in Chapter 15, "The Key to the Past." The discussion of paleogeography has been expanded. A brief treatment of paleoclimatology as well as a summary of organic evolution and correlation have been added. In Chapters 16 to 19 (on the Precambrian, Paleozoic, Mesozoic, and Cenozoic eras) the traditional chronologic order of geologic events and the development of life is followed, period by period. The orogenic histories for North America are summarized by geologic eras.

Many illustrations are new. We particularly thank Tad Nichols of Tucson, Arizona for providing outstanding photographs of geological phenomena. Sources are given for all photographic illustrations except the illustrations that were photographed by us. We also thank the National Geographic Society for permission to use parts of the colored maps of the Atlantic and Pacific ocean floors, which appear here as Plates II and III.

Derwin Bell of the Department of Geology at the University of Michigan transformed our crude sketches into excellent illustrations that will help the instructor in his teaching and the student in his comprehension.

We also thank many other people who helped in the preparation of this edition. Our colleagues in the College of Earth Sciences at the University of Arizona, in the Department of Geology at the University of California, Los Angeles, and elsewhere, generously contributed time and professional knowledge. J. W. Anthony, B. F. Buie, W. K. Hartmann, T. L. Péwé, R. W. Simonson, E. C. Stoever, S. R. Titley, and K. D. Watson were particularly helpful.

Donald Deneck of Wiley exercised extreme forbearance during the time that the manuscript was in its final stages.

Mrs. Shirley Talbert and Mrs. Vicki Jones typed the early manuscript pages, and Miss Rose Samardzich typed the manuscript in final form.

Finally, we express our gratitude to our respective families for their patience and understanding while this book was in preparation.

James H. Zumberge
January 1972   **Clemens A. Nelson**

Contents

List of Tables

ELEMENTS OF GEOLOGY

1 Introduction

Coal Canyon, Arizona (Photograph by Tad Nichols.)

There is nothing constant in the Universe,
All ebb and flow, and every shape that's born
Bears in its womb the seeds of change.
Ovid

During the past century, and most particularly in the decades following World War II there has been a tremendous expansion of the impact of science on everyday life. Explanations of scientific phenomena, particularly those of catastrophic effect on society, are given at all levels of education, science dominates the news media, most households make use of products that are the result of applied scientific research, and politicians and industrialists pay close attention to scientifically conducted polls. Heads of state have panels of scientific advisers, and international awards are given for outstanding scientific achievement.

In a society where science appears to rule our existence, some think of science in terms of tremendous technical achievement such as placing men on the moon or prospective visits to the planets, making better products for human consumption, or building 100 megaton bombs. Increasingly, science is looked on as the means by which we can rescue ourselves from a deteriorating environment. Or for some, science may appear to be merely a collection of facts that are discovered by a few for the benefit of many.

In view of the worldwide interest in science, it may be of value to stand back and take a hard look at this thing that so dominates our modern world. What is science?

In simple terms, science is the discovery of generalizations, which we call natural laws, through observation and experimentation. Science is not an organized collection of facts or a catalog of information about the physical universe, although it cannot be viable without the continuing collection of facts and cataloging of data. It is a dynamic development of concepts based on the logic of deduction and reasoning. The search for scientific truth is never ended or satisfied. It is not absolute; accepted scientific facts of one generation may be altered or discarded by the next. True science builds on what has gone before to predict what lies ahead.

Viewed in this perspective, science becomes more than the road to better living and an easier life, although one of its major aims must be the betterment of the human condition. It takes on a fuller meaning as the pursuit of knowledge for the sake of establishing the truth about the world in which we live and the universe to which we belong. The truths established by scientific endeavors may be beneficial or detrimental to mankind. The latter can be prevented, providing mankind can learn to regard himself as a part of the physical and biological environment and not its master.

The Scientific Method

The acquisition of scientific knowledge is a slow and laborious process. Accidental discoveries are often spectacular because they are unexpected. But the investigator in the pursuit of scientific truth employs the *scientific method.*

The scientific method involves the framing of a concept or idea to explain certain observed phenomena. Such a concept is termed an *hypothesis.* Next follows the prediction of the consequences of the hypothesis and the devising of methods of testing these predictions.

When, as is generally the case, the data initially available for framing an hypothesis are scanty, several *working hypotheses* may be framed, each of which must stand the test of more observation as more pertinent facts are accumulated. Only one hypothesis can eventually satisfy all conditions consequent to it. The surviving hypothesis, when elevated to the rank of a *theory,* may have been one of the original working hypotheses or, what is more likely, may have evolved during the course of testing the others.

In the early days of the Greeks, science as we know it today did not exist. Instead, scientific observations were unorganized and intertwined with mythology. The scientific method had not been applied and rigorous testing of ideas was not common. Gradually, however, detailed observation, and later, experimentation, provided for real advances in various branches of science.

As scientific endeavor expanded, certain broad fields such as physical science, biological science, and medical science developed. Today these broad fields are not as clearly defined as they were fifty years ago. New hybridized sciences such as *biophysics* and *biochemistry* indicate the overlapping of what formerly were distinct fields. The space age has ushered in the *planetary sciences,* and we may expect others to appear before the current century has ended.

All scientists, however, in whatever field they function have one thing in common — they employ the scientific method. That is to say, they carry out experiments and make observations to test some working hypothesis toward establishing a general law as to how natural forces act under certain conditions.

In fields such as physics, chemistry, and biology, experiments can be devised whereby the results can be duplicated time and time again, providing the conditions of the experiment can be controlled.

In other fields, especially the earth sciences, laboratory experimentation is not always possible, and the scientist must rely on his ability to observe natural phenomena or relationships in the field.

The Science of Geology

Geology is the science of the earth. Although many aspects of geology, such as the study of minerals and rocks and the conditions of temperature and pressure under which they were formed, can be approached by laboratory investigation, a geologist is first and foremost a field observer. The field geologist is concerned with observations of natural phenomena and how they combine to produce surface features, and with the observations of rock and mineral relationships and how they can be used to deduce phenomena that have operated in the past. Geology is thus a dynamic subject because it involves modern physical and biological processes and changes in the earth through enormous spans of time.

HISTORY OF GEOLOGY AS A SCIENCE

The concepts and principles embodied in modern geology evolved over a long period of history. The Greeks indulged in speculative philosophy about the earth and used mythological explanations to account for many phenomena such as earthquakes, volcanoes, and entombed organic remains (fossils). Aristotle (384–322 B.C.) held that all matter could be broken down into four basic elements: air, fire, soil, and water.

During the 600-year period of the Roman Empire advances were made in geological thought. In the reign of Tiberius, a widely traveled Greek geographer, Strabo (born 63 B.C.), recognized that the sea had once covered parts of the land. Pliny the Elder (A.D. 23–79), a learned historian and naturalist under the emperor, Vespasian, wrote voluminously on all aspects of natural science. Ironically, he died prematurely in the eruption of Vesuvius, which destroyed Pompeii and Herculaneum on the Bay of Naples in A.D. 79. The 37 volumes of Pliny's *Natural History* covered a wide range of subjects, not all of which were scientific in scope, and had a great influence on those who followed him. Of greater value to the domain of geology is the vivid description of the eruption of Vesuvius and accompanying earthquakes by Pliny's nephew and adopted son, Pliny the Younger (A.D. 61–113), in two letters to the Roman emperor, Tacitus.

The Dark Ages retarded the acquisition of scientific knowledge until the end of the Middle Ages when, in the fifteenth century, the spark of knowledge was rekindled throughout all Europe. In Italy, Leonardo da Vinci (1452–1519) recognized the true origin of fossils as remains of marine organisms that had accumulated on the floors of ancient seas in northern Italy. Somewhat later, George Bauer (1494–1555), a German who wrote in Latin under the name, Georgius Agricola, by which he is better known, published six books on geological subjects. The two best known, *De natura fossilium* (1546) and *De re metallica* (posthumously, 1556), provided the foundations for the fields of mineralogy and mining geology.

During the seventeenth century, Nicolaus Steno (1638–1687), a Dane who studied medicine in Copenhagen and spent ten years of his life from 1665 to 1675 in Florence as house physician to the Grand Duke Ferdinand II, was one of the most enlightened geologists of his time. He was the first to realize that the lower layers in a series of strata must be older than the upper, a principle we now call *superposition*. Moreover, Steno recognized that strata were originally deposited in a more or less horizontal position, but that they might become tilted or otherwise distorted through subsequent earth movements. This significant contribution is now referred to as the principle of *original horizontality*. Together, these ideas assure Steno a significant position in the history of geology.

Although the basic principles of geology were slowly emerging, they still lacked a unifying concept. Geology could not claim the status of a science until the middle of the eighteenth century. Many of the learned men who took up the study of earth history in the seventeenth and eighteenth centuries were theologians who hoped to find proof of the Noachian deluge in strata of the earth's crust. They were committed to a literal interpretation of Holy Writ, which included the

Mosaic account of creation and the Flood. Whether from personal convictions of the scientific authority of Scripture or because of fear of church censure, most of the natural historians of that time tried to mix geology with theology, an unfortunate circumstance that greatly retarded the development of geology as a science. For the few who dared to take a more liberal view of Genesis the church pronounced judgment. In England the Reverend Thomas Burnet (1635–1715), who wrote the *Sacred Theory of the Earth* (1681), was dismissed from a court appointment because he treated the Mosaic account of the fall of man as an allegory. Another Englishman, William Whitson (1666–1753), was deprived of a professorship in 1701 because of his heterodox views of scripture expressed in *The New Theory of the Earth* (1696). And in Paris, the first great naturalist to present a comprehensive work on the theory of the earth, George Buffon (1707–1788), was forced to recant his views before the Faculty of Theology at the Sorbonne.

It remained for James Hutton (1726–1797), educated in medicine at Edinburgh, Paris, and Leyden, to put into one volume, *Theory of the Earth* (1795), the first modern approach to geology. In the British Isles Hutton founded the so-called *Vulcanist* (or Plutonist) group who were opposed to the *Neptunists* led by Abraham Gottlob Werner (1749–1817) in Freiberg, Germany. Werner, an eloquent speaker and inspired teacher, held that all rocks were formed in water, even granite and basalt which are now known to have originated by cooling from a molten state. Hutton, as leader of the opposing Plutonist group, proved the true origin of granite and basalt. But more than that, Hutton established for the first time a unifying principle which geology so badly needed. He authored the famous *doctrine of uniformitarianism*. This concept teaches that geological phenomena of past geological time can be explained through the understanding of the operations of modern geological (physical, chemical, biological) processes. In essence, this doctrine states that the *present is the key to the past.*

The logic of Hutton's geology was so lucid that he gained many followers and collaborators, among which were several who previously supported the Wernerian School of Neptunism. One of Hutton's most ardent supporters and enthusiastic followers was John Playfair (1748–1819). His *Illustrations of the Huttonian Theory of the Earth,* published in 1802, is a further elaboration on the principles laid down by James Hutton.

The basic thought expressed in the uniformitarian concept was that worldwide catastrophes were not needed to explain past earth history. But catastrophism did not perish immediately. In France, Baron Georges Cuvier (1769–1832) made great strides in studies of fossil animals and was widely renowned throughout Europe for his work. But he taught that practically all the animals that populated the earth in past geological periods were wiped out catastrophically from time to time, and only a few survivors remained from which the next population was developed.

Strong opposition to this theory existed in Germany, France, and England in the early nineteenth century. Sir Charles Lyell (1797–1875) (see frontispiece) did more toward dispelling the catastrophists than any other man. He traveled widely in Europe and North America and wrote voluminously of his travels. He wrote two widely read books, *Principles of Geology* and *Elements of Geology,* both of which are today considered classics in earth science. Lyell was a contemporary of Charles Darwin, whose *Origin of Species* (1859) is commonly regarded as one of the most significant contributions to science to date. Together, their observations, Lyell's largely concerned with the physical world, Darwin's with the biological world, provided the framework on which our ideas of the physical evolution of the earth and the biological evolution of the earth's inhabitants are based. Lest the reader should get the idea that Lyell was "antichurch," it should be noted that he is one of three geologists buried in Westminster Abbey (see inscription beneath Lyell's picture on frontispiece).

The history of geological thought does not stop with Lyell, of course, but to continue the thread further would be anticlimactic. Lyell's influence gradually displaced the catastrophic school, and a controversy of similar magnitude has never arisen among geologists since that time. What Hutton started, Playfair promoted, and Lyell finished. Modern geological thought and practice all start with the foundations laid by these men.

Although catastrophism in the nineteenth century sense was laid to rest by Hutton, Playfair, and Lyell, we must not close our minds completely in this regard. The extremely short period of time during which man has observed geologic processes in action is less than a mere flick of an eyelash compared to the total length of geologic time. The possibility of catastrophic events of large magnitude in past earth history, especially in the very early stages, cannot be ruled out on statistical grounds alone. We refer here not to volcanic eruptions, violent "tidal waves," or destructive earthquakes, but rather to exceptional activity in the earth's crust of a kind not observed or known to exist. Many geologic features cannot be adequately explained by any known physical process. Although the first attempts at understanding them are, of necessity, based on the extrapolation of natural laws involving the interaction between matter and energy, it is possible and quite probable either that some past geologic events may have been related to forces and processes that are no longer operative, or that man's period of observation has been too short to encounter them.

AIMS OF GEOLOGY

Geology involves the knowledge of what is happening on and within the earth today, and the earth's past history. This requires not only an understanding of the materials involved, such as the rocks and minerals, but also a thorough understanding of the various geologic processes, for example, the way a river moves its load of sediment, or how petroleum moves beneath the ground, or the mechanisms involved in earthquakes. The first aim in geology is to develop an understanding of earth materials and how these materials are changed and modified through the action of natural forces over a period of time. Part One of this book deals with earth materials and processes.

When this aim has been accomplished, the student is ready for a second major aim, namely, the reconstruction of the geologic history of the earth. Even the elementary student will soon discover that the various features of the earth were not formed at the same time. Studying the forces that acted at different places on the earth at different times and deciphering the geologic record according to a time reference is an exciting experience.

Essentially, the geologist must answer three basic questions in his study of any geologic feature. First he wants to know, *what happened?* To find out he draws from the reservoir of knowledge that deals with the work of natural forces and agents. Many geologic processes are contemporary and can be viewed in action today. Phenomena such as the growth of a volcano, the movement of a glacier, the development of sinuous river channels, or the way in which rocks are distorted by the application of external forces, are just a few of the many geologic processes that man has studied in minute detail to learn how natural forces modify the earth.

The experimental phase of geology is a rather recent development that has placed the science on a firmer factual basis. There is no question that it will progress in future years through greater emphasis on experimental schemes whereby the forces of nature can be studied under controlled laboratory conditions, thereby placing geology on a quantitative rather than a purely qualitative basis.

The second major question that needs answering by the geologist when he studies some geologic feature, be it a mountain range or a coal bed is, *why did it happen?* In other words, having discovered the *effect*, the geologist then turns to the *cause*. Cause and effect relationships are fundamental in the study of geology, and it must be quite clear even to the casual observer that when the cause of a geologic phenomenon is clearly understood, it is possible to predict future geologic events. Geologists are now on the threshold of predicting the time of future earthquakes, and eruptions of Mauna Loa on Hawaii have been predicted within a few months time of the actual event. With the recognition that man's relation to his environment is of vital importance to his survival, geology as an environmental science can provide much of the basic information required for the solution of many problems related to man's future on earth.

The inability to recognize the difference between cause and effect led early geologists astray. For instance, it was generally assumed by many of the early philosophers that rivers flowed in valleys because the valleys represented a rift or crack in the earth's surface, thereby allowing water to flow in a ready-made channel. Modern geology, on the other hand, has shown that valleys are cut by the erosive power of running water, and from the standpoint of cause and effect relationship, the rivers caused the valleys.

Throughout Part One of this book, cause and effect relationships will be stressed repeatedly. Bear in mind, however, that we can observe the *effects* of past geologic processes today, whereas the *causes* are quite often based on assumptions rather than on known facts. The science will progress as we reduce the number of assumptions and increase our factual knowledge about the cause of geologic processes.

Consider, as an example, the origin of a common rock known as granite. Granite consists of interlocking particles called minerals which are natural chemical compounds. The individual mineral grains are clearly visible to the naked eye. Two hundred years ago, the Neptunists led by Werner argued that all rocks, including granite, were precipitated from seawater. It is true that certain rock types are precipitated in the seas, but this is not true of granite. The assumption that granite was a product of seawater precipitation was nullified when it was deduced that granite has characteristics similar to those rocks that are formed by the cooling of a molten material such as lava. This deduction was based largely on the manner of occurrence of granite masses in the earth's crust, since it was observed that granite rocks do not generally occur in layers as would be expected if they had formed at the bottom of a widespread sea. Rather, granitic bodies are irregular in shape and actually engulf other rock masses. Such discoveries eventually led to the formulation of the concept that granite was originally a molten mass of material called *magma* that formed beneath the surface of the ground and was later squeezed forcefully into the outer parts of the crust where it changed from a liquid to a solid by cooling. Sometimes the magma reached the earth's surface and poured out as lava.

Today even the magmatic origin of granite is challenged. A large number of geologists believe that although some granites are truly of magmatic origin, some form through the alteration of other rock types by the addition of heat and chemical constituents not present in the original rock.

Table 1.a Basic Geologic Time Scale Used in North America

Cenozoic Era
    Quaternary period (includes Pleistocene and Holocene)
    Tertiary period—began 65 m.y. ago
Mesozoic Era
    Cretaceous period
    Jurassic period
    Triassic period—began 230 m.y. ago
Paleozoic Era
    Permian period
    Pennsylvanian period
    Mississippian period
    Devonian period
    Silurian period
    Ordovician period
    Cambrian period—began 600 m.y. ago
Precambrian Era(s)—began about 3.5 billion years ago
Age of the earth—4.5 to 5 billion years old

Cause and effect relationships thus are not always obvious. As more and more observational facts are collected about a certain geologic feature, the ideas concerning its origin may have to be altered. And now that model studies are in widespread use in the geologic fields, the possibility of increasing our understanding of the geologic processes and the geologic features that they produce is extremely good. Like all sciences, geology is dynamic; it succumbs to new discoveries. Old ideas give way to new concepts as fact gradually replaces guesswork.

The geologist first asks *what* (effect), and *why* (cause), and finally *when?* Thus the third element in basic geologic reasoning involves the placing of cause and effect into a time scale—a sequence of geologic events worked out by field observation.

In Part One of this book, frequent reference is made to the various periods of earth history. The standard but simplified geologic time scale is shown in Table 1.a as it is used in North America. The reader should become familiar with the terminology in this table, even though a more detailed discussion of its origin and development is not presented until Chapter 15.

Geologic time starts with the origin of the earth and extends to the present. It covers 4 or 5 billion years according to the most recent methods of calculation discussed in Chapter 15. The most recent geologic periods are better understood simply because the geologic features that were formed then have not been erased by subsequent geologic forces. Part Two of this book is a brief history of the earth as inferred from the rocks of the earth's crust.

References

* Adams, Frank D., 1938, *The birth and development of the geological sciences,* Dover, Baltimore, 506 pp.

* Albritton, C. C., Jr., 1963, *The fabric of geology,* Addison-Wesley, Reading, Mass., 372 pp.

Bailey, Sir Edward, 1967, *James Hutton— The founder of modern geology,* American Elsevier Publishing Co., Inc., N.Y., 171 pp.

Davies, J. T., 1965, *The scientific approach,* Academic Press, New York, 100 pp.

* Eiseley, Loren C., 1959, Charles Lyell, *Scientific American,* V. 201, No. 2 (August), pp. 98-106. (Offprint 846, W. H. Freeman and Co., San Francisco).

* Greene, John C., 1959, *The death of Adam,* Iowa State University Press, Ames, Iowa, 388 pp.

Mather, K. (ed.), 1967, *Source book in geology, 1900–1950,* Harvard University Press, Cambridge, 435 pp.

Playfair, John, 1802, *Illustrations of the Huttonian Theory of the earth,* Edinburgh (Facsimile reprint, University of Illinois Press, Urbana, 1956), 528 pp.

Schneer, Cecil J. (ed.), 1969, *Toward a history of geology,* Proceedings of a conference, Durham, N.H., Sept. 1967, M.I.T. Press, Cambridge, Mass., 474 pp.

Wager, Laurence R., 1964, The history of attempts to establish a quantitative time scale, *Quarterly Journal Geol. Soc. London,* V. 120 s, pp. 13-28. Also *in The Phanerozoic Time Scale,* 1964, pub. by the Geol. Soc. London, Burlington House, London, W.I., England.

White, J. F. (ed.), 1962, *Study of the earth: Readings in geological Science,* Prentice-Hall, Inc., 408 pp.

Woodward, Horace B., 1911, *History of geology,* Putnam and Sons, New York, 204 pp.

* Recommended for further reading.

Part One The Dynamic Earth

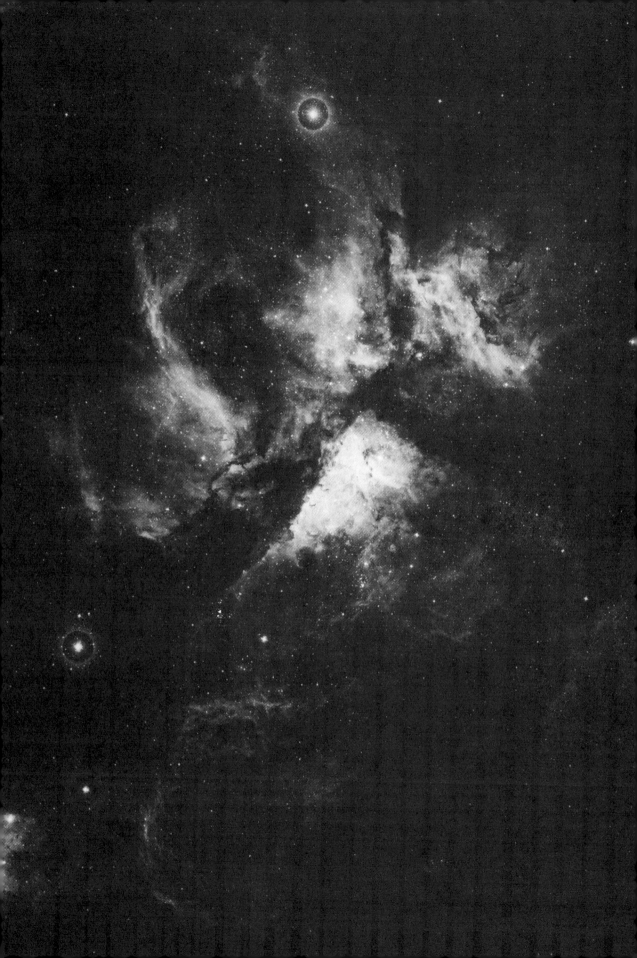

2 The Earth's Setting

Astronomy compels the soul to look upwards and leads us from this world to another.   Plato

It is appropriate to begin an elementary textbook on geology by considering some basic concepts about the universe and the solar system. It is possible to study geology without any knowledge of astronomy, but space exploration in the 1960's and 1970's has brought geology and astronomy closer together. Manned exploration of the moon and other planets in the solar system leads to the comparison of what is discovered there with what is known about the earth.

This chapter will deal first with the universe and solar system with special reference to some basic facts and known or inferred relationships. In most cases these facts will be presented without further explanation or proof since the aim is to develop a point of departure for the study of the earth rather than to pursue astronomy as a science. Some ideas about the origin of the universe will also be given.

The second part of the chapter deals with the earth as a planet with special emphasis on the two main features of its surface, the continents and ocean basins. Here too, the

Eta Carina Nebula, a large gas cloud of the southern Milky Way, is composed principally of hydrogen and helium. The Carina Nebula marks the heart of the region in the Milky Way where star formation, and possibly the formation of solar systems, is now in progress. The Carina Nebula is about 8500 light-years from the sun. (Courtesy of Bart J. Bok.)

material will be presented without exhaustive proof because in later chapters evidence in support of our existing state of knowledge about the gross features of the earth will be amplified.

The Universe

Astronomers use two units of measurement. One is the *astronomical unit,* the mean distance from the earth to the sun or about 93 million miles. The other is the *light-year,* which is the distance traveled by light in one year at a velocity of 186,290 miles per second. A light-year is equal to the number of seconds in one year (60 × 60 × 24 × 365) multiplied by 186,290. The product is nearly 6 million million miles. A pair of sharp eyes can see stars almost 2 million light-years away, and telescopes provide the means for extending this range to 2 billion light-years. An earthling who views a star in the sky is really looking back in time. For example, the earth's nearest star other than the sun is Alpha Centauri, 4.4 light-years away.

The light-year is a necessary unit for measuring the distances encountered in the universe because the standard earth units of the mile or kilometer are much too small for this purpose. To state the distance between the stars in miles or kilometers would be analogous to giving the distance from New York to London in millimeters or tenths of an inch.

Some stellar objects emit radio waves as well as visible light. Quasars[1] are such objects, and some astronomers consider them to be the most distant objects known. They were first detected by radio astronomers and later identified by optical telescopes. Although opinions differ among astronomers, quasars may be 4 to 8 *billion* light-years distant.

Galaxies. Most of the early Greeks regarded the universe as a hollow sphere, the inside of which was studded with stars. This heavenly sphere rotated on an inclined axis and the earth was its stationary center. With few exceptions this view was held until the sixteenth century when Copernicus (1473–

1543) propounded the revolutionary idea that the earth revolved around the sun, thereby making the sun the center of the universe. For nearly two centuries this view prevailed until it became apparent that the sun was only one of millions of stars in the disc-shaped galaxy of stars now called the Milky Way or our galaxy. Further studies showed our galaxy to have a diameter of about 80,000 light-years and the sun to be situated in a position about 26,000 light-years from the center. The Milky Way appears as a faint band of illumination stretching across the sky and radiating the combined light of millions of stars.

The Milky Way was considered to be the entire universe until 1923 when the American astronomer, E. P. Hubble (1889–1953), discovered indisputable evidence that other galaxies existed outside the Milky Way. These were named "island universes" in recognition of the fact that they represented separate assemblages of stars beyond what, until that time, was considered the known universe. An estimated 1 billion galaxies exist within the range of the 200-inch Hale telescope on Mount Palomar, and they too occur in clusters.

Galaxies lying 150,000 to 1,500,000 light-years from earth range in size from 2000 to 120,000 light-years in their longest

Figure 2.1 A spiral nebula 6 million light-years away. This galaxy is composed of billions of stars. The white specks are stars in the Milky Way. (Photograph by E. W. Dennison, University of Michigan Observatory.)

---

[1]A word coined from "quasi-stellar radio sources."

dimension. All galaxies are either elliptical, irregular, or discoidal in shape. The latter are especially intriguing because from the appearance of their spiral arms they seem to be rotating like giant pinwheels (Fig. 2.1).

The galaxies nearest the Milky Way are two irregular shaped assemblages of stars known as the Magellanic Clouds, which are actually visible to the naked eye in the Southern Hemisphere. They are 150,000 light-years distant and have diameters of 20,000 and 30,000 light-years, respectively. The Andromeda galaxy is another neighbor visible to the naked eye as a faint glowing patch in the Northern Hemisphere. A single galaxy may contain a *hundred billion* stars, some of which are distinctly visible by special photography through large telescopes. Most galaxies are so far away or are so densely populated with stars and glowing gas clouds (Fig. 2.2) that they appear only as a diffused zone of brightness.

The Expanding Universe. Probably the most intriguing thing about the universe in addition to its incomprehensible size is the apparent fact that it is expanding at a tremendous rate. The evidence for expansion is based on the fact that the light coming from distant galaxies shifts towards the red (long-wave) end of the light spectrum. This phenomenon is known as the Doppler effect or ''red shift'' and is caused by the apparent stretching of light waves that are received by an observer on earth from a receding galactic source. This is analogous to the lowering in pitch of a sounding horn from a fast-moving automobile as it speeds away from a stationary listener. The recessional velocity of galaxies increases in all directions away from the earth in almost direct proportion to the distance of each galaxy from earth. No expansion appears to be going on within the galaxies themselves, however.

The theory of an expanding universe poses some interesting questions and suggests some obvious corollaries. When did the expansion begin and how long will it continue? In answer to the first question it can be

Figure 2.2 Diffuse gases, mostly hydrogen, in our own galaxy, about 7500 light-years away. (Photograph by Albert Boggess III, University of Michigan Observatory.)

shown that, at the present rate of expansion, the universe would have had a "beginning" around 5 to 20 billion years ago. Greater precision in determining the "beginning" is not possible because the outer limits of the universe are unknown. According to this concept the birth of the universe must have been caused by a tremendous explosion, or "big bang," as the astronomers call it. Expansion has been going on ever since the initial explosion.

As to how long the expansion will continue, only educated guesses have been proposed. One of them is that the rate of expansion is diminishing and will continue to decrease until contraction begins. Contraction will continue until the galaxies get so close that expansion will begin again. A completely different idea is that the universe is in a steady-state condition. According to this view, expansion is caused by the continuous emergence of new matter so that the density of galactic clusters is kept constant in the universe. This hypothesis does not require a unique beginning of the universe as required in the "big bang" theory.

In the late 1960's the discovery of a radiation background, known by astronomers as 3° Kelvin radiation, gave the "big bang" hypothesis greater credibility. Calculations based on the physical behavior of matter created during the initial explosive start of the universe predicted that the 3° K radiation ought to be present, and so its discovery obviously encouraged the supporters of the "big bang" idea. It would be premature at

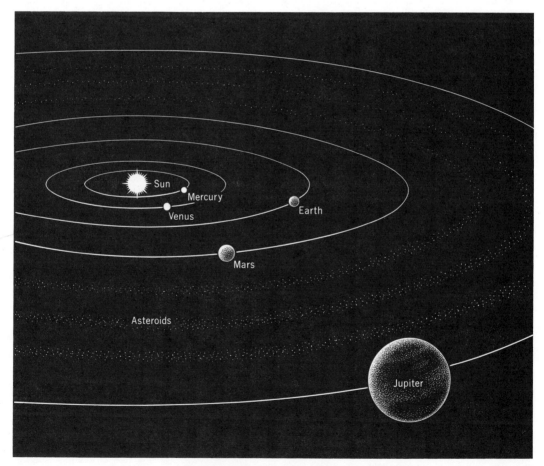

Figure 2.3 Sketch of the solar system in which the observer is located beyond Jupiter. (After Zim, Herbert S. and R. H. Baker, 1956, *Stars,* Simon and Schuster, New York, p. 123.)

THE DYNAMIC EARTH

Table 2.a Facts About the Solar System*

	Average Distance from Sun (Millions of Miles)	Mean Diameter (Miles)	Density (gm/cm³)	Length of Time for One Trip Around the Sun (Earth Units)
Mercury	36	3,190	5.61	88 days
Venus	67	7,842	5.16	225 days
Earth	93†	7,926	5.52	365 days
Mars	142	4,263	3.95	1.9 years
Jupiter	483	89,229	1.34	11.9 years
Saturn	886	78,293	0.69	29.5 years
Uranus	1,783	33,812	1.36	84 years
Neptune	2,793	30,821	1.30	164.8 years
Pluto	3,666	7,891(?)	(?)	248.8 years
Sun	—	864,327	1.42	—
Moon	93	2,160	3.36	—

* Data from *Handbook of Chemistry and Physics,* Chemical Rubber Publishing Co., 50th ed., 1969.
† The average distance from earth to sun, 93 million miles, is one *astronomical unit.*

this stage, however, to suggest that the question on the origin of the universe is answered once and for all. Indeed, it is unlikely that it will ever be answered with the same degree of certainty that other scientific questions are resolved.

The Solar System

The solar system occupies only a miniscule fraction of the Milky Way Galaxy. The sun is the center of the solar system. Bound to the sun by gravitational force are the other elements of the solar system, including nine planets and their 31 moons or satellites, myriads of asteroids, thousands of comets, and innumerable meteoroids[2] (Fig. 2.3). The farthest planet, Pluto, is about forty times more distant from the sun than the earth, but compared to distances between stars in our own galaxy, to say nothing of the universe, the solar system is exceedingly small. Indeed, an observer on the nearest star, Alpha Centauri, using the largest known telescope,

[2] *Meteoroids* are small solid bodies orbiting the sun, whereas *meteorites* are fragments of planets or meteoroids that have struck the earth.

could not discern a single planet in the solar system. Hence, planetary systems around other stars would be difficult to detect with existing instruments. Table 2.a summarizes the known facts about our solar system.

The Sun. The sun is a star in the Milky Way. Its diameter is 860,000 miles and it rotates on its axis once in every 25 days. Chemically, the sun is a gaseous mass of hydrogen that is continuously being changed to helium at extremely high temperatures. This reaction produces energy that radiates in the form of sunlight. Part of this energy reaches the earth, and were it not for the sun's energy the earth would be a lifeless planet. Even though the sun is entirely gaseous in composition, it contains 99 percent of the mass of the solar system.

The Planets. Moving in nearly the same plane around the sun are nine planets and their satellites. Planets are different from stars in that they are mostly crystalline or liquid masses rather than hot incandescent gas. Moreover, nuclear reactions in planets are not major energy sources as they are in stars.

The smaller inner four planets are known as the *terrestrial planets,* and in order of increasing distance from the sun they are Mercury, Venus, Earth, and Mars. The outer five or *Jovian planets* consist of Jupiter, Saturn, Uranus, Neptune, and Pluto. Pluto, the most distant planet, was the last to be discovered. Its position was predicted by astronomers more than a decade before its actual discovery in 1930.

No satellites are known for Mercury, Venus, or Pluto. Earth has one, Mars and Neptune each have two, Uranus has five, Saturn nine,[3] and Jupiter twelve.

There has been much speculation about the possibility of life on the other planets. A remote possibility is that Mars may be inhabited by extremely simple forms of vegetation as inferred from telescopic observations of the Martian surface. Subtle changes in color seen at different times of the Martian year have been interpreted by some astronomers to be due to a seasonal change in vegetational cover. However, the Mariner series of spacecraft that photographed the surface of Mars in 1965 and 1969 did not detect life of any kind.

The Martian atmosphere consists mainly of carbon dioxide with only traces of oxygen and water. Polar "ice caps" of dry ice (solid CO_2) expand and contract with the seasons on Mars, but temperatures in the equatorial region reach 50° F. As for the alleged "canals" of Mars, they are undoubtedly an illusion caused by the merging of complicated details and vague streaky markings when viewed with inadequate magnification. The Mariner spacecraft showed the surface of Mars to be dominated by a crater-pocked terrain not much different from the moon.

Planets other than Mars and Earth are even less likely to contain life as we know it. Their surface temperatures are either too hot or too cold, and they contain either no atmosphere at all or atmospheres consisting of gases that would be lethal to all forms of life on earth.

The Origin of the Solar System

Man's speculations concerning the origin of the solar system have been no less intense than his efforts to determine the origin of life. During the past 200 years three major hypotheses evolved, each of which has serious defects in the light of modern astronomy. Data from meteorites and from space flights have led to a newer concept of the way in which the sun and the planets came into being. The three early concepts are now of historical interest only. They will be described briefly in the paragraphs that follow, after which the modern view on this intriguing question will be presented.

Conditions that any Hypothesis Must Satisfy

All hypotheses that attempt to explain the origin of the solar system must be compatible with the known facts about the system and must be in accord with known physical and chemical laws. Although it is possible that some of the processes in operation during the formation of the solar system are not functioning today, it must be assumed that those processes were controlled by the laws of physics and chemistry as they are known today.

Generally speaking, the solar system has all the earmarks of an orderly system rather than a haphazard arrangement of bodies. This condition suggests a common origin for all units in the system and requires one to view the elements in the system as an organized whole rather than as individual components.

The following statements summarize the factual data about the relationships of the planets to the sun, and can be regarded as the basic elements that need explaining in any theory of origin.

1. All planets revolve around the sun in the same direction in nearly circular orbits, most of which lie almost in the same plane (Fig. 2.3).
2. The satellites of the planets exhibit the same regularities except for a few outer satellites that move in the opposite direction (their orbits are *retrograde*).
3. The sun's rotation is in the same direction as the planetary orbits (Fig. 2.4).
4. The sun has only 2 percent of the total angular momentum of the solar system.

---

[3] A tenth satellite, Janus, was observed in 1967, but as of this writing has not been verified (personal communication from G. Kuiper).

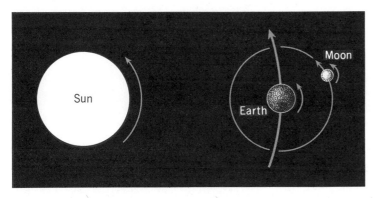

Figure 2.4 Sketch showing the rotation of the earth, sun, and moon. (Not to scale.)

The uniform relationship of the planets to the sun is contrasted with the apparent eccentricity of comets. Comets move in highly elliptical orbits that are strongly inclined to the orbital plane of the planets (Fig. 2.5). It is quite logical to assume that because comets do not bear the same orderly relationship to the sun that the planets do, they must have a different although not unrelated origin.

Kant-Laplace Nebular Hypothesis. This idea was developed by the independent efforts of a German philosopher, Immanuel Kant (1724–1804), and a French mathematician, Pierre Simon Laplace (1749–1827), in the late eighteenth century. The hypothesis assumed the existence of a large globular mass of gas (gaseous nebula) that was slowly rotating. The gravitational pull inherent in the initial mass caused contraction, which in turn caused an increase in rotational velocity according to the law of conservation of angular momentum. Eventually the gaseous mass became flattened into a discoidal shape. Gaseous rings near the outer rim of this system became detached or separated from the shrinking mass from time to time, each ring eventually condensing to form a planet. The central mass contracted and became the sun, whereas the rings shed during contraction went through the same process of contracting and shedding of rings to form the satellites of the planets.

The fatal objection to this hypothesis is that the sun is rotating too slowly in comparison to the planets for this mechanism to have worked. If the Kant-Laplace mechanism actually did take place, the sun should have the greatest angular momentum because it is the most massive element of the solar system and would be the fastest spin-

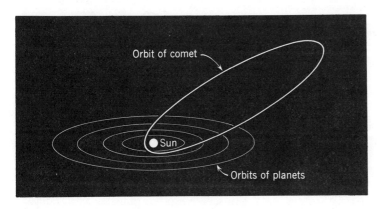

Figure 2.5 Sketch showing the orbits of a comet as compared to planetary orbits.

ning body. It has already been noted that the sun has only 2 percent of the total angular momentum.

The Chamberlin-Moulton Hypothesis. In this theory, proposed by an American geologist, T. C. Chamberlin (1843–1928), and an astronomer, F. R. Moulton (1872–1952), the sun was regarded as a star that existed before the planets were formed. Another star, passing close to the sun, exerted such a strong gravitational pull that material was torn from the sun and dragged along in the path of the passing star. The disrupted material pulled from the sun condensed into solid particles or "planetesimals," each pursuing its own orbit around the sun. Larger clusters of planetesimals were nuclei to which other planetesimals were welded by collision, eventually forming the planets. Still other knots or clusters of planetesimals, fortuitously located near the clusters from which the planets formed, became the satellites.

The major flaw in this hypothesis is that the proper distribution of angular momentum could not be imparted to the planetesimals by the passing star. According to this hypothesis, the angular momentum of the planets was derived from the pull of the passing star. But the major difficulty with that idea is that such a "sideways" motion of planetesimals could not be produced by the passing star, because when the star was about 100 million miles away from the sun it would be moving almost directly away from the sun. In this case the disrupted material would be dragged toward the passing star instead of being set into faster orbit around the sun.

The Jeans-Jeffreys Tidal Filament Hypothesis. This hypothesis also involves an original sun that had a close encounter with a passing star. The idea was conceived by Sir James Jeans (1877–1946), a physicist, and Sir Harold Jeffreys (1891–   ), an astronomer. Tidal action of the passing star pulled from the sun a gaseous filament that later broke into individual gaseous globules, each of which eventually condensed into a planet. Some of these newly formed planets were further disrupted during their first trip around the sun in an eccentric orbit, thus providing material for the eventual development of the satellites.

Like the planetesimal hypothesis, the tidal filament concept suffers from the inability of the passing star to impart the proper angular momentum to the gaseous filament. Furthermore, astrophysicists eventually showed that a hot filament pulled from the sun would not form solid planets but would simply diffuse into space.

The Solar Nebular Hypothesis (the Modern View). This hypothesis is based on observations of gas and dust around young stars. It supposes that when the sun was young, it was surrounded by a nebula of gas, called the "solar nebula," similar to the nebula proposed by Kant and Laplace. The nebula extended out as far as the planet Neptune. Astrophysical theory shows that instead of splitting up into rings as Kant and Laplace suggested, this nebula would have had a more complex history.

The nebula was probably initially hot, and as it cooled elements and compounds of different compositions would condense to form dust grains. Physical chemists have shown that grains of iron and various silicates would probably be the first to appear, a prediction which is borne out by certain chemical observations in meteorites. The nebula was probably in turbulent motion, causing a swirling of the gas and low-velocity collisions between the grains. It is assumed that the grains would stick together and grow into planetary bodies. The process might take about 100,000,000 years.

Angular momentum, the flaw in all three historical theories, is accounted for in the modern theory in the following way. Newly-forming stars seem to be the site of intense magnetic activity. The young sun is likewise assumed to have had a strong magnetic field that could "freeze" into the ionized material close to the fast-spinning sun. This means that the initially fast rotation of the sun would be slowed down and the nebula correspondingly speeded up. This explains how the sun can have very little angular momentum now while the planets have a large amount.

The massive outer planets, with their low temperatures and strong gravitational fields, have been able to retain the abundant light gases hydrogen and helium. This explains why Jupiter and the other major planets have massive hydrogen-rich atmospheres. The earth's atmosphere, on the other hand, is principally a *secondary* atmosphere of gases that escaped from the evolving earth.

Some satellites may have been formed along with the planets, while others may be examples of independently growing planetesimals that were later captured into orbits around the larger planets. This is borne out by certain outer satellites of Jupiter, Saturn, and Neptune, which have retrograde orbits, suggesting that they were captured.

Comets, according to current hypotheses, are "leftovers" formed from the scattered innermost parts of the original cloud. They were probably ejected from the inner solar system by gravitational interactions with the major planets shortly after the formation of the solar system. Their extremely eccentric orbits (Fig. 2.4) are due to the attraction of passing stars.

The Origin of the Earth's Atmosphere and Hydrosphere

After the earth became a solid body it is very likely that its atmosphere consisted chiefly of methane (CH_4) and ammonia (NH_4). The present atmosphere of the earth, therefore, is not a residue of primordial gases but rather a mixture of gases that have accumulated gradually by "degassing" of the earth's interior. Such gases as oxygen, carbon dioxide, nitrogen, and water vapor escaped from the earth's interior through volcanic eruptions to become part of the present atmosphere. Much of the water vapor condensed to become the water of the oceans. The lack of oxygen in the earth's atmosphere during the first billion years of earth history thus precluded any development of oxygen-breathing organisms until after free oxygen became available.

Geology and Space Exploration

With the development of the Apollo lunar program and unmanned space probes to the planets, a new discipline has emerged that is founded on traditional geology and modern astronomy. Variously known as planetary geology, "astrogeology," or planetology, it is the study of the rocks, surface features, and atmospheres of the moon and the planets, exclusive of the earth. This new scientific field provides an excellent opportunity of extending our knowledge out into space and back into time. Rock samples of lava flows on the moon have been dated at 3.7 billion years, nearly half a billion years older than the oldest rocks from earth. Meteorites, which are 4.6 billion years old, may well be samples of planetesimals similar to those that formed the earth and moon. Photographs of Mars and the moon reveal ancient surface craters resulting from countless impacts that have recorded events since the beginning of the solar system. Further exploration of the moon, closeup photography of Mercury, Venus, and Jupiter and, eventually, manned exploration of Mars will yield new and important data needed to understand the origin of the solar system and the origin of life.

The Earth's Major Features

The two major features of the earth are the continents and ocean basins. Each has distinguishing characteristics that will be covered in more detail in other chapters of this book. At this point, however, only the gross aspects of the earth will be considered.

The total area of the earth is 196.5 million square miles of which 57.5 million square miles or 29 percent is land, and 139.0 million square miles or 71 percent is ocean. The continents vary greatly in size from Eurasia (20.9 million square miles) to Australia (2.9 million square miles). The continental masses are unevenly distributed over the surface of the globe, with more than two-thirds north of the equator.

In the writings of many authors, reference is commonly made to the "seven seas," but in detail it is difficult to draw completely natural boundaries between the oceans, since they are all interconnected. The three largest, Pacific, Atlantic, and Indian converge around the Antarctic continent, and the ice-covered

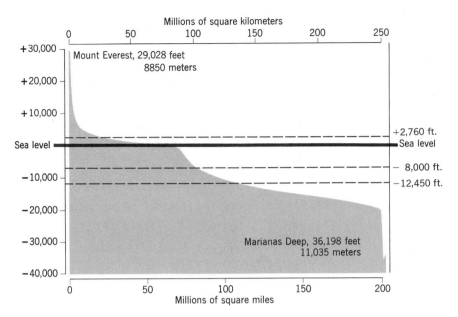

Figure 2.6 Graph showing cumulative area of the earth's surface for all elevations between Mt. Everest and the Marianas Deep. Average land elevation: 2760 feet above sea level. Average elevation of the crust of the earth: 8000 feet below sea level. Average ocean depth: 12,450 feet below sea level. (Data in part from *Encyclopaedia Britannica.*)

Arctic Ocean is connected to the Atlantic and Pacific by relatively narrow straits.

Besides the nonuniform distribution of land and sea, there exists also considerable difference in elevation of the earth's solid surface. The highest point, Mt. Everest in the Himalayas of Nepal, is 29,028 feet above sea level, and the deepest ocean sounding of 36,198 feet below sea level occurs in the Mariana Trench south of the island of Guam in the western Pacific.

The average elevation of the continents is 2760 feet above sea level, a little more than ½ mile, and the average ocean depth is 12,450 feet or roughly 2⅓ miles. Figure 2.6 portrays graphically the cumulative areas of the continents and ocean bottoms above and below sea level, the standard reference plane of all vertical measurements on the earth.

The more-or-less rigid shell covering the earth constitutes the earth's crust. The crust includes the rocks of the continents as well as the rocks beneath the ocean floor. The lower boundary of the crust is 20 to 25 miles (30 to 40 km) beneath the surface of the conti-

nents and only about 3 miles beneath the ocean floor (Fig. 2.7). This boundary is known as the *Moho* or *M-discontinuity*; both terms stand for Mohorovičić, a Yugoslavian geophysicist who first demonstrated the existence of the boundary in 1909. The Moho separates the earth's crust from the *mantle,* which is composed of denser rock, the details of which are discussed in Chapter 6.

The fundamental difference between the continental blocks and ocean basins is *not* to be found in the fact that the former generally lie above sea level and the latter are submerged but, instead, in the fact that the two are composed of different crustal materials. The continents are underlain by a rock type known as *granite* (density 2.7), which generally is absent in the oceans. Instead a heavier rock type, *basalt* (density 3.0), is the most common material underlying the floor of all ocean basins. Inspection of Figure 2.7 shows that the lighter granitic continental block is "floating" on the heavier basaltic "sea."

The light granitic crustal materials are gen-

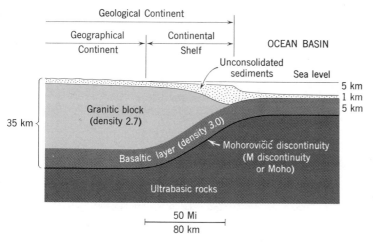

Figure 2.7 Diagrammatic cross section of a segment of the earth's crust showing the difference in thickness and composition of the crust beneath the continents and ocean basins. The base of the crust is defined by the Mohorovičić or Moho discontinuity. [Based on a diagram by J. Tuzo Wilson in Kuiper, G. P. (ed.), 1954, *The solar system,* v. II, *The earth as a planet,* University of Chicago Press, Chicago.]

erally referred to as the *sial,* a coined word taken from the first two letters of the words *si*licon and *al*uminum, two elements that occur in great abundance in granitic rocks. The basaltic crustal material beneath the ocean basins is identified by the word *sima,* a word derived from *si*licon and *ma*gnesium, two of the chief constituents of basaltic rocks. The terms sialic crust and simatic crust refer to the continents and ocean floors, respectively.

The granitic rock is not exposed everywhere on the earth's surface because it is covered by other rock types and soil that came into being after the continental masses were formed. The basaltic material beneath the oceans is covered with about a half mile (one-kilometer) thick layer of oceanic muds, silts, and oozes, but in some places in the ocean basins, notably in the Pacific and Atlantic, volcanic peaks composed of lava rise above the ocean floor. The Hawaiian Islands are outstanding examples of volcanic islands that rise 35,000 feet above the surrounding ocean floor. More details of the ocean floor are given in Chapter 14.

Origin of Continents and the Ocean Basins

There is really no satisfactory explanation to account for the origin of the continents and ocean basins. The reason for this is that very little is known about the early history of the planet Earth. Even if it is assumed that the solar nebular hypothesis of earth origin is a close approximation of what actually happened, one can only speculate about the events that followed during the first 1.5 billion years of the earth's existence.

A reasonable hypothesis suggests that the original solid earth was of uniform composition throughout and that its temperature was raised by radioactive heating. When the temperature was sufficiently high, melting occurred. In the liquid state heavier elements, such as iron and nickel, would gradually migrate toward the center while the lighter elements would remain closer to the surface, a process that ultimately resulted in a "layered" earth. Each successive layer away from the center would have a lower density. The evidence for a layered earth is presented in Chapter 6.

THE EARTH'S SETTING   21

Figure 2.8 A series of diagrams showing the consequences of an expanding earth, assuming that the original "skin" of the earth was a granitic crust that completely covered the primordial planet. [From a diagram in Carey, S. W. (ed.), 1958, *Continental drift: a symposium,* University of Tasmania, Hobart, 300 pp., after O. C. Hilgenberg, 1933, *Vom Wachsenden Erdball,* Giessmann and Bartsch, Berlin.]

The gravitational separation of heavy and light constituents of the earth is called *differentiation,* and it explains why the continental plates are less dense than the material of the ocean floor. What it does not explain, however, is the fact that the light continental crustal material is not a continuous rind around the earth but rather a series of granitic slabs "floating" in a basaltic layer as shown in Figure 2.7.

One of the older hypotheses, widely supported by many geologists until the mid-twentieth century, held that the original crust of the earth was not homogeneous in composition, but was more granitic in some places than in others. This view supposes that the present distribution of continents is a reflection of a random distribution of granitic material in the original crust and that the present continents occupy more or less the same position that they did from the very beginning.

Another idea is that the original layered earth had a granitic skin that surrounded the entire globe. Radioactive heating of deeper material resulted in expansion of the earth, which in turn caused the granitic skin to crack into separate fragments that became the continents. The gaps between these granitic blocks became the ocean basins as expansion continued (Fig. 2.8). In order for this mechanism to have operated, the original earth would have been a sphere with a radius of about 0.6 the present radius, as-

suming that the combined area of the present continents is equal to the area of the original granitic skin covering the whole earth. This idea is attractive, but at present it cannot be supported by any independent evidence that the earth has expanded the amount required to meet the conditions of the hypothesis.

Still another hypothesis claims that the original earth was molten and that it has been *shrinking* due to cooling. The idea of a shrinking earth was generated by geolosists attempting to explain "wrinkles" or folds in rock layers lying on top of the continental blocks. It was never intended as an explanation of the origin of either continents or ocean basins. It is mentioned here to demonstrate the diversity of ideas that have been invoked to explain certain geological phenomena, and to point out that an hypothesis that explains one aspect of earth history may be wholly incompatible with another. (The idea of a shrinking earth is considered further in Chapter 7.)

The concept of the permanency of continents and ocean basins has been badly eroded in recent decades because of the theory of continental drift and spreading ocean floors. Continental drift is an idea introduced independently by Frank B. Taylor (1860–1938), an American, and Alfred Wegener (1880–1930), a German, early in the twentieth century. They hypothesized that only one large continent existed when the earth was formed, but that it later ruptured into smaller pieces that have been shifting or "drifting" about ever since. Although the hypothesis of continental drift did not gain many adherents when it was first proposed, it was resurrected in the 1960's because of new evidence derived from geophysical studies of the ocean floors and continental masses. This matter is treated at greater length in Chapter 7, Global Tectonics.

The idea of permanently fixed continents from the beginning of earth history does not explain the present unsymmetrical distribution of them. Continental drift accounts for the present distribution of continents but does not address itself to the question of how the granitic material that now undergirds the continents was produced in the first place. The possibility that there were no large granitic crustal segments at all when the earth was first formed cannot be eliminated. This view supposes that the continents grew by the accretion of sediments around continental nuclei composed of volcanic materials that were later transformed into granite. The idea of continental accretion to explain the origin of continents is not without supporting evidence. North America, according to some geologists, has been growing in size since about 3.5 billion years ago.

From the foregoing it is clear that no satisfactory explanation is yet available to account for the origin of the continental blocks and the intervening ocean basins. The evidence at hand, however, does *not* support the idea that continents and ocean basins have always existed in their present relative positions and with their present sizes and configurations since the earth was born. Quite the contrary now appears to be true. The present distribution of continents and ocean basins is not a relict of the earth's original geography but rather the result of crustal evolution over a long period of earth history.

References

Brancazio, P. J., and A. G. W. Cameron (eds.), 1964, *The origin and evolution of atmospheres and oceans,* John Wiley and Sons, New York, 326 pp.

Cloud, Preston E., 1968, Atmospheric and hydrospheric evolution on the primitive earth, *Science,* V. 160, pp. 729–736.

Ewing, Maurice, and Mark Landisman, 1961, Shape and structure of ocean basins, in *Oceanography,* edited by Mary Sears, Pub. 67, American Association for the Advancement of Sciences, Washington, D.C. pp. 3–38.

Hoyle, Fred, 1966, *Galaxies, nuclei, and quasars,* Heinemann, London, 160 pp.

* Jastrow, R., and A. G. W. Cameron (eds.), 1963, *Origin of the solar system,* Academic Press, New York, 176 pp.

Larimer, J. W., 1967, Chemical fractionations in meteorites, *Geochim. et Cosmochim Acta* V. 31, pp. 1215–1238.

Mason, Brian, 1967, Meteorites, *American Scientist,* V. 55, No. 4, pp. 429–455.

Poldervaart, A., 1955, *The crust of the earth,* Geological Society of America, Special Paper 62, 762 pp.

* Reynolds, J. H., 1960, The age of the elements in the solar system, *Scientific American,* V. 203, No. 5 (November), pp. 171–182. (Offprint 253, W. H. Freeman and Co., San Francisco.)

* Urey, H. C., 1954, The origin of the earth, in *Nuclear Geology* (H. Faul, ed.), pp. 355–371, John Wiley and Sons, Inc., New York, 414 pp. Also reprinted in White, J. F. (ed.), *Study of the earth,* Prentice-Hall, Inc., Englewood Cliffs, N.J., pp. 286–401.

Wood, J. A., 1968, *Meteorites and the origin of the solar system,* McGraw-Hill, New York, 117 pp.

* Recommended for further reading.

3 The Earth's Crust

To a person uninstructed in natural history, his country or seaside stroll is a walk through a gallery filled with wonderful works of art, nine-tenths of which have their faces turned to the wall.   Thomas Henry Huxley

The crust of the earth is the outer shell of rock material that ranges in thickness from about 6 miles (10 km) beneath the oceans to 25 miles (35 km) beneath the continents. The base of the earth's crust is known technically as the *Moho* or *M-discontinuity* which is discussed in greater detail in Chapter 6. This chapter deals with the chemical composition of the crust and the structure and occurrence of the natural crustal constituents, minerals, and rocks.

All of the information about the chemistry of the earth's crust comes from direct sampling of rock masses that (1) occur at the earth's surface, (2) are extracted from deep mines and drill holes, or are (3) recovered from beneath the floor of the oceans by shipboard drilling techniques. Even though no samples from near the base of the crust have ever been recovered directly, enough samples from the upper part of it have been analyzed to permit some generalizations about crustal chemistry and physical properties.

Hallet Cove, Adelaide, South Australia.

25

Elements and Compounds

The ancient philosophers believed that four fundamental "elements" existed in all of nature: earth, water, air, and fire. The emergence of the science of chemistry was responsible for dispelling this mythical view of the nature of matter. The *Handbook of Physics and Chemistry* (1969) lists 105 elements of which more than 90 occur naturally. However, only 8 chemical elements occur in amounts greater than 2 percent in the earth's crust (Table 3.a), and the first 10 in order of abundance make up more than 99 percent of the entire crust.

An element is a substance that cannot be separated into simpler forms of matter by ordinary chemical means. The elements of the earth's crust usually occur in chemical union with one another, and only rarely are they found in their uncombined form. An element that does not occur in chemical combination with one or more other elements is called a *native element*. Gold, silver, copper, and carbon (graphite or diamond) are examples.

Two or more elements that are combined chemically form a *compound*. Some compounds contain only two elements such as the sodium and chlorine in sodium chloride (common table salt) while other compounds may contain a half-dozen or more elements, as for example hornblende (Table 3.b). Regardless of the number of elements in a given compound, however, the constituent elements are always in a definite proportion. Sodium chloride thus always has one atom of sodium to one atom of chlorine, a fact which is reflected in its chemical formula, NaCl. Quartz, one of the most common compounds in the crust, consists of one atom of the element silicon (Si) and two atoms of the element oxygen (O). Its formula is thus SiO_2. Water, H_2O, is also a compound because the ratio of hydrogen atoms to oxygen atoms is always two to one.

On the other hand, air is not a compound but rather a *mixture* of elements and compounds that do not occur in exactly the same proportions. Whereas a compound can be expressed in terms of a chemical formula, a mixture cannot be so represented. Water, ice, or steam, all of which are different forms of the same compound, can be defined by the chemical formula, H_2O, but air can only be defined as a mixture consisting of oxygen, nitrogen, carbon dioxide, water vapor, and other natural or artificial constituents that depend on local atmospheric conditions.

MINERALS: NATURAL ELEMENTS AND COMPOUNDS

A *mineral* is an inorganic element or compound that occurs naturally. A substance that fulfills the following four conditions will qualify as a mineral:

1. It occurs naturally as an inorganic substance.
2. Its composition can be expressed in terms of a chemical formula.
3. It has a definite geometric arrangement of its constituent atoms (that is, it is a *crystalline solid*).
4. It has a definite set of physical properties that are fixed within certain limits.

The external appearance of some minerals is evidence that some orderly arrangement of the constituents exists in the mineral. For example, Figure 3.1 shows several different minerals, each of which occurs naturally in some precise solid geometric form called a *crystal*. Not only does each of these crystals have a distinct *external* shape and configuration, but the *internal* arrangement of the constituent atoms is different for each. Or, stated differently, every mineral species consists of a distinct geometric arrangement or latticework of its atomic constituents, and this orderly three-dimensional internal arrangement controls the external shape of the crystal. But to say that a mineral is crystalline does not imply that it must necessarily have a perfect external geometric form; it simply means that the internal position or arrangement of the atoms of the mineral is the same for all specimens of that particular mineral species, and depending on the conditions under which it formed, it may or may not exhibit a definite external form.

The geometric pattern of atoms for a given

| Quartz | Feldspar | Muscovite |
| Garnet | Calcite | Fluorspar |

Figure 3.1 Crystals of some common minerals. The natural external shape of each crystal is determined by the internal arrangement of its constituent chemical elements.

mineral can be deduced by the use of two discoveries made more than 200 years apart. The first involved the measurement of the angle at which the sides or faces of a crystal meet. It was found that these *interfacial angles* were constant in specimens of the same mineral species. Using table salt again as an example, one can observe that the sides (faces) of the boxlike salt crystals always form an angle of 90°, a right angle. From this law of the *constancy of interfacial angles,* workers of the early eighteenth century deduced that the external shape of a given mineral must be a reflection of the internal geometric arrangement of the constituent atoms.

This deduction was verified in the early part of the twentieth century when the use of X-rays was introduced into the study of minerals. X-rays beamed through a crystalline solid are diffracted by the latticework of atoms in the crystal. The magnitude and direction of diffraction can be measured and are systematically related to the internal structure of the mineral. Every mineral spe-

cies has a characteristic X-ray diffraction pattern that can be used to calculate the size of individual atoms and the spacing between atoms in a particular mineral. These X-ray patterns are also useful in the identification of minerals that cannot be identified by simpler means.

In nature, two or more elements will combine to form a compound only if certain requirements are fulfilled. First, the sizes of the combining elements must be compatible, and second, the positive and negative charges must balance. For example, in the crystal structure of NaCl (halite) shown in Figure 3.2, it can be seen that the relative sizes of the sodium and chlorine permit a closely packed arrangement. A crystalline substance is composed of a very large number of these three-dimensional building blocks or *unit cells* stacked together in an orderly and repetitive way. This orderly arrangement is the most obvious and diagnostic characteristic of the crystalline state of matter, which is in contrast to the less orderly arrangements of constituent elements in gases, liquids, and

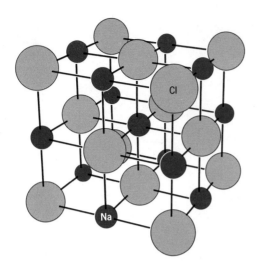

Figure 3.2 Internal structure of the mineral halite (sodium chloride). Notice the difference in size between the sodium and chlorine ions, and the boxlike external form. The diagram at left is expanded to show the geometry more clearly, whereas the diagram at the right is a truer picture of the actual physical arrangement of the sodium and chlorine ions.

amorphous solids. Most, but not all, minerals are crystalline solids and display the repetitive arrangement of their constituent atoms. Halite possesses one of the simplest geometric forms in the whole mineral kingdom.

The unit cell is the smallest three-dimensional unit in which the chemical composition and crystalline structure of that mineral can be seen. Using halite again as an example, each sodium atom has one positive electrical charge and each chlorine atom has one negative charge. Atoms with electrical charges are more precisely called *ions,* and the electrical charge is called the *valence.* The sodium ion has a valence of positive one (Na^+) and the chlorine ion (called chloride) has a valence of negative one (Cl^-). A unit cell of halite is electrically neutral because all of the positive charges of the sodium ions are neutralized by the negative charges of the chloride ions. In the crystal structure of halite, note that one sodium ion is surrounded by six chloride ions, each of which neutralizes one-sixth of the one positive charge of the sodium ion. In turn, one negatively charged chloride ion is surrounded

by six sodium ions, each of which neutralizes one-sixth of the one negative charge of the chloride ion. A unit cell of halite, therefore, satisfies the chemical formula of NaCl in the fact that the ratio of Na^+ to Cl^- is 1:1.

The size of an atom (or ion) is measured in *angstroms* (originally Ångström); one angstrom (abbreviated Å) is 1/100,000,000 centimeters (10^{-8} cm or 4×10^{-9} inches) in length. Figure 3.3 shows the relative size of 11 common ions and their valences. The occurrence of these substances in nature, therefore, is governed by physical and chemical laws that makes for an orderly and systematic association of elements with each other. Figure 3.3 shows also that ions of about the same size can proxy for each other in a crystal lattice. For example, because calcium has an ionic radius of 0.99 Å and sodium is 0.97 Å, these two elements are commonly associated with each other in an important group of minerals known as *feldspars,* different varieties of which have the same crystal structure but different percentages of sodium and calcium.

Crystal form or structure is thus a very use-

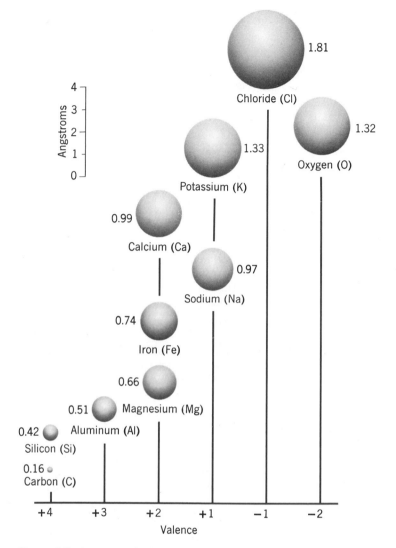

Figure 3.3 The sizes of some common ions and their valences. The size of each ion is called the *ionic radius* and is given in angstroms (Å). One angstrom = one hundred millionth of a centimeter or 10^{-8} cm. (Ionic radii from *Handbook of Chemistry and Physics* 50th ed., 1969, Table F-152, Chemical Rubber Co., Cleveland.)

ful means of identifying a givern mineral specimen. Because mineral specimens do not always occur in a form in which crystal faces and interfacial angles are well developed, other physical properties must be utilized in the identification of common minerals. Some of these properties are discussed in the following section.

Physical Properties of Minerals

The physical properties of a given mineral species are constant, or at most, variable within well-defined limits. Physical properties include not only crystal habit but also color, luster, specific gravity, cleavage, fracture, and hardness (resistance to abrasion) to name some of the most important ones.[1]

Crystal Habit. The crystal habit (external form) of a crystal is an excellent aid to the identification of a given mineral. Mineralogists have devised a classification of crystal forms that accommodates all possible geometric shapes in the mineral kingdom. A mineral specimen that occurs as a well-developed crystal can be identified by measuring its interfacial angles, and determining the shapes and spatial orientations of the various crystal faces. If no crystal faces are present in a specimen, the precise geometric form and crystal habit can be ascertained by X-ray techniques because the external shape of a crystal is controlled by the internal arrangement of the constituent atoms.

Color. Minerals of the same species may vary considerably in color because of small impurities. Quartz (SiO_2), for example, may be colorless, white, pink, violet, and even black. Color, by itself, therefore is not a reliable physical property insofar as mineral identification is concerned.

Luster. This property is the appearance of a fresh mineral surface in reflected light. Minerals with a metallic appearance have a metallic luster. Other adjectives used to describe various kinds of lusters include vitreous (glassy), dull (earthy), and greasy (oily).

Specific Gravity. The specific gravity of a substance is the numerical ratio between the weight of that substance and the weight of an equal volume of water at 4°C. Gold, for example, has a specific gravity of 19.3, quartz 2.65, and halite 2.16. When used in conjunction with other physical properties, specific gravity is a useful property for mineral identification.

Cleavage and Fracture. Cleavage is the property of a mineral to break along planes of weakness. These planes are related to the internal arrangements of the constituent atoms. Halite, for example, has three planes of cleavage at mutual right angles, as would be expected from the boxlike crystal lattice of this natural compound. Some minerals have only one plane of cleavage, and some have no cleavage at all because the crystal structure does not contain any planes that define incipient planes of weakness along which breakage will tend to occur. A mineral that breaks along a surface other than a cleavage plane is said to fracture. Minerals of an individual species may possess diagnostic fracture characteristics that are useful aids in mineral identification.

Hardness. As used in mineralogy, hardness is a property that reflects a mineral's resistance to abrasion. The hardness of a mineral is determined by comparing its abrasive capability with a standard hardness scale. The hardness scale, known as Mohs' Scale of Hardness,[2] consists of ten representative minerals arranged in order of increasing hardness from 1 (talc) to 10 (diamond).[3]

Other Physical Properties. The physical properties just described can be applied to most of the common minerals. Other physical properties are associated with only a few species, however, and they include *magnetic susceptibility, odor, taste,* and certain optical properties such as *double refraction.* The latter property is best illustrated in the common mineral calcite ($CaCO_3$).

Classification of Minerals

Whereas physical properties serve as a means for mineral identification, the systematic classification of minerals is based on

---

[1] Mineralogical museums generally exhibit mineral specimens in which most of the common physical properties are displayed. The users of this text will enhance their understanding and appreciation of minerals by viewing such exhibits. The comments on physical properties of minerals that are given on these pages are for general information only, and are not intended as a guide to mineral identification. More complete information on that subject is available in a number of physical geology laboratory manuals designed specifically for that purpose.

[2] Friedrich Mohs (1773–1839) was a protege of the famous geologist-mineralogist, Abraham Werner (1750–1817), of Freiburg University in Germany.

[3] Mohs' Scale of Hardness: 1, talc; 2, gypsum; 3, calcite; 4, fluorite; 5, apatite; 6, orthoclase; 7, quartz; 8, topaz; 9, corundum (sapphire, ruby); 10, diamond.

Table 3.a. **The Ten Most Abundant Elements of the Earth's Crust***

Rank	Element	Percent by Weight
1.	Oxygen	46.60
2.	Silicon	27.72
3.	Aluminum	8.13
4.	Iron	5.00
5.	Calcium	3.63
6.	Sodium	2.83
7.	Potassium	2.59
8.	Magnesium	2.09
9.	Titanium	.44
10.	Hydrogen	.14
	Total of first 10	99.17
	All others	0.83
		100.00

* From Brian Mason, 1958, *Principles of Geochemistry* (3rd ed.), John Wiley and Sons, New York, p. 48.

their chemical composition. Out of the approximately 2000 known mineral species only slightly more than two dozen are *major* constituents of the rocks of the earth's crust. These common *rock-forming* minerals are listed in Table 3.b, and it can be seen that the frequency of occurrence of the chemical elements that form these minerals bears some relationship to the most abundant elements of the earth's crust as listed in Table 3.a.

There is no systematic means of naming minerals. The name of a mineral may be based on its chemical composition, some characteristic physical property, a geographic locality, or the name of a person.

The main categories of minerals are listed in the following section, and some common examples are given. Special properties and characteristics of some of them are amplified where appropriate.

Native Elements. Among this group of minerals are the precious metals, *gold* (Au) and *silver* (Ag), and the important industrial metal, *copper* (Cu).

Two naturally occurring forms of carbon (C) are *diamond* and *graphite*. Chemically, these two minerals are identical, but in terms of crystallographic structure and physical properties, they are very different. In dia-

mond, the carbon atoms are closely packed and strongly bound to each other, a feature that explains the characteristic hardness of this gemstone and industrial abrasive. In graphite, the carbon atoms all occur in a layered structure, which accounts for the slippery and flaky property of that mineral, and explains why graphite is useful as a dry lubricant and why it is the main constituent of the common "lead" pencil. Different structural arrangements of the same chemical element thus produce two different minerals. This is known as *polymorphism.*

Oxides. An element combined with oxygen is an oxide. Not all elements are chemically or physically able to unite with oxygen. Perhaps the most ubiquitous oxides in nature are those containing iron. *Magnetite,* Fe_3O_4, and *hematite,* Fe_2O_3, are two common iron oxides. *Limonite,* $Fe_2O_3 \cdot nH_2O$, is a name given to a mixture of hydrous iron oxides, and is really not a mineral in the strictest sense because it cannot be represented by a fixed chemical formula. All three iron oxides are common rock-forming constituents, and both magnetite and hematite are the major sources of iron in iron ore. *Quartz,* SiO_2, is the most common of all oxides, but because of its internal structure, it is discussed under the silicate group.

Sulfates. Minerals containing sulfur and oxygen plus another element are known as sulfates. The sulfur and oxygen combine to form SO_4^{-2}, which is chemically known as a *radical.* A radical is a group of two or more atoms that acts as a single ion in chemical changes. The most common sulfate mineral is *gypsum,* $CaSO_4 \cdot 2H_2O$, which is widely used for making plasters and plasterboard. It is formed by precipitation from natural waters. The dehydrated form of this compound is *anhydrite,* $CaSO_4$.

Carbonates. These minerals contain the CO_3^{-2} radical and are important rock-forming materials that characteristically occur in rocks known as *limestone. Calcite,* $CaCO_3$, and *dolomite,* $CaMg(CO_3)_2$, are the two most common carbonates. Commercially these minerals and the rocks containing them are valuable sources of calcium and magnesium carbonates which are used in the cement industry and also in the iron and steel refin-

Table 3.b Some Common Rock-Forming Minerals and Their Chemical Composition

Chemical Group	Mineral Name	Chemical Compositions*
Oxides	Hematite	Fe_2O_3
	Magnetite	$FeFe_2O_4$
	Limonite	$Fe_2O_3 \cdot nH_2O$
	Quartz†	SiO_2
Sulfides	Pyrite	FeS_2
	Chalcopyrite	$CuFeS_2$
	Galena	PbS
Sulfates	Gypsum	$Ca(SO_4) \cdot 2H_2O$
	Anhydrite	$Ca(SO_4)$
Carbonates	Calcite	$Ca(CO_3)$
	Dolomite	$CaMg(CO_3)_2$
Halides	Halite	$NaCl$
	Fluorite	CaF_2
Silicates	Quartz	SiO_2
	Olivine	$(Mg,Fe)_2SiO_4$
Pyroxenes	Augite	$Ca(MgFeAl)(AlSi)_2O_6$
Amphiboles	Hornblende	$Ca_4Na_2(MgFe)_8(AlFe)_2(Al_4Si_{12}O_{44})(OH,F)_4$
Micas	Muscovite	$KAl_2(AlSi_3O_{10})(OH)_2$
	Biotite	$K(Mg,Fe)_3(AlSi_3O_{10})(OH)_2$
	Chlorite	$(MgFeAl)_6(AlSi_4O_{10})(OH)_8$
	Talc	$Mg_3(Si_4O_{10})(OH)_2$
	Kaolinite	$Al_2(Si_2O_5)(OH)_4$
Feldspars	Orthoclase	$K(AlSi_3O_8)$
	Microcline	$K(AlSi_3O_8)$
	Plagioclase	Isomorphous mixture of Ab and An
	Albite (Ab)	$Na(AlSi_3O_8)$
	Anorthite (An)	$Ca(Al_2Si_2O_8)$

*Chemical formulas based on Wm. H. Dennen, 1960, *Principles of Mineralogy,* revised printing, New York, Ronald Press Co., 429 pp.
†Chemically, an oxide; structurally, a silicate.

ing process. Like the sulfates, they are precipitated from natural waters.

Halides. The name halide comes from halogen, meaning "salt producer." The chief minerals of this group are *halite,* NaCl; *sylvite,* KCl; and *fluorite,* CaF₂. Halite is common table salt and occurs in large quantities as discrete layers of rock. Its importance to mankind is obvious. Fluorite is not as widespread in occurrence as halite, but it is found in veins and in association with calcite and dolomite deposits.

Silicates. Members of this group are by far the most important rock-forming minerals,

and it is necessary to go into some detail to get a better understanding of them. Structurally, they are considerably more complex than halite with its simple arrangement of sodium and chloride ions in a boxlike pattern.

The silicate group of minerals is subdivided on the basis of the different internal structures revealed by X-ray analysis. The basic "building block" of all silicate minerals is the SiO₄ tetrahedron shown in Figure 3.4. It consists of one atom of silicon, Si, surrounded by four atoms of oxygen equidistant from the silicon atom. Notice from Figure 3.3 that the

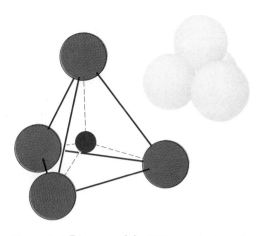

Figure 3.4 Diagram of the SiO₄ tetrahedron. Four oxygen atoms are symmetrically spaced around one silicon atom. The silica tetrahedron is the basic building block of all silicate minerals.

oxygen atoms are about three times larger than the silicon atom, so that in a diagram showing the SiO_4 tetrahedron to scale, the silicon would be hidden from view.

Quartz. Although *quartz* is chemically an oxide, it is structurally a silicate because it contains the SiO_4 tetrahedron. In quartz the SiO_4 tetrahedrons are grouped in a three-dimensional framework in which the ratio of silicon to oxygen is 1:2. In a single SiO_4 tetrahedron the ratio of silicon to oxygen is 1:4, but in quartz, all the oxygen atoms are shared by adjacent tetrahedrons, hence the formula for quartz is SiO_2.

Pyroxene Group. The SiO_4 tetrahedrons are arranged in single chains (Fig. 3.5A) that are held together by other ions such as calcium, magnesium, and iron. A very common example of a pyroxene is *augite,* a black mineral found in many igneous rocks.

Amphibole Group. The silicates of this group have a doube-chain arrangement of the SiO_4 tetrahedrons (Fig. 3.5B). Aluminum substitutes for silicon in some of the SiO_4 tetrahedrons, and the double chains are linked together by calcium, magnesium, or iron. An important difference between the amphiboles and pyroxenes is that the amphiboles contain the OH^{-1} radical, which makes them *hydrous.* The most common amphibole mineral is hornblende, a dark green to black mineral that is an important constituent of some rocks.

Mica Group. Chemically the micas contain a wide range of component elements, but structurally they have one thing in common. They consist of sheets or layers of linked tetrahedrons; hence, the term *sheet structure* is applied to all micas. This sheet structure explains why the micas have such excellent cleavage along a single plane. The thin flexible layers that can be peeled from a specimen of a mica mineral are actually cleavage fragments.

Two important rock-forming minerals of the mica group are *muscovite* and *biotite.* The first is transparent with thin cleavage plates, and the second is brown to black. Muscovite is a hydrated potassium aluminum silicate, and biotite contains all these constituents plus iron and magnesium.

Feldspar Group. No other group of minerals is as abundant in the rocks of the earth's crust as the feldspars. This fact is brought out when it is noted that these minerals contain a high proportion of the eight most abundant elements of the earth's crust (Table 3.a). *Orthoclase* feldspar, $KAlSi_3O_8$, is usually pink to flesh colored and is the chief mineral in granites. *Plagioclase* feldspar (varying mixtures of $NaAlSi_3O_8$ and $CaAl_2Si_2O_8$) also occurs in granites and rocks of similar origin. Feldspars are extremely important as rock-forming minerals. In fact, the classification of one whole group of rocks (igneous rocks) is partially based on the kind and amount of feldspar present.

The plagioclase feldspars are interesting from another point of view because they form a continuous series of different minerals depending on the amounts of the two end members, *albite,* $NaAlSi_3O_8$, and *anorthite,* $CaAl_2Si_2O_8$, present in the specimens. Such a series is called *isomorphous,* meaning similar in form. A plagioclase mineral can have any proportion of albite to anorthite, ranging from pure albite to pure anorthite. The unlimited variations of amounts of albite and anorthite in a plagioclase mineral are possible because the sodium ions of the albite and calcium ions of the anorthite are nearly the same size (Fig. 3.3), thereby allowing one to proxy for the other in the structural framework.

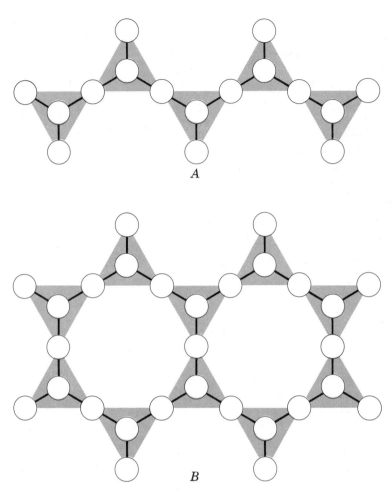

Figure 3.5 Two common geometrical patterns of the SiO$_4$ tetrahedron. *A:* single chain of tetrahedra characteristic of the pyroxene silicates such as augite; *B:* double chain of tetrahedra characteristic of amphibole silicates such as hornblende.

This *isomorphous substitution* explains why the plagioclase feldspars are rarely found in pure form (that is, albite or anorthite), but more commonly consist of an intermediate solid solution of the two end members of the series. Isomorphism is common in other mineral groups, but it is especially important in the plagioclase feldspars from the standpoint of igneous rock classification.

Rocks

A rock is a naturally occurring mass of solid inorganic or organic material that forms a significant part of the earth's crust. Strictly speaking, this definition includes hard granite as well as soft mud, but in the ordinary everyday sense the term rock is restricted to "hard" parts of the earth's crust. Most rocks are aggregates of minerals, but some important rock types such as coal and volcanic glass (obsidian) contain only insignificant amounts of minerals (that is, crystalline solids).

Rocks are the major units studied by the field geologist. He differentiates one rock type from another and plots the boundaries or contacts between the different kinds of rocks on a map, thereby constructing a geologic map. When the map is complete the geologist can draw general conclusions about the relationship between the rock units insofar as their time of origin and potential use to mankind are concerned.

Three main categories of rocks exist, and each is further subdivided on the basis of distinguishing characteristics such as mineralogical composition and texture. The three major groups are *igneous, sedimentary,* and *metamorphic.* Igneous rocks generally form from a molten mass such as lava; sedimentary rocks form by the accumulation of detrital mineral grains, from chemical precipitation, or by the buildup of organic substances; and metamorphic rocks are produced by the effect of heat and pressure on other rocks.

Igneous Rocks

Magmatic Origin. Igneous rocks are by far the most abundant of the earth's crust. Their chief distinction is that they were formed from a hot molten mass called *magma.* Magma that has reached the earth's surface through cracks and fissures is called *lava.*

Magma is a complex high-temperature solution of silicates containing water and other gases. Magma originates below the surface of the earth, probably in the upper mantle. Since rocks at these depths are under high pressure, they can exist in the solid state at very high temperatures. If the pressure is lowered by any means, however, the very hot rocks will change to the liquid state and become mobile. The accumulation of heat from a local radioactive source in the crust might also produce magma.

Once in the liquid state, the newly formed magma works its way toward the surface, either by the melting away of the overlying rocks (assimilation) or by forcing them aside. During the process of forcing its way into the surrounding and overlying hard rock, a process called *intrusion,* the magma cools. Initially the magma may have a temperature in the 1000 to 2000° F (500 to 1000° C) range, but eventually it will cool to the temperature of the enclosing medium, either rock or atmosphere.

The cooling rate of the magma is highly important in terms of the physical appearance of the igneous rock. Slow cooling permits the growth of megascopic crystals, that is, crystals large enough to be identified with the naked eye. Such rocks possess a *coarse* or *phaneritic texture.* More rapid cooling, on the other hand, results in microscopic crystals, which are clearly discernible only under a microscope or hand lens (magnifying glass). These rocks have a *fine-grained* or *aphanitic* texture. Furthermore, if the magma should break through to the surface and cool under atmospheric conditions, it literally freezes so quickly that the various atoms cannot arrange themselves into the different structural arrangements of the silicate minerals; hence, no crystals are formed and the rock is said to have a *glassy* texture.

Some igneous rocks show evidence of two stages of cooling; large crystals, indicative of slow cooling, are imbedded in a matrix of microscopic crystals, indicating more rapid cooling. The large crystals are called *phenocrysts* and the crystalline aggregate in which they are imbedded is called the *groundmass.* The rock itself is termed a *porphyry.* Such a relationship suggests that the magma was injected into a cooler environment after the first crystals formed.

The cooling of a magma is a complex chemical process, but generally speaking the various silicate minerals are precipitated in a definite sequence. The *ferromagnesian* minerals (iron and magnesium silicates) such as olivine and augite, which reach saturation early (at high temperatures), are among the first to be crystallized. These are followed by hornblende and biotite, then by the feldspars (plagioclase before orthoclase), and finally quartz. This sequence of crystallization is called *Bowen's reaction series,* after the man who discovered the concept.

The resulting igneous rock is thus a function of both the original composition (mineral constituents) of the parent magma and the rate of cooling (texture). It is on this basis that a general classification of igneous rocks can be made (Table 3.c).

A rock with a high proportion of ferromagnesian minerals has two general characteristics. It is usually dark-colored and has a specific gravity of about 3. In contrast, an igneous rock with a high quartz and ortho-

Table 3.c Simplified Igneous Rock Classification Chart.*

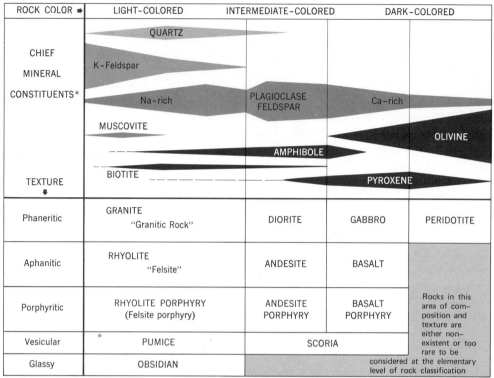

ROCK COLOR ➡	LIGHT–COLORED	INTERMEDIATE–COLORED	DARK–COLORED	
CHIEF MINERAL CONSTITUENTS*	*(QUARTZ, K–Feldspar, Na–rich, MUSCOVITE, AMPHIBOLE, BIOTITE)*	*(PLAGIOCLASE FELDSPAR)*	*(Ca–rich, PYROXENE)*	*(OLIVINE)*
TEXTURE ↓				
Phaneritic	GRANITE "Granitic Rock"	DIORITE	GABBRO	PERIDOTITE
Aphanitic	RHYOLITE "Felsite"	ANDESITE	BASALT	
Porphyritic	RHYOLITE PORPHYRY (Felsite porphyry)	ANDESITE PORPHYRY	BASALT PORPHYRY	Rocks in this area of composition and texture are either non-existent or too rare to be
Vesicular	PUMICE	SCORIA		considered at the elementary
Glassy	OBSIDIAN			level of rock classification

* Muscovite and biotite are accessory minerals and are not essential to the classes of rocks containing them. Amphiboles and pyroxenes are accessory minerals where shown as a thin dashed line or in the granite group.

*Adapted from Zumberge, J. H., 1967, *Laboratory Manual for Physical Geology*, W. C. Brown, Dubuque, Iowa.

clase feldspar content is commonly light colored and has a specific gravity of about 2.7. Because of these specific gravity differences, several kinds of igneous rocks can be generated from the same parent magma. For example, in an original magma the ferromagnesian silicates are the first minerals to become solid crystals as the magma cools. And, because they are also heavier than the remaining still-molten magma, they sink to the bottom of the magma chamber, leaving a residual magmatic solution of a different composition which, when completely crystallized, produces a light-colored igneous rock of high quartz and orthoclase content. The process whereby two distinct rock types are produced from a common magma is called *differentiation*.

Granitization. The magmatic theory is not the only explanation proposed for the origin of granitic rocks. The chemical composition of granite is such that all of its ingredients are present in some sedimentary rocks, that, if sufficiently heated, could be transformed into granite. This process is called *granitization,* the means whereby a rock that looks like granite is produced from preexisting rocks without the need for a molten phase. Granitization is thus an *in situ* process of extreme metamorphism that does not involve magma or the intrusive process of igneous rock emplacement.

In this sense, granite would be classified as a metamorphic rock rather than as an igneous rock. Most geologists agree that granite can be formed by magmatic differentiation as well as by granitization, but no general agreement exists as to the proportion of gran-

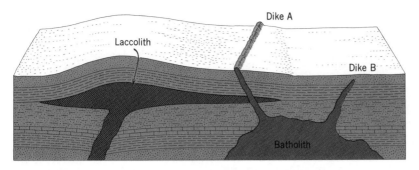

Figure 3.6 Diagrammatic cross section showing three kinds of intrusive rock masses. Dike A has been exposed at the surface by erosion. Dike B, the laccolith, and the batholith will crop out at the surface when the overlying rock has been removed by erosion.

itic bodies formed by each process. Some of the disagreement can be traced to definitions. For example, one school of "granitizers" would require that granitization is a process in which no liquid, or at most a small amount of water vapor, was involved during the change from, say a shale to granite. Laboratory studies with quartz and feldspar under high temperature and pressure tend to negate this concept and to support the idea that granitization requires "appreciable" melting in order for a granite to be produced. If this evidence is applicable, then the distinction between granite formed by granitization or by igneous activity becomes somewhat academic. Probably the only distinction is then one of how far the melted rock or magma has moved between the time it became molten and the time it cooled sufficiently to become solid.

Occurrence of Igneous Rocks

In general, igneous rocks are formed in the earth's crust in two ways. Either they crystallize beneath the surface, in which case they are *intrusive,* or they solidify at the surface and are termed *extrusive.* Intrusive rocks have a coarse texture as opposed to the fine grain or glassy texture of extrusive rocks.

Intrusive Masses. Although intrusive rocks were formed at a considerable depth below the surface in past geologic time, some of them are now exposed at the surface because the rock into which they were intruded has been removed by the slow but persistent processes of erosion. The geologist, therefore, has the opportunity to study the intrusive rocks now exposed at the surface even though they were formed in the depths. Studies of many different intrusive units have revealed the presence of several different shapes and sizes of intrusive rock masses.

Figure 3.6 shows a diagrammatic cross section of some intrusive rock masses. The *batholith* is the largest of all intrusive bodies. In plan view it may be circular, elliptical, or quite irregular. By definition its area of surface exposure must be greater than 40 square miles. Batholiths tend to increase in size with depth. Six miles (10 km) appears to be the maximum depth of a batholith. The three-dimensional picture of some batholiths is well known from their occurrence in mountain ranges. A *stock* is a small irregular shaped intrusive body that differs from a batholith only in the size of exposed surface area, less than 40 square miles.

Dikes are tabular intrusive masses ranging in thickness from a few inches to more than 100 feet (Fig. 3.6). They probably represent a crustal fracture into which magma was injected (Fig. 3.7). Where they are exposed at the surface, dikes commonly form ridges that can be traced for miles across the countryside. Dikes are *discordant* because they

THE EARTH'S CRUST   37

Figure 3.7 A vertical basaltic dike intruded into granite along the north shore of Lake Superior near Marathon, Ontario.

cut across trends of rocks they intrude. Dikes are commonly basaltic in composition, but some are extremely rich in quartz, ortho- clase, and muscovite, and are very coarse- grained in texture. These *pegmatite* dikes are important sources of commercial-grade mica, feldspar, and other minerals.

Another tabular intrusive rock mass is the *sill.* It is a layer of cooled magma between two preexisting rock layers, and hence is *concordant* in contradistinction to a dike. If the magma forces the overlying rock into a dome or arched shape, the resultant intru- sive mass is called a *laccolith* (Fig. 3.6).

Extrusive Rocks. The most common type of extrusive rock is the lava flow. Thick lava flows, like sills, are tabular in shape and may range in thickness from a few feet to several hundred feet. Although lavas are usually associated with volcanoes, it does not follow that all lavas are ejected from volcanic

cones. Some apparently oozed or welled out of crustal fractures without the spectac- ular "fireworks" attending volcanic erup- tions. Lava flows display a characteristic pattern of cracks or fractures when exposed in cross section. The fractures or joints form a columnar pattern (usually hexagonal) that develops during the cooling process and is very likely the result of shrinkage (Fig. 3.8).

Successive lava flows have accumulated in aggregate thicknesses of many thousands of feet, as for example, the mile-thick Colum- bia Lava Plateau of the northwestern United States and the Deccan Lava Plateau of India.

Yellowstone Park in northwestern Wyo- ming is underlain by extensive lava flows, some of which are still hot as indicated by the hot springs and geysers that originate when surface waters seep into the joints of the flows and are heated by the still hot rock.

Differences between sills and flows. Some sills are intruded so close to the surface that they develop a hexagonal joint pattern and other similarities of lava flows. Certain con- ditions, developed at the time of origin of a sill or flow, aid the geologist in distinguish- ing between them in the field. The possi- bility of confusion is readily appreciated when one considers the fact that both flows and sills are tabular, both may have identical textures and composition, and both may exhibit a well-developed hexagonal joint pattern. The possibility of mistaking a flow for a sill is increased still further if only part of the rock mass is visible to the observer.

The problem is usually resolved by a thor- ough examination of the *zone of contact* between the igneous mass and the adjoining rock *above and below* the sill or flow. A sill bakes the part of the rock it intrudes and may actually tear off parts of the intruded rocks and incorporate them into the still molten magma as fragments which are called *in- clusions* (Fig. 3.9). These baked contacts as well as inclusions from the overlying and underlying rock layers characterize both the top and bottom portions of the sill.

A flow also bakes the surface over which it moved while in the liquid state, and some of the fragments of the underlying rock could occur as inclusions near the base of the flow. The *surface* of a lava flow, however,

Figure 3.8 The vertical basalt columns in Devil's Postpile National Monument (Madera County, California) are defined by joints that formed during the cooling of a lava flow. The individual columns are about 2 feet in diameter, and most of them are six-sided although some have four, five, or seven sides.

is not in contact with any other rock when it forms (Fig. 3.10). Some flows, however, are buried at a later date by another deposit. In this case, the younger sedimentary deposit on top of the flow may contain fragments of the flow rock in it, thereby revealing the true sequence of events and relative age relationships of the various rock units. Figure 3.11 shows a series of diagrams illustrating a hypothetical situation in which

both a sill and lava flow bear a distinct relationship to the sedimentary rocks that existed before and after the igneous rocks were formed.

This digression on the field relationships of flows and sills with adjoining rock units is given as an example to show how important field investigations are in the study of geology. A single hand specimen of a fine-grained igneous rock would not reveal its field rela-

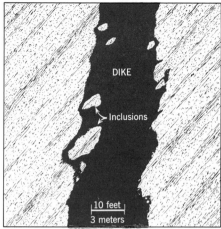

Figure 3.9 Left: the dark-colored angular fragments are inclusions in an igneous rock matrix (Inyo Mountains, California). Right: a diagram of the relationship of inclusions to an intrusive igneous body.

Figure 3.10 Lava flow in the Pinacate Mountains of northwestern Sonora, Mexico. Older lava flows and small volcanic cones are also visible in the photograph.

A

Undisturbed rock layers, limestone above, shale below

B

Sill intruded between shale and limestone. Note alteration of limestone and shale near top and bottom of sill.

C

Lava flow extruded on limestone. Note alteration of limestone at base of lava flow

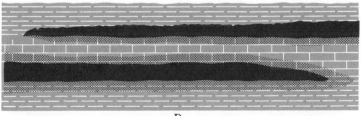

D

Shale deposited on top of lava flow. Note absence of alteration at shale–flow contact

Figure 3.11 Sequence of cross-sectional diagrams showing the difference between a sill and a lava flow as deduced from field relationships shown in diagram *D*. The stippled coloring denotes contact baking. Diagrams *A* to *C* show the sequence of geologic events leading up to the relationships that exist in diagram *D*.

Figure 3.12 This volcanic bomb was a molten clot of lava ejected from a volcano. While still aloft and spinning, the mass solidified into the typical shape shown here. (Photograph by Robert Logan.)

tionships. The importance of field studies in working out the relative sequence of geologic events is obvious from this example.

Another group of extrusive rocks consists of the *pyroclastics*. These are all associated with volcanism and represent rock fragments of various size blown out of volcanoes during eruptions. They are classified according to size and include *volcanic bombs* (Fig. 3.12), *cinders*, and *volcanic ash*. Collectively, these materials ejected from a volcano and transported through the air are known as *tephra*. Some volcanic cones consist almost wholly of volcanic ash, as for example, Paricutin in Mexico or SP Crater in Arizona (Fig. 3.13). Lava was extruded from the lower flanks of SP Crater some months after eruption commenced, but the cone itself is mainly of tephra.

Sedimentary Rocks

Sediment is material that settles out of air or water. Sedimentary rocks form from sedi-ments that have accumulated on dry land or in water. Sediments may accumulate in the bottom of a lake, in a riverbed, on a desert plain, or at the bottom of the sea, and the sedimentary particles range in size from microscopic to large rock fragments.

Some sediment is produced by chemical action called *precipitation*, the process whereby solid particles form from a liquid solution. The temperature of the solution and the *solubility* of the precipitate are important factors in this process. Other sedimentary rocks consist almost wholly of the remains of plants and animals.

During the transformation from sediments to sedimentary rocks, a number of physical and chemical changes take place, either during the process of sedimentation or shortly thereafter. Chief among these changes is *lithification*, the process whereby sedimentary rock-forming materials are changed from the loose or soft state into a hard or *indurated* rock. Lithification is usually accomplished by *cementation* of the individual particles. For example, loose sand grains will be lithified to sandstone when the individual sand particles are bonded together by some mineral material such as calcium carbonate (calcite), or iron oxide (limonite).

The process of sedimentation, although more susceptible to direct observation than the formation of intrusive igneous rocks, is exceedingly complex. Clues about the exact manner in which sedimentary rocks originated are obtained by direct observation and study of the chemical and physical environmental conditions that prevail in places where sediments are accumulating today. Some of the sedimentary processes that account for different kinds of sedimentary rocks will be noted in the simplified classification of sedimentary rocks. This classification does not include all the many different kinds of sedimentary rocks, but it will suffice to show what major categories exist and how they differ in terms of origin.

Three major categories of sedimentary rocks are recognized: *detrital, chemical,* and *organic.*

Detrital Sediments. The word detrital means fragmental, and detrital sediments are derived from fragments or individual minerals of other rocks. The size of the con-

Figure 3.13 SP Crater in Arizona is a cinder cone of rather recent geologic age. The lava flow extending from the base of the cone represents a late stage in the eruptive sequence, and is typical of these small volcanic events. (Photograph by Tad Nichols.)

stituent fragments, or *texture,* is related to the conditions under which the sediment accumulated. Textural terms such as sand, silt, and clay are commonplace in the vocabularies of civil engineers, soil scientists, and geologists, and these terms have definite meanings in terms of the size of particles in various detrital sediments. Table 3.d shows the terms used in geology for the size ranges of the various particles found in detrital sediments.

The three major kinds of detrital rocks are, in order of decreasing size of constituent particles, *conglomerate, sandstone,* and *shale.*

Table 3.d. Size Range of Particles According to the Wentworth Scale*

Name of Particle	Diameter	
	Millimeters	Inches (Approximate)
Boulder	larger than 256	larger than 10
Cobble	64 – 256	2.5 – 10
Pebble	4 – 64	0.15 – 2.5
Granule	2 – 4	0.07 – 0.15
Sand	$\frac{1}{16} - 2$	0.0025 – 0.07
Silt	$\frac{1}{256} - \frac{1}{16}$	0.00015 – 0.00025
Clay	smaller than $\frac{1}{256}$	smaller than 0.00015

* After Wentworth, C. K., 1922, A scale of grade and class terms for clastic sediments, *The Journal of Geology,* V. 30, p. 381.

Conglomerates contain particles of pebble and cobble size (Fig. 3.14*B*). Sandstones (Fig. 3.14*A*) consist of cemented sand grains that range from very fine to very coarse. Shale is compacted silt and clay, and is the most abundant of all sedimentary rocks.

Arkose is a coarse detrital sedimentary rock that contains quartz and feldspar as its major mineralogical constituents. This fact suggests that arkoses were derived from minerals that originated in granites.

Breccia (Fig. 3.14*C*) is a coarse detrital rock in which the individual particles are angular in shape rather than rounded like the pebbles of a conglomerate. The differences lie in the fact that the pebbles in a conglomerate were rounded by rolling on a stream bed or by wave agitation on a seacoast. The sharp edges of the coarse fragments in a breccia imply that no such rounding process was involved and that the fragments, however derived, must have become incorporated in a fine-grain matrix without an intervening stage of wear and abrasion.

Chemical Sediments. These rocks consist mainly of material precipitated from bodies of fresh- or saltwater, and include *limestone, dolomite, rock gypsum,* and *rock salt.*

True limestones consist predominantly of

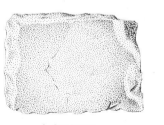
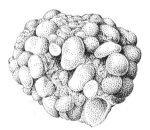

A. Sandstone   *B.* Conglomerate   *C.* Breccia

Figure 3.14 Three kinds of detrital rocks: *A,* sandstone; *B,* conglomerate; *C,* breccia.

the mineral calcite, $CaCO_3$, but impurities in the form of SiO_2 (chert) or $CaMg(CO_3)_2$ may be present. Some limestones consist of fragments of other limestones; some limestones are also derived from organisms, and this group is considered under organic sediments.

Dolomite is the name of a mineral or rock consisting of calcium magnesium carbonate. No satisfactory explanation exists for the origin of dolomite, because at the present time it is not known to form directly from seawater. A large proportion of dolomites probably originate from limestones that are altered to dolomite through a reaction between magnesium ions (Mg^{++}) in seawater and the originally deposited $CaCO_3$.

Rock gypsum, $CaSO_4 \cdot 2H_2O$, and rock salt, $NaCl$, belong to a special class of sediments known as evaporites. The name stems from the concept that sediments of this type develop by evaporation of sea water, or in special cases, salt water lakes. Restricted bays or arms of the seas in past geologic time were sites of the deposition of hundreds of feet of evaporites.

Organic Sediments. Organic limestone and coal make up the bulk of organic sediments. They are derived from plant or animal fragments cemented or compacted into rock layers. Some organic limestones are formed through the secretion of calcium carbonate by coral-reef builders, which inhabit the warm shallow seas. Coral reefs flourish in clear water in a depth not greater than 150 feet. The atolls of the southern Pacific Ocean are fine examples of organic limestones in the process of formation. Most coral reefs contain not only the main coral mass in place but also the broken fragments of coral and seashells washed in by wave action. Some organic limestones consist of calcareous (rich in calcite) shell fragments that accumulate on the shallow sea floor and become cemented together by calcite. One of the best examples of a sedimentary rock of this origin is *coquina* found extensively along the Florida coast.

Coal represents the accumulation of vegetation that originated in a swamp environment. The original organic material is transformed to coal by increase in pressure which expels the moisture and gaseous constituents, and increases the percentage of fixed carbon. *Lignite, bituminous coal* (soft coal), and *anthracite* (hard coal) are varieties or *ranks* of coal that contain, respectively, higher percentages of fixed carbon.

Fossils. Fossils are the preserved remains or traces of prehistoric plants or animals that have been preserved by natural causes in the earth's crust. The study of fossils is called *paleontology*. Fossils occur most commonly in sandstone, shale, and limestone. Rocks containing conspicuous shells, shell fragments, bones, teeth, or other animal or plant remains are *fossiliferous*. The general subject of fossils and their geologic significance is treated more fully in Chapter 15.

Occurrence of Sedimentary Rocks. By volume, sedimentary rocks or their metamorphic equivalents comprise about 5 percent of the rocks in the earth's continental crust; but when it comes to the rocks exposed on the earth's land surface, sedimentary rocks or sediments, as they are sometimes called, cover nearly three-fourths of the land surface. A knowledge of the extent and character of layers of sediment provides the geologist with a means of identifying the existence and extent of areas in which sediments accumulated in past geologic time, including shallow saltwater seas that formerly covered parts of the present dry land.

All sedimentary rocks occur in nature as layers called *strata* or *beds* (Fig. 3.15). Most strata were deposited in a nearly horizontal position. Earth movements, after lithification, moved the strata so that the original flat position was altered in one way or another. Some strata are simply uplifted (elevated) with no change in their horizontal position. Other sedimentary beds are warped or folded in broad flexures so that any part of the original horizontal layering is tilted or inclined. These distorted strata are referred to as *geologic structures* and are discussed later in this chapter.

Metamorphic Rocks

The most complex of all rock types are classed as metamorphic, a word which lit-

Figure 3.15 Bedded layers of limestone, Culbertson County, Texas. (Photograph by P. B. King, U. S. Geological Survey.)

erally means "change in form." Metamorphic rocks are formed by the chemical and physical alteration of other rocks under conditions of high pressure and temperature associated with depths of many thousand feet beneath the surface. Also, the formation of metamorphic rocks takes place essentially in the *solid state,* although some kinds of metamorphic processes take place in the presence of hot liquids and gases.

Metamorphic Processes

Because metamorphism occurs at great depths, it is impossible for geologists to observe the various metamorphic processes *in situ.* Laboratory experiments with chemical reactions between minerals under conditions of high temperatures and pressures provide some insight into the various reactions that are thought to occur at moderate depths in the earth's crust. Some rocks, such as limestone, have been deformed in the laboratory by squeezing or stretching under high pressure, but neither chemical experiments nor

mechanical deformation can duplicate the complex conditions found deep within the crust where most metamorphic processes occur. The immense spans of geologic time involved in the production of metamorphic rocks cannot be reproduced in the laboratory, either. Consequently, laboratory experiments are somewhat limited in the information they provide about metamorphism. Metamorphic processes, therefore, must be inferred in part from the evidence found in the metamorphic rocks themselves. These rocks formed at considerable depth, but they are exposed in outcrops where the overlying rocks have been removed by long periods of erosion.

Metamorphic processes are very complex but, generally speaking, they can be divided into four categories: (1) mechanical deformation, (2) recrystallization, (3) chemical recombination, and (4) chemical replacement.

Mechanical Deformation. All metamorphic processes take place at considerable depth beneath the surface where temperatures are high and pressures are great. Under these conditions, rocks under stress do not behave in the same manner as they do at or near the surface under low temperature-pressure conditions. At depth, some coarse-grained rocks such as granite may be crushed by purely mechanical means so that the original rock texture is completely destroyed. Other rocks such as shale are deformed plastically so that the micas and other platy minerals are aligned parallel to each other. The original bedding planes in a shale that has been metamorphosed are obscured by the new orientation of the platy minerals. Metamorphism of this kind imparts a cleavage to the rock known as *slaty cleavage,* and the resulting metamorphic rock is a *slate.*

Recrystallization. Recrystallization is the process whereby the minerals that exist in a rock prior to metamorphism are transformed to larger crystals during metamorphism. Recrystallization has its best manifestation in rocks that contain a single mineral species. *Marble,* for example, consists mainly of interlocking crystals of calcite, which were originally calcite particles in limestone. The metamorphism of limestone to marble causes the calcite grains to be reorganized into fewer, but larger crystals during the transformation from sedimentary limestone to marble. Similarly, a ''pure'' quartz sandstone is changed to *quartzite* during metamorphism. The quartz grains of the sandstone are recrystallized in the process.

Chemical Recombination. A rock that contains more than one mineral species may be metamorphosed into a new rock by a recombination of the chemical constituents in the original rock. New mineral species are thus produced solely from the minerals in the original rock without the addition of any new material. For example, a sedimentary rock may contain both quartz (SiO_2) and calcite ($CaCO_3$). When these two minerals are exposed to high temperature and pressure, a chemical recombination takes place to produce the mineral wollastonite ($CaSiO_3$) and the gas carbon dioxide (CO_2). If the mineral dolomite, $CaMg(CO_3)_2$, is present in the original limestone, then the quartz-dolomite reaction may produce another metamorphic mineral, diopside, $CaMg(Si_2O_6)$. The key to chemical recombination, therefore, lies in the mineralogic composition of the original rock and the intensity of heat and pressure to which it is subjected during metamorphism.

Chemical Replacement. Rock masses lying several thousands of feet beneath the surface may be invaded by gases and liquids which, because of the very high temperature and pressures, are able to penetrate minute fractures and intergrain boundaries. These gaseous solutions react chemically with the host rock, dissolving some of the ionic constituents of the original mineral components and replacing them with new ions brought in by the solutions. New minerals are formed in this replacement process. Laboratory experiments under high temperatures and pressures have revealed the temperature-pressure conditions under which these replacement reactions occur. Certain mineral species, for example, can be produced only at these elevated temperatures and pressures. Some minerals are produced in the 300°C. range while others require temperatures in the 600°C. range. Depending on the rock type, temperatures much higher than 700°C. will cause the rock to melt, in which case we are no longer dealing with metamorphism but with a magmatic process.

Types of Metamorphism

Two general types of metamorphism are recognized, *contact metamorphism,* and *regional metamorphism.*

Contact Metamorphism. This type of metamorphism is associated with the effects of a magmatic intrusion on the rock that it invades. The invaded or *host rock* is affected both by the heat of the magma and the chemical constituents emanated from it. Contact metamorphism is most intense near the zone of contact between the magma and the host rock, and with increasing distance away from the contact zone, the metamorphic effects are less pronounced. Contact metamorphism may produce recrystallization, chemical replacement, and even some mechanical changes in the host rock. The degree to which one or all of these metamorphic processes will be operative in contact metamorphism depends on the composition of the magma and its emanations, the temperature and pressure conditions in the contact zone, and the mineralogy and texture of the host rock. Not infrequently, the host rock receives such heavy concentrations of metallic substances from the magma that it can be a source of valuable minerals.

Regional Metamorphism. During the course of geologic time, earth movements cause the deformation of crustal rocks over belts hundreds of miles wide and thousands of miles long. These crustally deformed belts are generally associated with mountain chains such as the Andes, Rockies, Urals, Alps, and others. Rocks that occur in the deeper zones of these deformed belts are subjected to mechanical stresses and some elevated temperatures. Regional metamorphism results in the production of highly deformed rocks with slaty cleavage and other manifestations of plastic deformation. Recrystallization is also an effect of regional metamorphism so that, for example, sandstone is changed to quartzite, and limestone and dolomite to marble. The process of granitization (page 36) is also associated with regional metamorphism.

Classification of Metamorphic Rocks

Two general categories of metamorphic rocks are recognized, (1) *foliated,* and (2) *nonfoliated.*

Foliated Metamorphic Rocks. This group is characterized by the parallel or subparallel arrangement of platy minerals, such as the micas, or needlelike crystals, such as hornblende. Foliated metamorphic rocks generally are produced during regional metamorphism. Variations of the type and intensity of foliation are indicators of the intensity of metamorphism to which the original rock was subjected. The best illustration of this relationship is a shale that has been subjected to varying degrees of regional metamorphism.

The first step in the metamorphic evolution of a shale is slate. Further deformation produces a *schist,* and the final metamorphic product is a *gneiss* (nice). The difference between a slate and a schist is that the former is finer-grained than the latter. In a slate, the platy minerals are so small that they are not visible to the naked eye, but in a schist the micaceous flakes and other *folia* are distinctly visible without the aid of a magnifying glass. A gneiss is coarser grained than a schist and usually consists of alternating bands of light- and dark-colored minerals.

Foliated metamorphic rocks are named according to their predominant mineral assemblages as, for example, *garnet-mica schist.* Some gneisses contain all the minerals found in a granite, in which case they are called *granite gneisses.*

Nonfoliated Metamorphic Rocks. The mineral grains in nonfoliated rocks are commonly equidimensional so that no preferred orientation is possible. Two examples of this class of metamorphic rocks are quartzite and marble, which are the respective metamorphic equivalents of sandstone and limestone or dolomite. If, however, inequidimensional grains such as micas or hornblende are present, they lack the parallel or subparallel arrangement characteristic of foliated metamorphic rocks. *Hornfels* is a dense, hard, nonfoliated metamorphic rock commonly produced from shale in a contact-metamorphic zone, and commonly containing high-temperature minerals that were formed by chemical recombination or recrystallization.

The metamorphic equivalents of some common sedimentary rocks are given in Table 3.e.

Table 3.e Metamorphic Equivalents of Some Common Sedimentary Rocks

Original Rock	Metamorphic Equivalent
Shale	Slate, schist, gneiss
Sandstone	Quartzite
Conglomerate	Quartzite conglomerate
Limestone	Marble

Occurrence of Metamorphic Rocks

Metamorphic rock masses tend to retain the approximate geometric shape of the rock body from which they were formed unless the metamorphism reaches such high intensity that melting takes place, which is what happens during granitization (see page 36). Slates, schists, gneisses, quartzites, and marbles, therefore, generally occur in layered form because the original rocks also occurred in layers. These metamorphosed rock layers may be highly contorted into complex folds as a result of regional metamorphism during the formation of the mountain chains previously mentioned. If foliation exists in these contorted rock layers, the foliation planes (for example, slaty cleavage) are independent of the original bedding planes, and generally cut across them, thereby proving that the metamorphism postdated the original sedimentary features.

Some rocks have been exposed to several episodes of metamorphism over long periods of geologic time. In such cases, the original rock mass may have suffered such physical and chemical changes that its original texture and composition can be inferred only by indirect means. This situation is best illustrated in the several metamorphic stages that a shale passes through in the shale→schist→gneiss→granite sequence.

Rocks that have been subjected to contact metamorphism occur in zones around the magmatic intrusive. The boundaries of these zones tend to be gradational and are usually identifiable only after considerable mineralogical studies have been made on many specimens at progressively greater distances from the intruded igneous rock. The host rock that has been subject to contact metamorphism can thus be divided into several metamorphic zones, each of which contains a characteristic suite of minerals that reflects the diminishing influence of the intrusive igneous mass.

Fundamentals of Structural Geology

In the broadest sense, the structure of the earth's crust may be described as the granitic continental blocks resting on the basaltic layer (Fig. 2.7). In a narrower sense, however, the term structure is applied to the geometric shapes and mutual relationships of rock masses that are much smaller than continental proportions. Structural geologists are also concerned with the mechanics of rock

deformation. Still another use of the word structure, in the geological sense, refers to the arrangement and orientation of mineral grains in a rock.

The geometric or structural relationships of adjacent rock masses is basic to the understanding of the origin and age of the various rock units. These relationships are not always apparent from the visual examination of isolated patches of rock that crop out at the earth's surface or are exposed on the walls of a canyon. Most rock units, such as a layer of sandstone or an igneous dike, have two of its three dimensions measurable in thousands of feet. Some rock layers are traceable as discrete units over hundreds of square miles.

Geologic Maps. To portray the relationships of these widely outcropping rock units, the geologist constructs a *geologic map*. Just as a political map shows the boundaries of different states, provinces, and countries, so a geologic map shows the boundaries between different rock masses. These rock boundaries are called *contacts*. The construction of geologic maps and their interpretation is a unique function of the geologist. Relationships between rock units that are difficult and unnecessarily wordy when described in prose become unambiguous and lucid when portrayed on a map. In many cases, the complex relationships among different rock masses can be unraveled only by the careful construction of an accurate geologic map.

The geologic map has one important additional use besides the delineation of contacts between rock units. It can be used to decipher *subsurface* geologic conditions in the upper part of the earth's crust. This is made possible by the fact that a geologic map is merely a two-dimensional view of a three-dimensional situation. In this respect, a geologic map has the characteristics of a house plan. Both are drawn on a flat piece of paper, and both are scale representations of a three-dimensional structure. One does not have to be a professional architect to visualize the layout of a house from floor plans, nor does one have to be a professional geologist to visualize the shape of rock units depicted on a geologic map.

There is a further analogy between architectural drawings and a geologic map. The designer of a house may wish to illustrate the view of the structure from the front, side, or back instead of only the floor plan or *plan view* as it is called. To do this he draws elevations of all sides of the house. The geologist does the same thing with his rock structures, only he calls a vertical view of the rock units a *cross section* instead of an elevation. The vertical wall of a canyon is a natural cross section of the earth's crust, but because canyons or deep chasms are not always available or accessible for visual inspection, the geologist must be able to construct a geologic cross section from a geologic map. By combining the geologic map with a cross section, a three-dimensional perspective drawing known as a *block diagram* can be sketched. These are very useful in depicting the geometric relationships of the geologic units and the terrain that they underlie.

This section deals mainly with the geometric shapes of rock units, their portrayal on geologic maps, and their appearance on geologic cross sections and block diagrams. Some basic definitions and concepts are treated first, after which a few simple geologic map patterns are introduced.

Geologic Contacts. A contact is the boundary between two adjacent rock units. On a map or cross section a contact appears as a line, but in nature, a contact is a surface. Two books standing side by side on a book-

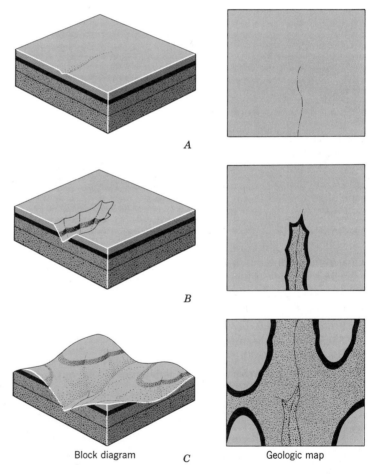

Block diagram   C   Geologic map

The geologic maps, *A, B,* and *C,* are all based on an area underlain by horizontal strata. The three maps are different because of the differences in topography as shown in the block diagram next to each.

shelf are in contact along the plane (surface) formed by their touching covers. Yet, the observer sees only the line between the two books since he is looking at the edge of the plane which separates them.

Geologic contacts may be very regular; the contact between two flat-lying sedimentary rock layers may be traced as a nearly straight line for many miles along the walls of a canyon, or the contact of an igneous dike may extend across the landscape in a single direction for thousands of yards. On the other hand, some contacts are extremely complex and very irregular.

The geometry of a contact is dependent not only on the shape of the two rock units in contact with each other but also on the configuration of the terrain that characterizes the area in question. Notice the influence of the topography on the contacts between the three horizontal rock layers of Figure 3.16 and the vertical dike of Figure 3.17. In both cases the geometry of the rock units remains the same. Only the surface topography is different in A, B, and C of both figures.

Attitude. Many rock units occur as layers or beds. These beds are rarely visible in their entirety over large areas because parts of them have been destroyed by erosion or are buried beneath the soil. In other cases the beds or layers are severely contorted so that the geometry of the whole rock mass is not

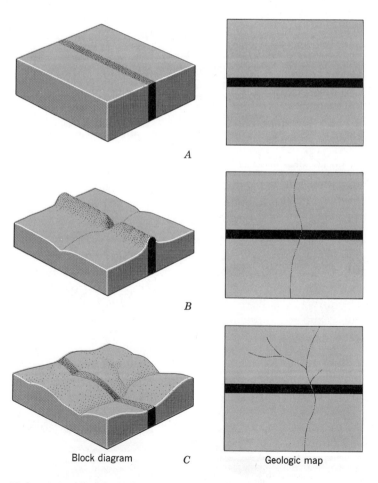

Block diagram   C   Geologic map

Figure 3.17 The block diagrams, A, B, and C depict three different topographic situations in which a vertical dike crops out at the earth's surface. The corresponding geological maps are identical, regardless of topography.

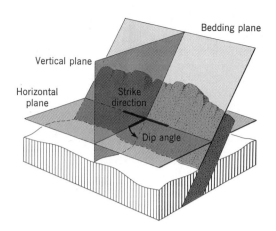

Figure 3.18 Three-dimensional view of a rock outcrop in which the attitude of a bedding plane is measured with respect to a vertical and horizontal plane. The strike and dip symbol is a common notation used by field geologists.

question. It consists of the angle between the plane to be measured and a horizontal reference plane parallel to sea level (Fig. 3.18). The strike and dip notation may be applied to any geologic plane such as a bedding plane or discrete layers in a single formation.

Figure 3.19 shows a geologic map of inclined strata. The attitude of the beds is shown by the appropriate strike and dip symbol. The number next to the symbol is the dip angle, the inclination of the beds in degrees from a horizontal reference plane.

immediately recognizable from a single vantage point. In such circumstances the geologist must decipher the geometry of the rock unit piecemeal until he has enough observations and measurements on isolated parts of the structure to reconstruct the gross geometry of the whole rock unit.

At an outcrop where a contact is visible, the position of the contact in space is called the *attitude*. The attitude consists of two parts, the *strike* and *dip*. The strike is the bearing or compass direction of a horizontal line in the plane of the contact. It is drawn as a line on the map and recorded as a direction (for example, north-south). The dip is a measure of the inclination of the plane in

Types of Geologic Structures

Layered Rocks. All sedimentary rocks or *strata* occur in layers. Three general types of layered structures are recognized: (1) horizontal, (2) monoclinal, and (3) folded. Figure 3.16 is a geologic map of an area underlain by horizontal beds. Figure 3.20*A* is a block diagram showing a horizontal layer modified by a single flexure. This is a *monoclinal* structure. Folded beds are shown in Figure 3.20*B* and *C* as they appear in a block diagram. Folds are of two geometric types, *anticlines* and *synclines*.

Parts of a Fold. The two sides of an anticline or a syncline are the *limbs* of the fold (Fig. 3.21). An anticline and an adjacent syncline share a common limb between them. An imaginary plane separating the two limbs of a fold is called the *axial plane* which, on a geologic map, appears as a line called the *fold axis*. If the axial plane has a vertical attitude, the fold is said to be *symmetrical*. If the axial plane is tilted so that the

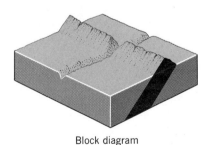

Block diagram

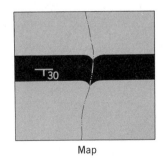

Map

Figure 3.19 Block diagram and corresponding geologic map of inclined strata.

THE DYNAMIC EARTH

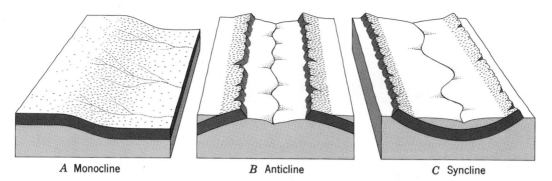

A Monocline   B Anticline   C Syncline

Figure 3.20 Block diagrams of three common types of folds in sedimentary strata.

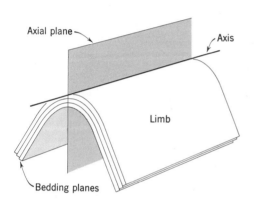

Axial plane

Axis

Limb

Bedding planes

Figure 3.21 Schematic diagram showing the terminology employed in describing folded strata.

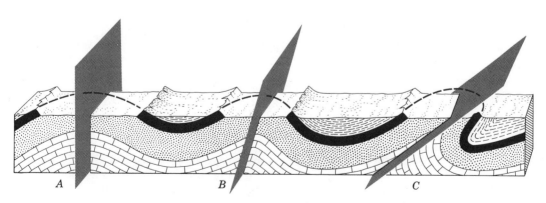

A   B   C

Figure 3.22 Block diagram in which three variations of folds are shown: A, symmetrical anticline; B, asymmetrical anticline; and C, overturned anticline. Notice the difference in attitudes of the three axial planes.

THE EARTH'S CRUST   53

Figure 3.23 Plunging anticline, Washington County, Maryland. (Photography by C. D. Walcott, U. S. Geological Survey.)

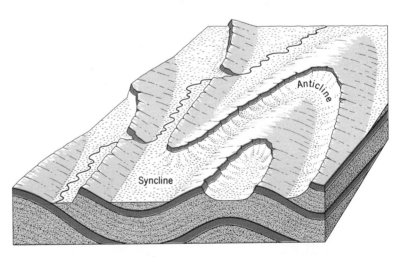

Figure 3.24 Block diagram of plunging folds. Compare with Figure 19.3, p. 000.

one limb is inclined more steeply than the other, the fold is *asymmetrical*. A fold in which the axial plane is inclined and both limbs dip in the same direction is an *overturned* fold. Figure 3.22 shows a series of folded sedimentary strata in which three different types of folds occur. Note that the strike of the axial plane is parallel to the strike of the beds in the limbs of the fold.

Plunging Folds. A further complication of folds occurs when the fold axis is not horizontal but is inclined. In this circumstance the strike of the beds forming the limbs of the fold is not parallel to the strike of the axial plane of the fold (Fig. 3.23). A fold of this type is a *plunging fold*. A geologic map or aerial photograph of plunging folds shows a diagnostic outcrop pattern (Fig. 3.24).

Figure 3.25 Aerial photograph of a structural dome near Rawlins, Wyoming. (From *Geology Illustrated* by John S. Shelton. Copyright © 1966 by W. H. Freeman and Company. All Rights reserved.)

Structural Domes and Basins. A sequence of sedimentary rock layers in which the individual beds form a roughly circular or elliptical banded outcrop pattern is a *structural dome* or *basin*. The term basin or dome in a structural sense does not refer to the shape of the surface topography but, instead, to the configuration of sedimentary layers. Figure 3.25 shows that the outcrop pattern of a structural dome is produced by beds dipping outward from the central area. A similar appearing outcrop pattern will result from beds that are dipping toward the center of the circular pattern, in which case the geometrical shape of the beds is a structural basin. One of the best-known structural basins in the United States is the Michigan Basin shown in Figure 3.26.

Faults. A *fault* is a break or fracture in the earth's crust along which relative movement of the rocks on either side of the plane of

fracture has occurred. A fault is a planar feature; the actual break or fracture is called the *fault plane*. Its attitude can be determined and recorded on a map by the use of strike and dip symbols.

Faults are found in all kinds of rock masses. Rocks on opposite sides of a fault plane may be displaced or offset a matter of a few inches or feet or greater distances, as in the case of the great San Andreas fault of California (Fig. 6.4), or the large Alpine fault of South Island, New Zealand (Fig. 3.27). The movement of rocks along either side of the fault plane in these two cases may have been more than 100 miles over a period of geological time.

It is a generally accepted fact that the energy released during faulting is a cause of earthquakes. This subject is treated in more detail in Chapter 6. We are more concerned here with faults as geologic structures. Al-

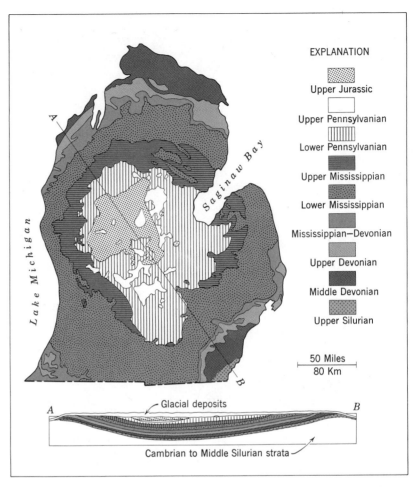

Figure 3.26 Generalized geologic map and cross section of the Michigan basin. The explanation is a conventional way of showing the relative ages of the sedimentary strata according to the standard geologic time scale. The youngest rocks (Upper Jurassic) are listed at the top of the list of names and map patterns, and the oldest (Upper Silurian) are shown at the bottom. (Glacial deposits of Pleistocene age are not shown on the map.) The geologic time scale as used in North America is shown in Table 1.a and Figure 15.9. (After Michigan Geological Survey, 1969.)

though a great number of faults are known and mapped on all continents, only a relatively small percentage are actually in an active state. Recurring movement along the San Andreas and related faults in California has been responsible for many of the devastating earthquakes in that state during historical times.

The majority of known faults are inactive, however. They usually are recognizable be-cause the continuity of a geologic structure is interrupted by the fault itself. A part of the surface of an inactive fault plane is occasionally visible in an outcrop, and these surfaces reveal the effect of rock masses sliding over each other in the form of *slickensides*. Usually, a fault is inferred from field evidence because of discontinuities in structural trends or outcrop patterns.

Faults may be classified according to any

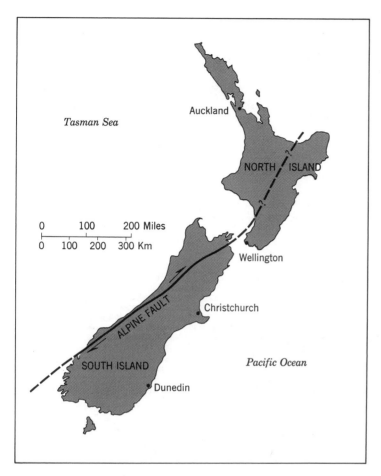

Figure 3.27 The Alpine fault of New Zealand. Note the relative movement on either side of the fault trace. The exact position of the Alpine fault on the North Island is not known. (From Wellman, H. W., 1956, *Structural Outline of New Zealand,* New Zealand Department of Scientific and Industrial Research, Bulletin 121, Figure 5.)

one of a number of systems, but here we need to consider only three general types, *normal, reverse,* and *strike-slip.* All three are defined on the basis of the relative movement of rocks on either side of the fault plane. If we consider the simple geologic situation in Figure 3.28*A*, it is apparent that the fault plane strikes east-west and dips toward the south. The block on the underside of the inclined fault plane is termed the *foot wall,* and the block on the upper side is the *hanging wall.* These terms are derived from old mining terminology and relate to the underground min-

ing tunnels constructed along a fault plane or other planar bodies.

A *normal fault* is one in which the hanging wall has moved down with respect to the foot wall. A *reverse fault* is one in which the hanging wall has moved up with respect to the foot wall (Fig. 3.28*B*). A *thrust fault* is a reverse fault in which the dip of the fault plane is very small—say 5 to 10 degrees.

In some faults, the hanging wall has not moved either up or down relative to the foot wall. Instead, the two blocks have moved in a horizontal direction parallel to the strike of

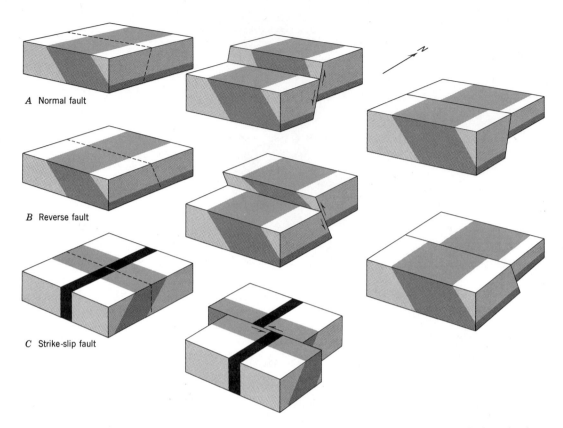

Figure 3.28 Block diagrams showing three kinds of faults. The first diagram in each sequence (from left to right) shows an incipient fault plane (dashed line) cutting across sedimentary strata. The second diagram in each sequence shows the movement along the respective fault planes, and the third diagram in *A* and *B* shows the outcrop pattern after the up-faulted (up-throw) side has been eroded to the same level as the down-faulted (down-throw) side. *A:* normal fault. The hanging wall has moved *down* with respect to the foot wall. *B:* reverse fault. The hanging wall has moved *up* with respect to the foot wall. *C:* strike-slip fault. The movement has been in a horizontal direction only, parallel to the strike of the fault plane.

A Normal fault

B Reverse fault

C Strike-slip fault

the fault plane. Such a fault is a *strike-slip* fault (Fig. 3.28C).

References

Billings, Marland P., 1959, *Structural geology,* 2nd ed., Prentice-Hall, New York, 514 pp.

deSitter, L. V., 1956, *Structural geology,* McGraw-Hill Book Co., New York, 552 pp.

Krauskopf, K. B., 1967, *Introduction to geochemistry,* McGraw-Hill Book Co., New York, 721 pp.

* Mason, Brian, and L. G. Berry, 1968, *Elements of mineralogy,* W. H. Freeman and Company, San Francisco, 550 pp.

Poldervaart, A., 1956, *The crust of the earth,* Geological Society of America, Special Paper 62, 762 pp.

Raguim, Eugene, 1965, *The geology of granite* (translation of the 2nd French edition

* Recommended for further reading.

by P. R. Eakins, E. R. Kranck, and Jean M. Eakins), Interscience Publishing Co., New York, 314 pp.

*Tuttle, O. F., 1955, The origin of granite, *Scientific American,* V. 192, No. 4 (April) pp. 77–82. (Offprint 819, W. H. Freeman and Co., San Francisco.)

Walton, M., 1960, Granite problems, *Science,* V. 131, pp. 635–645.

Whitten, E. H. T., 1966, *Structural geology of folded rocks,* Rand McNally, Chicago, 663 pp.

4 Geologic Forces

Observe always that everything is the result of change, and get used to thinking that there is nothing Nature loves so well as to change existing forms and to make new ones like them.   Marcus Aurelius

It is obvious from even quite casual observation of the world around us that our earth is a dynamic body. We can deduce from more detailed study that in its 5-billion-year history it has undergone significant change. Mountains have arisen from accumulations of marine sediments and have been eroded away; glaciers have come and gone; and various lowlands of the earth periodically have been inundated by the oceans. Today the dynamic character of the earth is manifested in erupting volcanoes, retreating and advancing glaciers, and violent earthquakes. What are the causes of this dynamism? Where does the energy come from?

These are old questions, asked by geologists of past generations, and modern geologists are still seeking the answers. Progress has been made, but much is still to be learned about the forces that keep the earth "wound up." In this chapter some ideas on the origin of geologic forces are presented and some of the older concepts are reviewed.

Entrenched meanders of the San Juan River, Utah. (Photograph by Tad Nichols.)

What Forces are Involved

The surface of the earth is in constant motion as a consequence of forces acting on the surface and forces acting within the earth. Dynamic *internal forces,* which generally tend to elevate the earth's surface and are sometimes collectively called *diastrophism,* are in constant battle against *external forces* that tend to wear away the land surface.

In a broad sense, the interaction of these two groups of forces determines what the configuration of the earth's surface will be at any one time at any one place. It is well to keep in mind the fact that the kinds of forces and their intensities vary with space and time. Different parts of the earth are dominated by certain forces during one segment of geologic time and by other forces at other times.

Internal Forces

The devastating consequences of an earthquake or the explosive violence of a volcanic eruption is convincing evidence that powerful forces are at work beneath the earth's surface. No less spectacular to the geologist are lofty mountain ranges containing rocks that were deposited below sea level. Our problem is to account for the forces that can uplift a mountain range or cause a volcano to explode. The more fundamental problems of volcanology, earthquake activity, and the origin of mountains will be treated in subsequent chapters. Here we will discuss briefly the evidences for uplift as a result of these internal forces.

Earth Movements. As we examine the present distribution of sedimentary rocks which originally were deposited in ancient marine seaways, we find them at various elevations above sea level. Some of these rock layers, for example, the ones along the Atlantic and Gulf Coastal Plains of the United States, are only a few tens of feet above sea level. Others, such as the strata exposed in Glacier National Park, have been uplifted a mile or more, although they are still in a nearly horizontal position. In central Illinois, coal beds known to have been formed in

near-sea-level terrestrial swamps are now well below sea level. A classic example of earth movement is the Roman Temple of Serapis at Naples, Italy. The columns of the temple ruins have marine clam borings 20 feet above their base, clearly signifying post-construction subsidence below sea level, followed by uplift to their present position. In many areas of the Pacific, coral reefs are found as much as 2000 to 3000 feet above sea level, and in some areas submergence of reefs below the wave zone in which they thrive has been documented. These vertical movements of segments of the earth are called *epeirogenic* movements.

In contrast to the vertically uplifted or downwarped crustal segments are the belts of folded and distorted stratified rocks that comprise the magnificent mountain chains of the earth. Not only have these mountainous areas been elevated high above sea level but they have been subjected to strong lateral compressive forces acting more or less parallel to the earth's surface. Such mountain-building forces, generally confined to elongate linear belts in contrast to the regional character of epeirogenic movements, are termed *orogenic movements.* This type of activity is discussed more fully in Chapter 7.

Volcanic Activity. An additional significant way in which the earth's surface may be built up is by volcanic activity. The outpourings of truly colossal quantities of volcanic lava have formed such high plateaus as the Columbia Plateau of eastern Washington and Oregon. Also, volcanic peaks like those that occur in the Cascade Mountains of the northwestern United States, the Hawaiian Islands, and Mt. Vesuvius represent built-up parts of the earth's surface. This process and others related to it are discussed in Chapter 5.

Volcanoes are, of course, the surface manifestation of deeper-seated igneous processes. The emplacement of large bodies of magma into the upper crustal layers may produce accompanying uplift of the surface and dislocation, distortion, and metamorphism of the older rocks into which the material is being intruded.

Orogeny, epeirogeny, volcanism, and intrusion are all manifestations of some force acting beneath the surface. A major problem

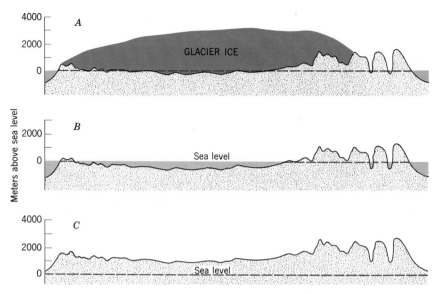

Figure 4.1 *A:* The land beneath Greenland is depressed below sea level because of the weight of the Greenland Ice Cap. *B:* If the ice melted, the earth's crust would begin to rise, but the rate of uplift would lag behind the rise in sea level caused by the return of melted glacier ice to the ocean. *C:* Final isostatic balance is reached long after the ice melted. All of the area formerly covered by glacier ice is now above sea level.

is whether these *effects* are *caused* by different forces, or whether they all can be explained by a single underlying force. This question will be raised again in subsequent chapters.

Isostasy. A special problem concerning movements of segments of the earth's crust is that of *isostasy.* Isostasy is neither a force nor a process, it is a condition of gravitational balance between crustal segments of different thickness, or a tendency toward restoration of balance once it has been disturbed by some other force or process. For example, ice-covered Greenland is now in isostatic equilibrium (Fig. 4.1*A*), but if the thick ice cap should melt (Fig. 4.1*B*), the result would be a gradual rise of the landmass (Fig. 4.1*C*). Such uplift, called *glacial rebound,* is postulated for parts of Scandinavia and North America that were covered by extensive ice sheets during the "Ice Age" (described more fully in Chapter 13).

Another example of isostasy is shown in Figure 4.2. Suppose that a mountain range is eroded and the sediments derived from it are transported to an adjacent basin of deposition. Evidence from earthquake waves indicates that the light crustal material beneath the mountains extends deeper into the subcrustal material than does the crustal material beneath lower-lying continental segments, just as thick icebergs extend to a greater depth below sea level than thinner icebergs. When material is removed from the mountain range by erosion, the crust below rises for the same reason that the base of an iceberg will rise if its top is melted away. Conversely, as the sediments derived from the mountain accumulate in a depositional basin, there will be a tendency for the basin to be depressed because of the additional weight added to it.

If one segment of the crust rises and another sinks, a logical conclusion is that a very slow subcrustal transfer of material from the sinking area to the rising area must take place as shown in Figure 4.2. It can be seen from the above that isostatic adjustments involve essentially vertical movements of the crust. Isostasy by itself, therefore, cannot be invoked to explain horizontal compression

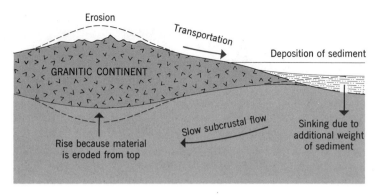

Figure 4.2 A continental landmass rises as material is eroded from its surface and transported to the ocean as sediment. This situation is analogous to the rising of an iceberg as ice is melted from its surface. (After Longwell, C. R. and R. F. Flint, 1962, *Introduction to Physical Geology,* John Wiley and Sons, New York, p. 409.)

so commonly displayed in folded sediments of mountain ranges, nor can it explain crustal thickening beneath mountains. The origin of forces causing those phenomena must lie elsewhere.

External Forces

Uplifted segments of the earth's crust, however they are formed, do not stand forever, nor do volcanic plateaus or peaks exist for geologically long periods of time. All mountains, no matter how lofty, eventually succumb to the relentless forces acting on the earth's surface. These forces act in opposition to the internal forces and, hence, tend to lower the uplifted parts of the continents to sea level. It is this constant battle that produces the tremendous topographic and scenic variety of the modern landscapes. The battle is clearly not over, as the earth is a dynamic, ever-changing planetary body; but it can be argued that for any single continental area the uplifting forces have the upper hand over those forces that tend to wear away the land surface. Studies of the erosion rates in the United States (Chapter 11) indicate that the surface is being reduced at an average rate of 1 foot in 5000 years. Since the continents are very old they should

have been worn down to sea level long ago, even at the average rate of erosion for North America. Because, at least, parts of the continental masses are still high above sea level, it is evident that the erosional agents, principally rivers, are failing in their attempt to reduce the level of the land. The external forces of erosion, however, can for a time succeed, because in parts of the northern Great Lakes region, ancient mountain ranges are now reduced to low plains.

What are these forces that can reduce lofty mountain ranges to low-lying hills and plains, or that can cut deep gashes into mountains and high plateaus? We see them in action every day, but they are so commonplace that many people fail to recognize them for what they are.

The Hydrologic Cycle. Water, covering 71 percent of the earth's surface, fills the ocean basins more than brim full. If it were not for the energy of solar radiation, the water would be confined to the ocean basins and would never fall as rain or snow on the dry land. Because of solar heating, however, water from the sea is evaporated and carried inland as clouds of water vapor. Precipitation in the form of rain, sleet, or snow releases the water from the atmosphere and returns it to the surface of the ground. Then, one of the earth's major forces, gravity,

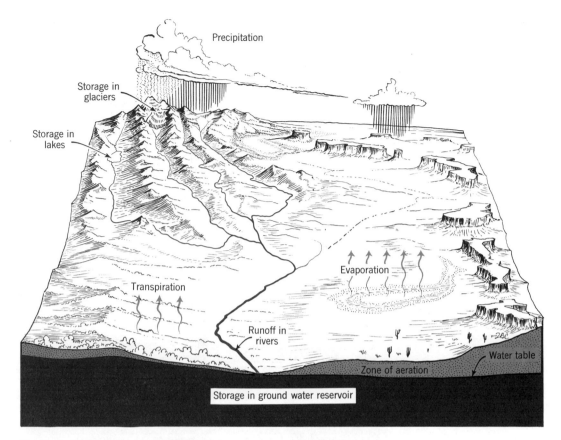

Precipitation

Storage in glaciers

Storage in lakes

Transpiration

Evaporation

Runoff in rivers

Water table

Zone of aeration

Storage in ground water reservoir

Figure 4.3 A diagram showing the major components of the hydrologic cycle on land. Precipitation on land originates by evaporation of seawater. Snow and rain that falls on the land surface is returned eventually to the ocean via direct or indirect routes. See text for details and Table 10.a for the estimated volume of water contained in each component of the hydrologic cycle.

comes into full play. The water returns to the sea, sometimes directly, more often indirectly. But whatever path it takes from high elevation to the ocean basin, the water expends energy along the way. It is this energy that wears away the land. Throughout the entire geological history of the earth, the hydrologic cycle, driven by solar energy and gravity, has drastically altered the face of the earth.

Elements of the Hydrologic Cycle. The path taken by a water particle from the time it is evaporated from the sea until it is returned again to the oceanic reservoir may be extremely short or extremely long. The many facets of the hydrologic cycle (Fig. 4.3) are worthy of examination in some detail, for

in them we find the reasons for many different land forms that impart characteristic forms to different parts of the earth's surface.

Consider first the precipitation that falls on the continents as rain. Some of it collects in natural surface water courses and flows directly back to the sea; this fraction is called *runoff*. The geologic work done by running water accounts for most of the erosive features on the earth's surface, and is the subject of Chapter 11.

Some of the rain collects in lakes where it is temporarily stored before being reevaporated into the atmosphere or discharged into a surface water course via the lake's outlet. Another portion of the rainfall soaks into the ground and infiltrates the pores of soil and rock where it remains temporarily stored as *groundwater*. This underground water dissolves some rock material and eventually returns to the sea carrying a huge load of dissolved mineral material with it. Chapter 10 deals with the groundwater part of the hydrologic cycle.

Plants, which use soil moisture in their life processes, return water to the atmosphere through the process of *transpiration*.

Precipitation falling as snow may accumulate during the winter but will be set free during the spring breakup. But in the polar regions and in high mountainous areas, where annual precipitation exceeds annual melting and evaporation, the snow builds

up year after year to form *glaciers*. Some glaciers move as rivers of ice and carve deep valleys. Others, in the form of great ice sheets thousands of feet thick, creep slowly over areas of continental proportions, scraping, grinding, and wearing away the land surface.

The hydrologic cycle, then, is the unifying thread that binds the surface geologic agents together. The geologic work accomplished by each agent will be considered in more detail in the chapters that follow. But for the moment, keep in mind that the geologic agents of running water, moving glaciers, groundwater, and others, operating over fantastically long periods of time, can reduce mountains to hills, and hills to plains.

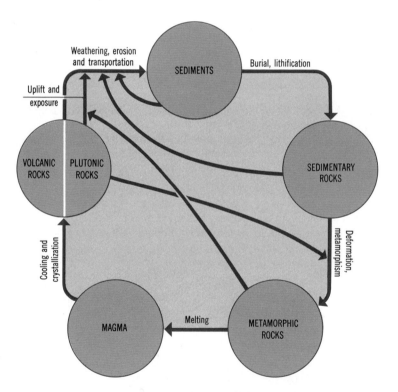

Figure 4.4 The rock cycle. Magma produces either plutonic (intrusive) or volcanic (extrusive) rocks. These, in turn, are converted to sediments that become sedimentary rocks. Deformation and metamorphism transform sedimentary rocks to metamorphic rocks which, when heated and further metamorphosed, may become molten magma. The complete rock cycle is shown by the arrows around the outside of the diagram. Other arrows show various interruptions, or short-circuits of the cycle.

The work accomplished by these agents is called *erosion*.

Unlike the forces from within, the external forces lend themselves to direct observational study. Even with this advantage we still have much to learn about these geologic agents and the way that they operate. We know enough about the workings of each to permit us to interpret past geologic events in the light of the hydrologic cycle, however, because it is through the study and understanding of the hydrologic cycle of today that we provide ourselves with some of the necessary information for the understanding and interpretation of the geologic past.

The Rock Cycle. Thus far we have discussed briefly the three major rock types, the forces that may bring them into contact with surface forces, and the actions of these forces on earth materials. An instructive way to describe these interrelationships is provided by the *rock cycle* (Fig. 4.4).

The primary source material from which all rock materials are derived is, of course, magma beneath the surface of the earth. This material may reach the surface and crystallize as hardened lava or may crystallize beneath the surface as igneous rock and be brought to the surface by uplift and subsequent erosion.

Whatever the mechanism of bringing the igneous material to the surface, it will undergo chemical and mechanical weathering, the products of which will be transported by some mechanism to a site of deposition where the materials will accumulate as sediments, eventually to be lithified into sedimentary rocks. These sedimentary rocks may be brought under intense heat and pressure (metamorphism) to form metamorphic rocks; these, in turn, under very high temperatures may be melted and returned to a magmatic state. This is the theoretically ideal and complete rock cycle. The cycle may be inter-rupted, however, by the insertion in the cycle of additional geologic events, out of order in regard to the ideal cycle. Thus, igneous rocks may be directly metamorphosed without passing through the weathering-transportation-sedimentation-sedimentary rock stage, and the metamorphic rocks may directly undergo weathering and thereby return to an earlier stage in the cycle. Figure 4.4 shows not only the ideal rock cycle but also the various "short circuits" that actually occur in nature.

References

Badgley, Peter C., 1965, *Structural and tectonic principles,* Harper and Row, New York, 521 pp.

Gaskell, T. F. (ed.), 1967, *The earth's mantle,* Academic Press, New York, 509 pp.

Paige, Sidney, 1955, Sources of energy responsible for the transportation and deformation of the earth's crust. *In:* A. Poldervaart, *The crust of the earth,* Geological Society of America, Special Paper 62, pp. 221–342.

* Penman, H. L., 1970, The water cycle, *Scientific American,* V. 223, No. 3 (September) pp. 98–108.

Phinney, R. A. (ed.), 1968, *History of the earth's crust,* Princeton University Press, Princeton, 244 pp.

* Takeuchi, H., S. Uyeda, and H. Kanamori, 1970, *Debate about the earth,* rev. ed. (translated from the Japanese by Keiko Kanamori), Freeman, Cooper and Company, San Francisco, 281 pp.

Umbgrove, J. H. F., 1947, *The pulse of the earth,* 2nd ed., Martimus Hijhoff, The Hague, 358 pp.

*Recommended for further reading.

5 Volcanoes and Volcanism

Night view of Paricutin Volcano, Mexico. (Photograph by Tad Nichols.)

. . . . the cinders, which grew more abundant and hotter the nearer he approached, fell into the ships, together with pumice stones, and black pieces of burning rock. . . . Pliny the Younger

Of all nature's violence, the eruption of a volcano is the most spectacular. The outpouring of large quantities of water vapor, volcanic gases, and molten lava provide the most startling of sights. The early Greeks and Romans identified volcanic eruptions with the activities of certain of their gods. Vulcan, the Roman god of fire, forged arrows for Apollo, a shield for Achilles, and the armored breast plate of Hercules. The eastern Mediterranean region, the seat of western civilization, provided a number of active volcanoes such as Vesuvius, Etna, and Stromboli, toward which the early writers could direct their imaginations.

Many peoples—the Aztecs of Mexico, the Polynesians of the Pacific, and the Japanese of Honshu—had rituals related to fire gods who dwelt in the volcanoes of their lands. This attraction and awe is understandable, for volcanoes have caused death and destruction on a large scale. It is estimated that some 500 volcanoes have been active during the past 400 years. It is further estimated that they have killed 190,000 people during that time, the largest scale devastation being that for Tamboro Volcano in the East Indies which killed 56,000 persons in 1815 during a single explosion. On the other hand, volcanic ash produces exceedingly rich soils, and the gaseous emanations have added tremendous quantities of carbon dioxide to the atmosphere.

The potential for harnessing volcanic energy is great. It is estimated that the hot springs and geysers of Yellowstone National Park give off 220,000 kilo-calories—enough to melt 3 tons of ice—every second. Very little of the potential energy of the world's volcanic centers is utilized, however. In Italy, natural volcanic steam has been used to generate electricity since 1904. In Iceland, volcanic steam is used to heat many buildings and to heat fields, allowing crops to be grown that are normally confined to more temperate latitudes.

Volcanological observatories, such as the Hawaiian Volcano Observatory of the United States Geological Survey on Hawaii and the Vesuvius Volcano Laboratory in Italy, provide detailed studies about the birth, life, and decline of volcanoes; they also have provided the data from which eruption predictions can be made, hopefully making further large scale loss of life less likely.

In Hawaii, for example, precise observations using tilt meters on the cone surface at many places indicate that the mountain "swells" a measurable amount as magma moves upward preceding an eruption.

DISTRIBUTION OF VOLCANOES

Of the 500 active volcanoes, the vast majority are concentrated in a few discrete zones. They occur along the circum-Pacific belt, the Atlas-Alpine-Caucasus-Himalayan belt, the East-African rift system, and a few selected portions of the Pacific and Indian oceans. The volcanoes of the circum-Pacific belt are near the continental margins or are situated on island arcs. These volcanoes and those of the Alpine-Himalayan chains occur also in a zone of earthquake activity; both regions are the sites of young mountain systems, suggesting that both volcanoes and earthquakes may be surface manifestations of the more fundamental process of mountain building.

The volcanoes of the Atlantic Ocean occur along the Mid-Atlantic Ridge (Plate II) and form the islands of the Azores, Cape Verde Islands, St. Paul Rock, and Iceland. Generally speaking, the borders of the Atlantic Ocean contain no volcanoes and are also relatively free of earthquakes, except for the Lesser Antilles (West Indies), a volcanic island arc system of the Caribbean. Geologically, the West Indies are a part of the Andes-North American mountain range rather than a part of the Atlantic basin.

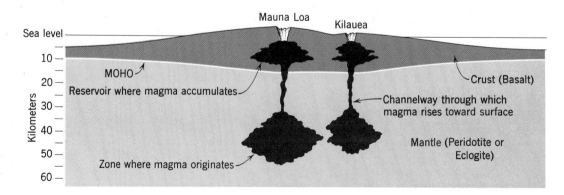

Figure 5.1 Cross-sectional diagram through the island of Hawaii. The distance from the crest of Mauna Loa to Kilauea is about 30 kilometers. (After Macdonald, G. A., 1961, Volcanology, *Science,* v. 133, p. 677.)

The Origin and
Source of Volcanic Materials

A volcano is an opening in the earth from which molten material, solid rock fragments, and gases are ejected. The molten material, or lava, issuing from a volcanic vent is derived from some distance beneath the surface of the earth where it formed. How the magmatic material forms is a significant problem in geology.

Geologists who have studied the Hawaiian volcanoes believe that the magma which feeds the Mauna Loa and Kilauea cones comes from a zone about 40 or 50 kilometers below the summit. This would place the magma generating zone about 35 kilometers below the Moho in that region (Fig. 5.1). The fact that volcanic tremors occur at that depth tends to confirm this depth as the zone of magma generation.

The temperature at 35 to 40 kilometers below the crust is not known with any degree of certainty, but it is generally believed on the basis of activity of earthquake waves that at this depth the mantle is significantly cooler than the measured temperatures of molten Hawaiian lavas of 1100° to 1200° C. This means, therefore, that magma is generated from the solid mantle by the addition of local heat.

The origin of this local heat is a difficult question. The material of the solid mantle may become fluid either by the reduction of pressure or by a rise in temperature, or both. The deformation attendant on earthquake activity or on a variety of orogenic forces may reduce the pressure; the temperature may be raised by localized radioactive "hot spots," although concentrations of radioactivity do not seem to accompany the outpouring of lava or volcanic clouds. Another possible source of heat is from deeper in the mantle where rising convection currents (Chapter 7) bring hotter mantle material upward toward the base of the crust. When the material becomes fluid, it is lighter than the surrounding solid rocks and tends to rise, providing that some avenue is available.

The composition of lava is significant in comparison with the composition generally attributed to the mantle. Solidified lava ranges in composition from the light-colored felsites (rhyolites) to the dark-colored andesites and basalts. The rhyolites generally contain more than two-thirds SiO_2 by weight and are the so-called "acidic" volcanics. Those containing between one-half and two-thirds SiO_2 are intermediate lavas, or andesite; and lavas with less than one-half their composition SiO_2 are the basalts.

By far the greatest volume of lava is basaltic in composition and is concentrated in the Pacific basin. Andesitic lavas also occur in great abundance, generally confined to the landward side, on both sides of the Pacific, of the "Andesite line," a line separating two distinct suites of lavas and presumably representing significant differences in their generating areas at depth.

The Hawaiian chain represents a volume of lava approaching 100,000 cubic miles. The Deccan region of India and the Columbia River Plateau in the northwestern United States, both more ancient accumulations, have volumes of lava of similar magnitudes. It is, therefore, apparent that large volumes of basaltic magma of rather uniform composition were generated in past geologic time.

Basaltic magma is thought to originate in the upper mantle. As yet, no one has sampled the mantle, although the feasibility of drilling down to the mantle is well within the technological capabilities of both the United States and the Soviet Union. Earthquake waves indicate a sharp difference between crustal rocks and the mantle. The boundary between crust and mantle is the M-discontinuity or Moho. Earthquake waves also suggest that the mantle has a density similar to the ultrabasic rock called *peridotite,* which contains much less SiO_2 than basalt; even so, many geologists believe that fractional melting of peridotite of the mantle would produce basaltic lava. This view suggests that the Moho represents a *compositional* change between the lower crust (sima) and the mantle.

Other geologists argue that the Moho represents a *phase* change rather than a compositional difference between the sima and upper mantle. This hypothesis holds that the chemical composition on either side of the Moho is the same but that the minerals normally found in basalt of the lower crust cannot exist below the Moho because of the

Table 5.a. Composition of Volcanic Gases from Hawaii*

Gas	Composition	Percent by Volume
Steam	H_2O	70.75
Carbon dioxide	CO_2	14.07
Sulfur dioxide	SO_2	6.40
Nitrogen	N_2	5.45
Sulfur trioxide	SO_3	1.92
Carbon monoxide	CO	0.40
Hydrogen	H_2	0.33
Argon	A	0.18
Sulfur	S_2	0.10
Chlorine	Cl_2	0.05

*Based on Jaggar, T. A., 1940, Magmatic gases, *American Journal Science,* V. 238, pp. 313–353.

higher pressures. The material of the mantle that is equivalent in composition to the basalt of the crust but that has the density of periodtite is *eclogite*. The higher density phase is a result of closer packing of the atoms as a consequence of the higher pressures at increased depth.

It appears, then, that basaltic magma could be generated from either peridotite or eclogite. If the former, melting would have to be only partial because peridotite, like any rock composed of different minerals, does not have a finite melting point but a melting range of several hundred degrees. Partial melting of peridotite would produce a basaltic magma that would migrate upward. In the case of eclogite, local complete melting would produce the basaltic magma that would migrate upward toward the volcanic vent.

Volcanic Gases

Gases of various types constitute the greatest volume of material produced from volcanic centers. Water vapor and carbon dioxide are the most abundant gases discharged in an eruption. While the magma is still confined under high pressure, gases are dissolved in the liquid melt. As the magma rises, however, the gases expand under lessening pressure and at the surface produce a frothy, or very porous rock. The basaltic lavas are quite fluid (low viscosity) so the expanding gases can escape quickly; the result usually is a relatively quiet outpouring over fairly large distances. The more silicic lavas such as rhyolite, on the other hand, are extruded at lower temperature than their basaltic counterparts and are highly viscous, a condition that keeps the gases entrapped for a longer time; the result is a more violent explosion and the resulting lava flows generally do not travel far before congealing.

Table 5.a gives the typical composition of gases emitted from Hawaiian volcanoes. The gaseous products of other volcanoes are generally similar. Although the major volume of water vapor analyzed in volcanic gases is from rainwater that is held in the rocks and soil, W. W. Rubey, an American geologist, has postulated that small amounts of "original water" discharged over long periods of geologic time at volcanic centers, including hot springs and fumaroles, have been sufficient to produce all the oceanic waters of the earth's surface.

Figure 5.2 Lava fountain, 1000 feet high, spouting from the eastern rift zone of Kilauea Volcano during the eruption of January 10, 1960 near Kapoho, Hawaii. Frothy pumice ejected into the air cools into dust which is blown over the countryside. The road (lower right) through the sugarcane field has been blocked by an *aa* lava flow. (Photograph by Wayne Ault.)

Types of Volcanoes and Their Eruptive Characteristics

A number of classifications of volcanic eruptions have been proposed. Generally, they are based on the nature of the eruptive sequence, the character of the extruded material, the nature and extent of the vent, and the resulting morphology of the volcanic accumulation. Volcanic materials that issue from a point source form the typical volcano; those issuing from elongate fissures produce what have been called fissure eruptions or lava plateaus. Volcanoes in this book are classified as (1) the Hawaiian type or shield volcanoes, (2) cinder cones, and (3) the Vesuvian type, also called composite cones, or stratovolcanoes.

Hawaiian Type. This type of eruption is characterized by the outpouring of tremendous quantities of basaltic lava that builds a broad based, convex upward, gently sloping cone of enormous height, hence the term *shield volcano*. Because the lava has a high fluidity, associated gases are liberated quickly with very little explosive violence, although *lava fountains* are sometimes projected upward to heights of 1000 feet or more during an eruptive sequence (Fig. 5.2).

Mauna Loa, the world's largest active volcano, is the most magnificent example of the Hawaiian type (Fig. 5.3). It rises 13,680 feet above sea level and about 30,000 feet above its base on the ocean floor. Even for a cone of this height, its maximum slope angle is only about 12 degrees. It has been built by thousands of individual lava flows that average about 10 feet thick. Since 1832 it has erupted lava intermittently for a total of more than 1300 days from the summit crater and has erupted lava for a similar length of time from

Figure 5.3 Mauna Loa, the largest active volcano in the world, rises to an elevation of 13, 680 feet above sea level as a broad shield cone. Rounded boulders in the foreground were deposited by glaciers in an earlier geologic period on the slopes of Mauna Kea from where this photograph was taken. The faint depression barely visible on the skyline is the summit caldera of Mauna Loa. The irregular black patches extending down the flanks are lava flows that originated from the summit caldera or a rift zone on the left flank. Some of these highly fluid basaltic lava flows reached the ocean, some 20 miles away. (Photograph by Wayne Ault.)

fissures on its flanks. During the last century, its eruptions have poured out a total of more than 3½ billion cubic yards of lava. The longest recorded eruption began on April 20, 1873 and lasted for about 1½ years. Mauna Loa is one of the great volcanoes comprising the island of Hawaii (Fig. 5.4). Its companion crater, Kilauea (Fig. 5.5) on the east flank of the mountain has a similar long history of eruptions, many of which have been intensively studied.

Kilauea differs from Mauna Loa in the presence of greater amounts of explosive debris in the area of the caldera and in the presence of a large lava lake, Halemaumau, which changes levels during the eruptive period due to the circulation of lava from the magma chamber. It is at Kilauea where tilt-

meter studies have best illustrated the modern activity of the Hawaiian-type eruption.

Shrinking and swelling in the crater region suggests the presence of a reservoir within the volcano, in which magma accumulates preceding an eruptive phase. The eruption of 1967 to 1968 was preceded by a detectable swelling of the volcano that was followed by volcanic tremors and sharp earthquake shocks, and eventually, the outbreak of lava in Halemaumau. By April 1968, the level of the lava lake had risen 320 feet above the preeruption floor (Fig. 5.6).

A typical eruption of the Hawaiian type begins with the opening of a fissure or fissures several miles long along the flanks of the volcano. In the initial stages, lava fountains erupt from these fissures, and they increase in

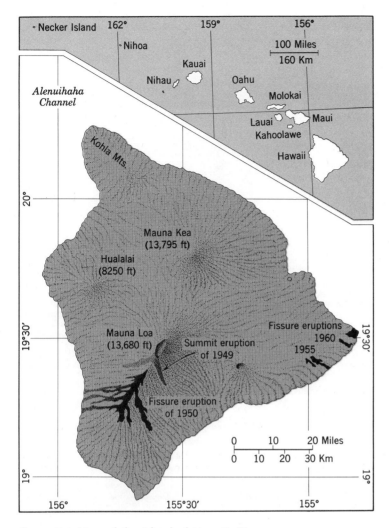

Figure 5.4 Map of the island of Hawaii. (From Macdonald, G. A., and D. H. Hubbard, 1961, *Volcanoes of the National Parks of Hawaii*, Hawaii Natural History Association.)

intensity until a maximum height of several hundred feet is reached. Large gas clouds rise several thousand feet above the fountains as lava pours down the volcano's flanks. Eventually the fountaining ceases as the gaseous emanations diminish and a short period of lava outpouring ends the eruptive sequence.

Some of the volcanic materials formed during a Hawaiian eruption consist of pumice and Pele's hair (natural spun glass), which rain down during the lava fountain stage. Two principal types of lava flows result from the Hawaiian eruptions, those called *aa* and

pahoehoe (Fig. 5.7). Aa lava is characteristically blocky and has a very rough and jagged surface. Pahoehoe lava is also known as *ropy lava* because of its appearance as a series of parallel strands of twisted rope. Aa and pahoehoe lavas may have closely similar compositions; their different physical appearances result from differences in the amount of enclosed gas at the time of solidification and minor chemical differences. The pahoehoe lava flow moves forward in a fluid state resembling the movement of warm molasses, and the aa flow, more viscous be-

Figure 5.5 Aerial view of Kilauea Crater on the east flank of Mauna Loa. Mauna Kea is in the distance. The large pit in the main crater is the lava lake of Halemaumau, and the dark irregular patch to the right of Halemaumau is a lava flow of a 1954 eruption. (U. S. Navy photograph.)

cause of higher silica content, moves as a mass of frozen blocks above fluid lava.

Cinder Cones. A cinder cone is an accumulation of solid volcanic materials in the form of cinders and ash. Cinder cones are typically steep-sided, owing to the angularity of the fragments, with slopes as high as 40°. The SP Crater in Arizona (Fig. 3.13) is an excellent example of this type. At Paricutin, Mexico, the eruptive period began suddenly with the ejection of large quantities of ash,

cinders, and bombs from a flat plain in a region of older volcanic activity. After one year a cone nearly 1500 feet high had developed (Chap. 5, opening photo), and volcanic activity continued for nearly 10 years. As is common in many cinder cones, the Paricutin crater was ultimately breached and large volumes of lava were extruded.

Vesuvian Type. The Vesuvian, or stratovolcano, owes its principal characteristics to alternations of ash, cinder, and bomb ejec-

Figure 5.6 Basaltic lava streaming from the Halemaumau lava lake during the 1967-1968 eruption of Kilauea Volcano, Hawaii. The white area to the upper right of the center of the photo is an incandescent lava fountain. (Photograph by Richard S. Fiske, U. S. Geological Survey.)

tions and quieter episodes of lava extrusion. This alternation produces a volcanic profile (Fig. 5.8) intermediate between that of the shield volcano and the cinder cone. Examples of this type are among the best-known volcanoes, and include Vesuvius, Stromboli, and Vulcano of the Mediterranean region, Fujiyama in Japan, and Mt. Rainier and Mt. Shasta of the Cascade Range.

Mt. Vesuvius, near Naples, Italy, is widely known for its historic eruptions (Fig. 5.9). The Volcano Observatory was established in 1845, but records of eruptive activity go back to the beginning of the Christian Era when in A.D. 79, Vesuvius erupted with great violence and destroyed the cities of Pompeii and Herculaneum. An eyewitness account of the eruption was recorded by Pliny the Younger, nephew of Pliny the Elder who, at the time of the eruption, was in command of the Roman fleet near Naples. Pliny the Elder lost his life during the eruption while directing the evacuation of refugees fleeing the holocaust.

The A.D. 79 eruption had been preceded by earthquake shocks several years earlier. The major eruptive phase consisted mainly of

Figure 5.7 Lava flows on the flank of Mauna Loa, Hawaii. The dark upper flow is *aa* lava, and the lower flow is composed of ropy or *pahoehoe* lava.

Figure 5.8 Mt. Bachelor in Oregon is one of a number of composite volcanic cones of the Cascade Range. Composite cones are built of interbedded layers of ash and lava and, hence, have steeper slopes than shield cones. Compare this with the photograph of Mauna Loa, a shield cone, in Figure 5.3. (Oregon State Highway Department Photograph.)

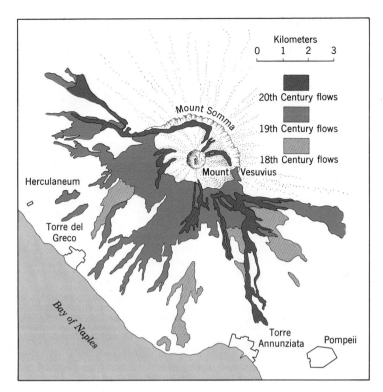

Figure 5.9 Map of Mt. Vesuvius and vicinity showing the distribution of lava flows of three centuries. Pompeii and Herculaneum were destroyed by the eruption of A.D. 79. (After Bullard, F. M., 1962, *Volcanoes in History, in Theory, in Eruption,* University of Texas Press, Austin, plate 10.)

pyroclastic debris exploded in vast quantities. It appears that Pompeii was buried by ash, while Herculaneum was inundated by mudflows moving down the slope of the mountain. These mudflows were the consequence of tremendous quantities of newly formed ash mixed with water produced during the eruption. Both cities lay buried until they were discovered in the sixteenth century. The speed with which the catastrophe overtook the cities is indicated in the exhumed ruins. Families had been trapped, along with food and table settings, and Roman sentries were found buried at their posts.

Stromboli is another Mediterranean volcano that rises from the waters of the Tyrrhenian Sea. It has been more or less continuously eruptive for more than 2500 years. Like Vesuvius, Stromboli ejects both ash and lava, but its distinguishing feature is a dense white cloud of steam that rises from the crater from time to time. The whiteness is attributed to the lack of ash during this phase of eruption.

Vulcano, 50 miles from Stromboli, is the cone from which the term volcano was derived. The style of activity here, as exemplified by the eruption of 1888 to 1889, has been termed by some the Vulcanian type, in which a central crater plug was blasted skyward in the form of volcanic bombs, ash, and scoria. Most of the material produced at Vulcano consists of white ash (pumice) and other incandescent bombs which create a bright glow during night eruptions.

Other Features Produced by Volcanism

Calderas. A caldera is a greatly enlarged volcanic crater. Some are formed by the ex-

Figure 5.10 Crater Lake, Oregon occupies a six-mile wide caldera produced by the collapse of Mt. Mazama in prehistoric time. Wizard Island is a small volcanic cone that formed during the final stages of eruptive activity after the collapse of Mt. Mazama. (Oregon State Highway Department photograph.)

plosion of a volcano summit, as in the case of the Bandai caldera in Japan, which was formed by explosion in 1888; some are formed by the collapse of the summit area owing to crustal weakening produced by transfer of lava and other volcanic material from a subcrustal zone to the surface.

Perhaps the best studied caldera is that of Crater Lake, Oregon (Fig. 5.10). The caldera is just over 5 miles in width and approximately 4000 feet deep. It is speculated that the caldera formed when the summit portion of a large volcanic cone, called Mt. Mazama, disappeared during a prehistoric eruption. An early theory held that the volcano had exploded to produce the caldera. In this event, the products of the explosion should be recognizable in the surrounding region. Howell Williams, an American geologist, has estimated, however, that of the approximately 17 cubic miles of material represented by the lost top of Mt. Mazama, only some 2 cubic miles of material can be accounted for by

ash, lava, and other volcanic ejecta in the surrounding region. The remainder, he believes, was lost by collapse into the underlying chamber. The sequence of events leading to the formation of Crater Lake is shown in Figure 5.11.

Figure 5.11 Evolution of Crater Lake, Oregon. Diagram *A* shows the filled magma chamber feeding the central orifice and several vents from which lava was extruded to build Mt. Mazama. Diagrams *B* and *C* show an increase in the eruption of pumice and volcanic ash and a reduction in lava outpourings as the magma sank. After the major eruptive activity had ceased, Mt. Mazama collapsed into the underlying chamber, forming a caldera (*D*). Diagram *E* shows Wizard Island as the result of a final eruption, and the filling of the caldera with water to form Crater Lake. (After Williams, Howell, 1951, Volcanoes, *Scientific American,* v. 185, No. 5, November, pp. 45–53.)

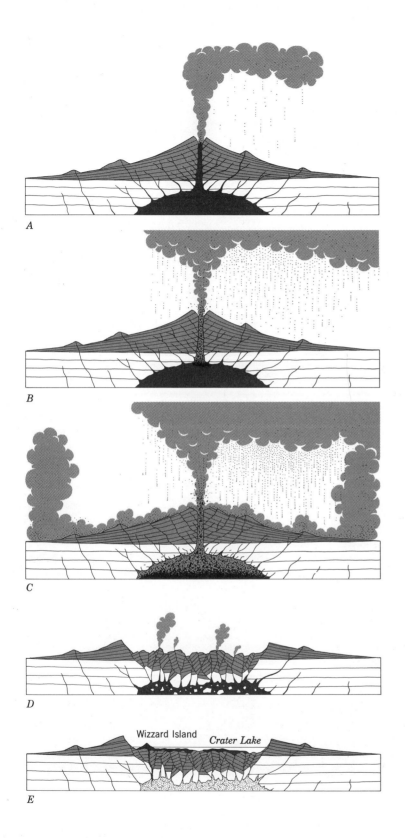

A

B

C

D

E

Wizzard Island   *Crater Lake*

Figure 5.12 Map showing the Columbia Lava Plateau. (From Kay, Marshall, 1964, *Geologic Map of North America,* Geological Society of America, Boulder, Colorado.)

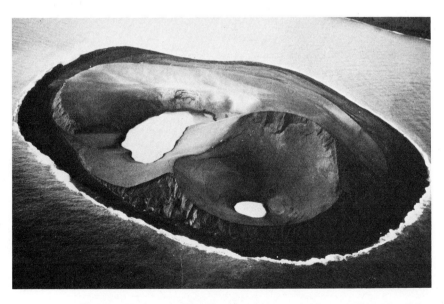

Figure 5.13 Aerial photograph of Jolnir, a satellite volcano of Surtsey composed entirely of volcanic ash. This island rose above the sea in late 1965, but all that remains today is a submarine shoal. (Photograph by R. S. Williams, Jr., Air Force Cambridge Research Laboratories.)

Nuée Ardente. On the island of Martinique in the West Indies is the volcano of Mt. Pelée. On May 8, 1902 its sudden eruption resulted in the death of all but two inhabitants of the city of St. Pierre, a seaport city five miles from the crest of the volcano. In a matter of minutes, 28,000 persons were killed by a dense cloud of ash ejected with explosive violence from the summit area. The crater itself was blocked by a plug of frozen lava and, as a consequence, the ash was directed horizontally, in part explaining the high velocities reported. The cloud was so dense that only part of it was disseminated in the atmosphere. The rest moved rapidly down the mountain side as a density current made up of incandescent gas, dust, and ash. Eyewitness reports estimate speeds up to nearly 100 miles per hour, a speed comparable to that of a hurricane. This glowing mass of gas and ash is called a *nuée ardente* ("fiery cloud").

The inhabitants of St. Pierre were killed by the inhalation of hot gases, suffocation from lack of oxygen, or by burns. The temperature of the cloud, based on its effect on glass and wood, is estimated at several hundred degrees Fahrenheit.

Five months after the expulsion of the *nuée ardente,* a mass of stiff lava was extruded from the crater of Mt. Pelée. It rose like a gigantic spine from the throat of the crater and eventually reached a height of more than 1000 feet above the crater floor. The spine continued to rise until August 1903, after which it disintegrated into a stump standing in its own debris.

Fissure Eruptions. Volcanic eruptions that issue from linear fissures and in which lava is extruded over large areas are termed *fissure eruptions.* They produce broad, for the most part, gently rolling terrains in which successive lava flows may accumulate to a thickness of several thousands of feet. The Columbia Plateau (Fig. 5.12), for example, is underlain by approximately 5000 feet of basaltic lavas interbedded with terrestrial lacustrine (lake) deposits, and covers an area of about 200,000 square miles. Similar outpourings of basaltic lava occurred on the Deccan Plateau of western India. Today, Iceland is the only place where modern fissure eruptions are taking place.

Iceland is also the site of explosive volcanic activity. This was magnificently displayed by the birth of a new volcanic island, Surtsey, just south of Iceland, beginning in November 1963. The visible eruption, which had been preceded by a series of earthquake shocks, began in mid-November with a violent explosive phase of eruption produced by the mixing of seawater with the volcanic material. Colossal columns of water vapor, sulphurous gases, and volcanic ash were thrown into the air. Within two months a cinder cone 670 feet in elevation had developed. This type of explosive activity continued until early April when the sea was blocked from entering the vent. After that, quiet Hawaiian type lava outpourings began and added to the dimensions of the island. This style of eruption continued for about 13 months. In May 1965, a second explosive phase began which formed an island nearby to a height of about 210 feet. This new island was later destroyed by wave action, so that by late October only an underwater shoal remained.

A third island, Jolnir, was born in late December 1965 (Fig. 5.13). From then until August 1966, the island was washed away by wave activity and rebuilt by eruptions five times. The sea won in the end because all that remains of Jolnir is a submarine shoal where the island had previously been.

The Surtsey eruptions afforded a firsthand study of volcanic processes located on a surface expression of the Mid-Atlantic Ridge, a feature that will be considered in more detail in Chapter 7.

Volcanic Hazards in the Western United States

The Cascade Range between northern California and the Canadian border is studded with a number of snowcapped volcanic peaks (Fig. 5.14). Air travelers flying the north-south routes along the west coast of the United States are treated to the spectacular view of these beautiful peaks on clear days. No *significant* eruptions of any of these volcanoes have occurred within historic times, so both the air passengers flying over them and the residents living within sight of

Figure 5.14 Map showing the distribution of major volcanic peaks of the Cascade Range. The only one that has erupted in historical time is Mt. Lassen. A major eruption of any one of these could have undesirable effects on populous areas nearby. (Adapted from Crandell, D. R. and H. H. Waldron, 1969, *Geologic hazards and public problems, conference proceedings,* Santa Rosa, California, p. 6. See references at end of chapter for full citation.)

them are inclined to overlook the fact that these volcanic spires are potentially hazardous.

Unfortunately, geologists who have studied the histories of these volcanoes are unable to do little more than speculate on the consequences of an eruption of one or more of them in terms of the effects on the surrounding population and man-made features. No geologist can predict the time of the next eruption of any one of the dozen major volcanoes in California, Oregon, or Washington, but because each of the peaks shown in Figure 5.14 could erupt with little or no warn-

ing, there is nothing to be lost and much to be gained in assessing the hazards to life and property in the event of future eruptions.

Past Eruptions. Geologic studies on the flanks and surrounding area of many of the volcanoes of the Cascade Range provide the basis for piecing together the past eruptive events of any single cone. The geological history of Crater Lake has already been given. Not all of the other volcanoes of the Cascades have been so thoroughly studied as Mt. Mazama, but some are better known than others in terms of their eruptive histories.

Mt. Rainier, for example, is known to have experienced 12 eruptions of volcanic ash, several hot avalanches of rock debris and, at least, one period of lava outpouring during the past 10,000 years. The last major eruption occurred about 2000 years ago when lava, pumice, and several large mudflows were produced. Mudflows are common byproducts of volcanic eruptions and are produced on the flanks of an active volcano when the loose volcanic debris becomes mixed with rain or melting snow. In the last 10,000 years more than 50 large mudflows have originated on the flanks of Mt. Rainier. One of them occurred 6000 years ago and contained about 800 million cubic yards of debris having the consistency of very wet concrete. It flowed down the valley of the White River, a distance of 30 miles. The last time Mt. Rainier showed any life was in 1894 when some steam and smoke was observed.

Mt. St. Helens in southern Washington erupted frequently in the nineteenth century, and during the last 10,000 years mudflows, lava, hot avalanches, and pumice have originated from this peak. About 3500 years ago, 2 feet of pumice was laid down 50 miles away from the peak and winds carried some of it north into Canada.

Other peaks in the volcanic chain from Canada to northern California have had similar histories. Lassen Peak in California experienced a series of eruptions during the period 1914 to 1917, the most violent of which took place on May 22, 1915 when a column of volcanic ash was blown 6 miles skyward.

For more than half a century the volcanic cones of the Cascade Range have remained silent. There is no reason to conclude from

Figure 5.15 Eruption cloud of volcanic ash from Taal Volcano in the Philippine Islands, 1965. (U. S. Navy photograph.)

this, however, that these majestic peaks will remain mute. On the contrary, from what is known about their past performance, long periods of quiescence have occurred between eruptive spasms over the last 10,000 years, a fact that should not be lost sight of. On the other hand, since it is impossible to predict the time and place of the next volcanic activity in the Cascades, much less the magnitude and nature of that activity, all that can be done is to assess the possible consequences of a major eruptive event if it should occur.

Types of Volcanic Hazards. Assuming, then, that future violence from one or more of the Cascade volcanoes is an inevitable eventuality, what are the likely consequences in terms of human life and works of man? Geologists of the United States Geological Survey consider ash eruptions, mudflows, and secondary effects of mudflows to be the greatest hazards. Volcanic ash can be car-

ried for great distances, depending on the size of the ash particles and the force of prevailing winds (Fig. 5.15). Ash falls may be accompanied by toxic fumes close to the volcano, and both fine ash and fumes may be deleterious to respiratory systems. Heavy ash falls in residential areas could cause collapse of house roofs, clogging of storm sewers, and contamination of surface water supplies.

Mudflows are of high potential hazard in the case of an eruption of a Cascade Range volcano because of the perennial snow cover. A volcanic event liberates considerable heat that could melt large volumes of snow or ice that, when mixed with ash and volcanic debris from previous eruptions, will produce mudflows. Mudflows move down valleys toward low areas at 20 to 30 miles per hour and are capable of destroying almost everything in their paths (see Chapter 9). If a mudflow of large volume should enter a reservoir behind a hydroelectric dam, the displaced waters could spill catastrophically over the dam and send floodwaters down the valley to do further damage. Several hydroelectric reservoirs exist along the pathways of potential mudflows from Mt. Baker, Mt. Rainier, and Mt. St. Helens. A large mudflow moving down the flanks of Mt. Shasta could ultimately enter the Sacramento River Valley and flow into Lake Shasta, causing the lake to overtop Shasta Dam with disastrous consequences.

It must be emphasized that, lacking the ability to make accurate predictions on the volcanoes, in the Cascades or elsewhere, the safest course to follow is one in which the residents of potentially dangerous areas are alerted to the nature of possible hazards should these events actually become a reality.

References

* Bullard, Fred M., 1962, *Volcanoes in history, in theory, in eruption,* University of Texas Press, Austin, 441 pp.

Coats, R. R. and others (eds.), 1968, *Studies in volcanology,* Geological Society of America, Memoir 116, Boulder, Colorado, 678 pp.

Crandell, D. R., and H. H. Waldron, 1969, Volcanic hazards in the Cascade Range, pp. 5–18, in *Geologic Hazards and Public Problems, Conference Proceedings,* May 27–28, 1969, R. A. Olson and M. M. Wallace (eds.), Office of Emergency Preparedness, Region 7, Federal Regional Center, Santa Rosa, California, 335 pp.

* Eaton, J. P., and K. J. Murata, 1960, How volcanoes grow, *Science,* V. 132, pp. 925–938.

* Macdonald, Gordon A., 1961, Volcanology, *Science,* V. 133, pp. 673–679.

* Macdonald, Gordon A., and A. T. Abbott, 1970, *The geology of Hawaii,* University of Hawaii Press, Honolulu, 442 pp.

Thorarinsson, Sigurdur, 1967, The Surtsey eruption and related scientific work, *The Polar Record,* V. 13, pp. 571–578.

* Williams, Howell, 1951, Volcanoes, *Scientific American,* V. 185, No. 5 (November) pp. 45–53 (Offprint 822, W. H. Freeman and Co., San Francisco.)

* Recommended for further reading.

6 Earthquakes and the Earth's Interior

Collapse of apartment building caused by the July 29, 1967 earthquake, Caracas, Venezuela. (Photograph by El Nacional, courtesy of Mary Hill.)

. . . oft the teeming earth is with a kind of colic pinch'd and vex'd by the imprisoning of unruly wind within her womb which for enlargement striving, shakes the old bedlam earth and topples down steeples and moss grown towers. Shakespeare

Similar to volcanic activity, earthquakes have been viewed with awe and fear because of their devastating effect on many populated regions. Earthquakes such as those in San Francisco in 1906, Tokyo and Yokohama in 1923, Anchorage, Alaska on Good Friday of 1964 and the San Fernando earthquake of 1971, serve as reminders of the restless nature of the earth beneath us. As in the case of volcanoes, some hope can be held out for earthquake prediction. Although it is doubtful whether seismologists will ever be able to predict the timing of an earthquake to the nearest day or week, or to the precise place of the severest shock, precise measurements with strain gauges and other instruments permit the identification of the areas most likely to be affected by earthquake activity.

Studies of this nature and the detailed analysis of the prehistoric earthquake history of an area can provide useful information to those responsible for site selection for human

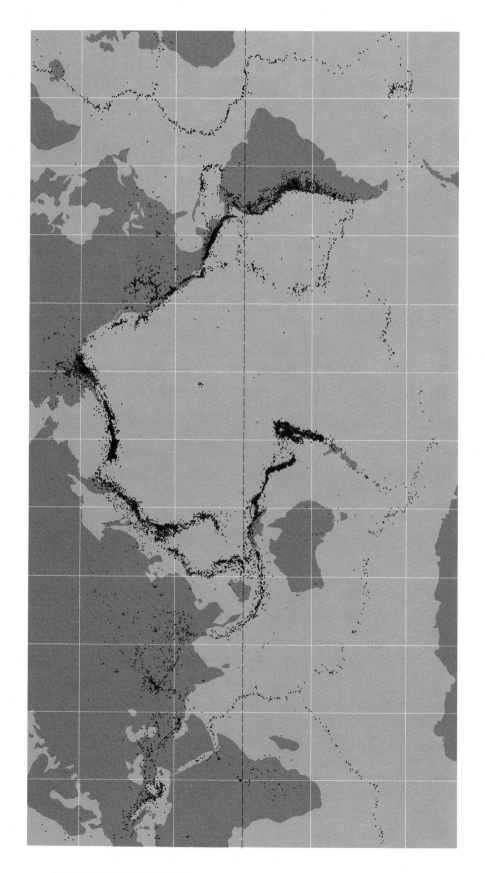

THE DYNAMIC EARTH

activity. The restless earth is a major factor of the environment of any area, and its characteristics must be understood and its potential for violent activity appreciated by not only scientists but by all concerned citizens.

Seismology: Facts about Earthquakes

Seismology is the science that deals with earthquakes. An earthquake is a natural vibration of the ground produced by the rupturing of large masses of rock beneath the surface. The intensities of earthquakes vary over a wide range from those perceptible only to delicate instruments to those that create widespread destruction of life and property. The energy released during the largest shocks is roughly equivalent to 10,000 times the energy of the atom bomb dropped on Hiroshima late in World War II.

The place beneath the earth's surface where an earthquake originates is called the *focus,* and the point on the surface vertically above the focus is the *epicenter.* Earthquake foci are distributed in three general depth

zones. Shallow earthquakes originate within 40 miles of the surface, intermediate earthquakes have foci between 40 and 200 miles depth, and deep focus earthquakes originate at depths between 200 and 500 miles. Most of the estimated one million earthquakes per year are of the shallow type, and these release the greatest energy.

Distribution of Earthquakes. Earthquakes are concentrated in belts around the earth, and their distribution is similar to the distribution of volcanoes. Approximately 80 to 90 percent of shallow and intermediate shocks and all deep focus quakes are concentrated in the circum-Pacific belt (Fig. 6.1). Other areas of concentration are the Alpine-Himalayan chain, the mid-ocean ridges, and the African rift-valley system.

Theoretically, no place on the earth's surface is entirely safe from earthquakes, although the map of Figure 6.1 indicates large areas where destructive earthquakes are not likely. Nevertheless, the eastern United States has experienced significant quakes, for example, in Boston (1775), Missouri (1811),

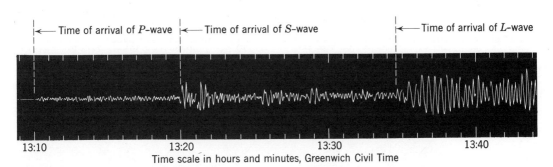

Figure 6.2 Seismogram of an earthquake recorded at Victoria, B. C. on May 24, 1944. The dashes along the lower margin give the time in hours and minutes at the recording station. The epicenter of this quake was located at 2.5° south latitude, 152° east longitude, near New Ireland. The time of the shock was fixed at 12 hours 58 minutes, Greenwich Civil Time. (Courtesy of James T. Wilson, University of Michigan.)

Figure 6.1 Map showing distribution of earthquake epicenters recorded between 1961 and 1967 with depths ranging from 0 to 700 kilometers. The earthquakes along the mid-ocean ridges in the Pacific, Indian, and Atlantic oceans all occur at depths less than 100 kilometers. (After Barazangi, M. and J. Dorman, 1969, *World Seismicity Maps compiled from ESSA, Coast and Geodetic Survey, Epicenter Data,* 1961-1967, Bulletin of the Seismological Society of America, v. 59, p. 369-380, Plate I, by permission.)

THE EARTH'S INTERIOR 91

South Carolina (1886), Maine (1904), New York City (1937), and Chicago (1938).

The California-Nevada region has had about five thousand earthquakes a year since the first human record of a California quake in 1769. This is one-half of one percent of the estimated million quakes per year for the entire world, and about 90 percent of all shocks felt in the United States, exclusive of Alaska and Hawaii. It has been estimated that an earthquake of sufficient intensity to be felt by a person somewhere in the California-Nevada region occurs on the average of once every hour and three-quarters. It is for this reason that *seismology* made its North American debut in California when the first earthquake recording instrument, a *seismograph,* was set up at Berkeley in 1887. A seismograph is nothing more than a pendulum device that records the motion of the earth beneath the pendulum and records the time and intensity of each motion. The graphic record produced by a seismograph is a *seismogram* (Fig. 6.2).

The San Francisco Earthquake of 1906

On April 18, 1906, San Francisco was partially destroyed by a severe earthquake in which 700 persons were killed. To this day it is probably one of the best documented quakes of California, and its study has made important contributions to our understanding of earthquakes and faulting. Oddly enough, direct damage to buildings and property amounted to only 5 percent of the estimated 400-million-dollar loss. Most of the havoc was caused by fire which broke out after the quake. The earthquake itself lasted only about 40 seconds, but the disruption of water mains, caused by the shock, drastically reduced the fire-fighting capacity of the city. The quake's intensity has been exceeded many times elsewhere, but the geologic circumstances surrounding it makes the San Francisco quake unique in seismology. More recent examples of well-documented earthquakes are the Hebgen Lake earthquake in Montana in 1959, the Alaskan earthquake of 1964, and the San Fernando earthquake of 1971. Reference to these and other quakes will be made throughout this chapter.

Causes of Earthquakes

Aristotle (384–322 B.C.) explained earthquakes as a result of entrapped air escaping

Figure 6.3 Fault scarp associated with the Fairview Peak, Nevada earthquake of December 16, 1954. (Photograph by James T. Wilson.)

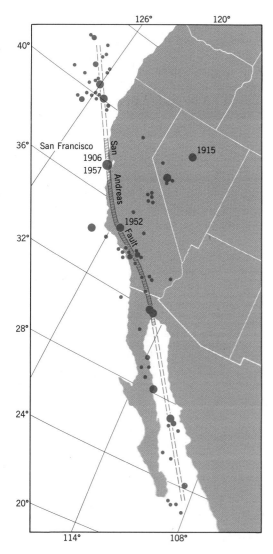

Figure 6.4 The black dots are earthquake epicenters in California and Nevada. Many of them are related to the San Andreas fault zone. Many other active faults exist in southern California, but they are not shown here. (After Benioff, H. and B. Gutenberg, 1955, *Earthquakes in Kern County, California during 1952,* California Department of Natural Resources, Division of Mines, Bulletin 171.)

from the earth's interior. No less fanciful were the writings of some theologians two thousand years later who taught that all earthquakes were the manifestations of God's wrath.

Earthquakes are now interpreted as the result of slippage of rock masses along a rupture or break called a *fault.* Where such faults intersect the surface, a *fault scarp* may be produced (Fig. 6.3).

One of the most intensely studied and best-known faults in the earth's crust is the San Andreas fault of California, which passes near San Francisco. The surface trace of this fault system extends from a point north of San Francisco to the lower reaches of the Gulf of California (Fig. 6.4), a distance of more than 1800 miles. Beyond its visible trace on the ground, its trend beneath the ocean has been established by the position of many earthquake epicenters. Some geologists believe that the total cumulative horizontal movement along the fault is more than 300 miles, the west side having moved to the north, relative to the east side. Sporadic movements along this fault have resulted in many earthquakes, including the San Francisco quakes of 1906 and 1957. Displacement along the fault zone in 1906 was essentially horizontal and amounted to 21 feet.

The Elastic Rebound Theory. Why did the slippage take place along the San Andreas fault in April 1906? The answer lies in a theory expounded as a result of analyzing the exact location of numerous geographic points before the quake and what had happened to these places as a result of the quake. The explanation gleaned from these data constitutes the *elastic rebound theory.* This theory does not account for the forces that produced slippage (faulting), however, but only the manner in which the rocks yielded to those forces. Ultimately, the forces that cause faulting are the same as those that produce mountains and other major structural features of the earth's crust.

When a solid is squeezed (by compression) or stretched (by tension) it is deformed according to physical laws that depend on the properties of the solid. The squeezing or stretching force is called a *stress,* and the deformation (change of shape or volume) of the solid yielding to the stress is called *strain.*

Elastic materials are those in which the stress is proportional to the strain. For example, to increase the length of a rubber band to twice its length, a certain pull (stress) is required. If the amount of stretching is to be doubled, twice the pull must be exerted. But the band cannot be stretched indefinitely because eventually it will break. The same analysis can be applied to a stick of wood bent across the knee. The stick can be bent without breaking, up to a certain point. If the bending stress is released, the stick returns to its original unstressed shape, but if the stress continues to increase, the stick snaps. At that instant the stress returns to zero.

In applying the foregoing to the San Francisco earthquake, the status of the area before faulting and immediately after the shock must be considered. It was observed that places located closest to the trace of the San Andreas fault experienced the greatest lateral displacement during the movement. Those places farther removed from the fault showed smaller amounts of movement; places at some distance from the fault had not moved at all. These relations are simply illustrated in Figure 6.5. The bottom diagram represents the circumstances before any stress had been applied; the middle diagram

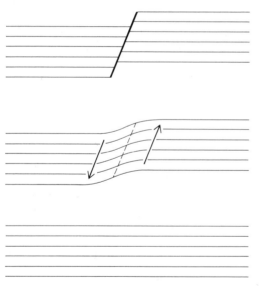

Figure 6.5 Diagram (map view) showing the development of an earthquake by faulting according to the elastic rebound theory. In the bottom diagram, rock strata are unstressed. In the middle diagram the elastic rock layers are strained due to a stress pattern indicated by the arrows. The incipient plane of failure or fault plane as shown by the dashed line. In the upper diagram, rupture has occurred and the rock layers have adjusted to a new unstrained condition. The rupturing and movement along the fault plane produces an earthquake. Not to scale.

shows the effect of a stress applied in the directions of the arrows, but before faulting. When the rocks failed along the fault, shown in the top diagram, they snapped back to a position of "rest." Thus, the differential movement of points according to their distance from the fault is explained.

In other words, surface rocks act elastically and store up the energy applied by the stresses involved until they reach an elastic limit, following which they fail, or break, and faulting is the result. Some breaking may occur before the principal shock, producing *foreshocks,* and adjustments along the fault zone after the principal shock produce *aftershocks.* After the principal shock occurs, most of the stress in the rocks is released. However, the ruptured zone is a plane of weakness along which further movement will take place when the stress becomes large enough to overcome the friction along the fault plane. Hence, recurrent movements along faults like the San Andreas are to be expected.

Earthquake Waves

The energy released during faulting is transmitted away from the focus of the earthquake through the rocks by wave motion. In a similar fashion, the energy imparted by dropping a stone on the surface of a still pond is transmitted outward by wave motion, represented by the concentric ripples produced. Earthquake waves are received and recorded at seismograph stations. A network of stations is necessary for the study of earthquake activity and the study of the interior of the earth. Several hundred stations are in operation all over the world, and seismic stations are in almost instant communication with one another, providing accurate location and intensity information for most earthquakes.

Types of Earthquake Waves. Earthquake waves are of two types. One group, known as *body waves,* are waves that penetrate deep into the body of the earth. The other type, known as *surface waves,* travel along the surface of the earth or along other surfaces within the crust, in this case the outer 20 to 30 miles of the crust. Body

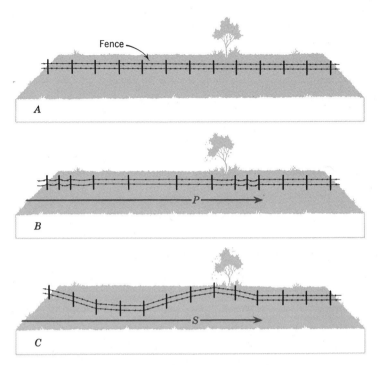

Figure 6.6 Earthquake waves are vibrations of the ground. These diagrams show the effect that *P* and *S* waves would have on a straight fence if the wave motion were greatly exaggerated. *A*, no vibrations; *B*, vibrations due to *P* waves; *C*, vibrations due to *S* waves. The arrow in *B* and *C* shows the direction of wave propagation.

waves consist of *Primary* (*P*) waves, which are of the longitudinal wave type, and *Secondary* (*S*) waves, which are of the transverse type. The terms longitudinal and transverse refer to the vibration direction in relation to the direction of propagation of the wave (Fig. 6.6). The *P* waves travel faster than the *S* waves and, therefore, arrive at a seismograph station first, hence the names, primary and secondary.

Important in the study of the deep interior of the earth is the fact that longitudinal (*P*) wave velocity increases with greater elasticity (rigidity and incompressibility) and, therefore, with greater depth in the earth, and that transverse (*S*) waves depend on resistance to shearing for their propagation. Fluids have no shear resistance and thus cannot transmit an *S* wave.

By tabulating the travel times of *P* and *S* waves from earthquakes of known sources, seismologists have constructed time-distance

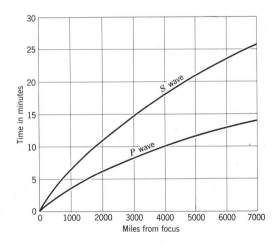

Figure 6.7 Graph of time versus distance traveled by *P* and *S* waves. (From Bullen, K. E., 1954, *Seismology*, Methuen and Co., London, p. 95.)

graphs from which the distance of a new quake can be determined (Fig. 6.7). These

Table 6.a Modified Mercalli Earthquake Intensity Scale*

I. Not generally felt.

II. Felt by persons resting, particularly on upper floors of buildings.

III. Felt indoors, especially on upper floors of buildings. May not be recognized as an earthquake. Vibrations resemble those of a passing light truck.

IV. Felt by some outdoors, by many indoors. Vibrations resemble those of a passing heavy truck, or a sensation of a heavy object striking the walls. Windows, dishes, and doors rattle or make creaking sounds. Hanging objects swing and stationary cars rock noticeably.

V. Felt outdoors; sleepers wakened. Liquids disturbed, some spilled. Small unstable objects displaced or upset. Doors swing, close, open. Shutters and pictures move, and pendulum clocks stop, start, or change rate.

VI. Felt by all, many frightened and run outdoors. People walk unsteadily. Windows, dishes, glassware broken. Objects fall off shelves and pictures fall off walls. Furniture moved or overturned. Weak plaster and poor masonry cracked. Small bells ring, and trees or bushes shaken visibly or are heard to rustle.

VII. Difficult to stand. Felt by drivers of cars. Hanging objects quiver. Furniture broken. Poor masonry damaged; fall of plaster, loose bricks, stones, tiles, cornices, and architectural ornaments. Weak chimneys broken at roof line. Waves on ponds; water turbid with mud. Sand and gravel banks cave or have small slides. Concrete irrigation ditches damaged. Large bells ring.

VIII. Difficult to steer motor cars. Ordinary unbraced masonry damaged or partially collapsed; some damage to reinforced masonry but no damage to masonry reinforced against horizontal displacement. Stucco and some masonry walls fall. Twisting or fall of chimneys, factory stacks, monuments, statues, towers, or elevated tanks. Frame houses move on foundations if not bolted in place, and loose panel walls dislocated. Weak piling broken. Branches broken from trees. Springs change flow, wells change level, and temperatures in both may change. Cracking in wet ground and on steep slopes.

IX. General panic. Poor masonry destroyed, good masonry damaged seriously. Frame structures, not bolted, shift off foundations; foundations generally damaged. Frames crack. Reservoirs seriously damaged. Conspicuous cracks in ground; underground pipes broken. In areas of loose sediment, sand and mud, and water ejected.

X. Most masonry and frame structures and foundations destroyed. Some well-built wooden structures and bridges destroyed. Serious damage to dams, dikes, embankments. Large landslides occur. Water thrown on banks of canals, rivers, lakes, and reservoirs. Flat areas of sand and mud shifted horizontally. Railroad tracks bent slightly.

XI. Few, if any masonry structures remain standing. Railroad tracks bent severely, underground pipes completely out of service, many bridges destroyed.

XII. Damage to man-made structures nearly total. Large rock masses displaced, lines of sight distorted, objects thrown into air.

*modified from Tocher, Don, 1964, Earthquakes and rating scales, *Geotimes,* v. 8, no. 8, p. 19: and Wood, H. O. and F. Neuman, 1931, Modified Mercalli intensity scale of 1931, *Bull. Seismological Society of America,* v. 21, pp. 277–283.

graphs show that the difference in time of arrival of *P* and *S* waves at a seismograph station is a function of the distance between the station and the focus of the earthquake. Since the speed with which *P* waves travel increases more rapidly with greater depth of penetration than the speed of *S* waves, it is clear that the farther away the station is from the shock, the greater will be the time lapse between the *P* and *S* waves. The seismic records indicate the distance but not the direction to the point of shock; therefore, seismograms from a minimum of three stations are necessary for precise location.

Earthquake Intensity. The energy released during an earthquake is a difficult thing to measure. Measurements of earthquake intensity or magnitude have been made in both quantitative and qualitative fashions. The widely used *Richter scale,* named after the California seismologist who developed it, describes the amplitude of the earthquake waves, and is related to the amount of energy released at the source of the earthquake. The scale describes magnitudes on a logarithmic base from 1 to 10 in which a magnitude of 7 indicates a disturbance 10 times as large as a magnitude of 6. On this scale an earthquake of magnitude 2.5 can be felt by persons nearby; a magnitude of 7 or over represents a major earthquake capable of extensive damage.

A qualitative earthquake intensity scale is illustrated in Table 6.a. It is the *Mercalli scale* and is used in situations where insufficient seismographs do not permit the more analytical approach to intensity determinations.

Earthquake intensity maps show the diminishing intensity of an earthquake with increasing distance from the epicenter. Lines connecting points of equal intensity are *isoseismal lines.* Figure 6.8 shows an example of such a map, constructed on the basis of the Hebgen Lake earthquake of 1959. In the epicenter area of this quake near the Wyoming-Montana border, an intensity of X on

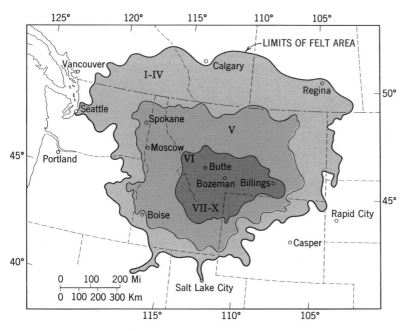

Figure 6.8 Earthquake intensity map of the Hebgen Lake Earthquake, August 17, 1959. The epicenter of this quake was in Yellowstone National Park in the northwestern corner of Wyoming near the Montana state line. Roman numerals on the map indicate earthquake intensities according to the Modified Mercalli scale (see Table 6.a for details). (After Eppley, R. A. and W. K. Cloud, 1961, *United States Earthquakes,* 1959, U. S. Coast and Geodetic Survey.)

Figure 6.9 Fault displacement associated with the 1964 Alaska earthquake. The white area in this photograph was formerly beneath the
sea in Hanning Bay on the west side of Montague Island in the Gulf
of Alaska. The right-hand side of the white area is the trace of the
Hanning Bay fault along which the movement took place. The white
area was uplifted 4 to 5 meters relative to the downthrow side of the
fault. The white material consists of the bleached coatings of calcareous algae and other marine organisms that lived below mean
tide level before faulting occurred. (Photograph by George Plafker,
U. S. Geological Survey, May 30, 1964.)

the modified Mercalli scale was assigned to it. The limits of the felt area extended over more than 500 miles from the epicenter. The quake was felt by persons as far away as Regina, Saskatchewan; Calgary, Alberta; Seattle, Washington; Salt Lake City, Utah; and Dickenson, North Dakota.

The Hebgen Lake earthquake caused considerable damage and loss of life. Surficial offsets along the fault plane were measured at 10 feet in a horizontal direction and 15 to 20 feet in a vertical direction. The most spectacular and disastrous effect of the earthquake was the dislocation of large masses of soil and rock that cascaded from the south wall of the Madison River Canyon. The debris roared down the steep canyon slope from 1300 feet above the canyon floor and formed a barrier across the Madison River, thereby creating a lake 175 feet deep and nearly 7 miles long. Most of the people killed as a result of the earthquake were engulfed in the Madison Canyon slide. The main shock occurred on August 17, 1959, but aftershocks continued almost daily until the end of the year.

For the Alaskan Good Friday earthquake of March 27, 1964, similar earthquake intensity maps have been prepared. The quake was centered near Anchorage, Alaska, and the limits of human perceptibility extended as far as Juneau and to the north of Fairbanks. The Good Friday quake had a Richter magnitude of 8.5 and created great havoc. It is estimated to have released, at least, twice the energy of the San Francisco earthquake of 1906, and it was felt over an area of nearly 500,000 square miles. The earthquake triggered large-scale avalanches, produced landslides (Fig. 9.3) and slumps causing great devastation, and sent tsunamis (seismic sea waves, see Chapter 14) as far as Japan, Hawaii, and southern California. Uplift and subsidence produced by the earthquake affected an area of approximately 34,000 square miles in south-central Alaska. Some areas were uplifted from a few feet to more than 30 feet (Fig. 6.9), while other tracts suffered as much as 5 feet of subsidence. Much of the damage was a result of horizontal and vertical movements produced by the failure of unstable alluvial deposits lying below the surficial

materials. Man-made structures on bedrock, although they sustained considerable damage, in general fared better than structures on alluvial materials, demonstrating that, in earthquake-prone areas, detailed geologic knowledge is essential in the selection of building sites. The effects of the Alaskan earthquake were intensively investigated and documented by a special task force of the United States Geological Survey. The published results of these studies are of great value in guiding the planning and land-use practices of earthquake-prone areas.

The San Fernando earthquake of February 9, 1971 served notice on a densely populated urban area of the potential for catastrophe that earthquakes represent. This earthquake had a magnitude of 6.4 on the Richter scale; its epicenter was located a few miles north of the San Fernando Valley on the northern extremity of Los Angeles. Although a shock of this magnitude is not regarded as a great earthquake, such as the one predicted to take place on the San Andreas fault just 33 miles northeast of Los Angeles, a considerable amount of damage was sustained. Sixty-four persons lost their lives, and property damage was estimated at about half a billion dollars.

In the San Fernando quake, the major loss of life was caused by the collapse of old structures that were built before rigid earthquake engineering standards were developed in California. Other similar structures, also located near the epicenter of the earthquake but which had been built to rigid specifications sustained great damage (Fig. 6.10), but loss of life was small. Highway and freeway damage (Fig. 6.11) was exceptionally severe, the estimated damage placed as high as 40 million dollars. Severe damage to a major reservoir (Fig. 6.12) required the temporary evacuation of 80,000 persons living below the dam while the water level in the reservoir was drawn down by pumping. Fortunately, no lives were lost, but the potential destruction from dam failures caused by an earthquake is obvious.

The San Fernando earthquake, with its very considerable devastation and relatively low Richter magnitude, illustrates the potential for a catastrophe of truly major proportions when a large earthquake is located near

Figure 6.10 Effects of February 9, 1971 earthquake on Olive View Hospital, San Fernando. Major features include right-handed shear of the main structure above the ground floor, collapse and 90° rotation of the large wing in left of the picture, and collapse of the ambulance port. (Photograph by Bradford S. Newman.)

Figure 6.11 Collapse of freeway overpass structure, San Fernando earthquake, February 9, 1971. (Photograph by Bradford S. Newman.)

Figure 6.12 Failure of the Van Norman reservoir dam, San Fernando earthquake of February 9, 1971. (Photograph by Bradford S. Newman.)

Table 6.b Earthquake Safety Rules (from United States Department of Commerce, Environmental Science Services Administration)

During the shaking:

1. Don't panic. The motion is frightening but, unless it shakes something down on top of you, it is harmless. The earth does not yawn open, gulp down a neighborhood, and slam shut. Keep calm and ride it out.

2. If it catches you indoors, stay indoors. Take cover under a desk, table, bench, or in doorways, halls, and against inside walls. Stay away from glass.

3. Don't use candles, matches, or other open flames, either during or after the tremor. Douse all fires.

4. If the earthquake catches you outside, move away from buildings and utility wires. Once in the open, stay there until the shaking stops.

5. Don't run through or near buildings. The greatest danger from falling debris is just outside doorways and close to outer walls.

6. If you are in a moving car, stop as quickly as safety permits, but stay in the vehicle. A car is an excellent seismometer, and will jiggle fearsomely on its springs during the earthquake; but it is a good place to stay until the shaking stops.

After the shaking:

1. Check your utilities, but do not turn them on. Earth movement may have cracked water, gas, and electrical conduits.

2. If you smell gas, open windows and shut off the main valve. Then leave the building and report gas leakage to authorities. Don't reenter the house until a utility official says it is safe.

3. If water pipes are damaged, shut off the supply at the main valve.

4. If electrical wiring is shorting out, shut off current at the main meter box.

5. Turn on your radio or television (if conditions permit) to get the latest emergency bulletins.

6. Stay off the telephone except to report an emergency.

7. Don't go sight-seeing.

8. Stay out of severely damaged buildings; aftershocks can shake them down.

urban centers. It obviously is essential that strict construction codes are established for all earthquake-prone areas and that funding for extensive research in earthquake prediction and prevention be continued and increased. If the San Fernando earthquake had been of a magnitude of the San Francisco or the Alaskan Good Friday earthquakes, it seems certain that deaths would have numbered in the hundreds, and property loss would have been several billions of dollars.

Persons living in earthquake-prone areas would do well to acquaint themselves with the earthquake safety rules recommended by the Environmental Science Services Administration of the United States Department of Commerce. Table 6.b summarizes what individuals caught in an earthquake should do during and after the shaking.

The Earth's Interior

In an earlier chapter the major subdivisions of the earth were defined as the crust, mantle, and core. The reader may wonder how such positive statements can be made about parts of the earth that have never been observed directly. The answers are provided by many different kinds of investigations, the most important of which are related to the analysis of earthquake waves.

Direct observations of the earth are limited to its outermost skin. The deepest mines are no more than 3 miles below the surface. Deep oil wells provide data from more than 25,000 feet in sedimentary basins, and some deep canyons expose rocks several thousand feet below the general surface. Very deep erosion in the ancient shield areas of some continents exposes rocks that can be inferred to have been formed 20 miles beneath the surface, and samples of volcanic materials and some diamond-bearing rocks provide us with "samples" of the mantle. These sources of information about the "depths of the earth" are too fragmental to provide general conclusions, so it is necessary to approach the study of the earth's interior by indirect means.

One of the indirect sources of information about the earth's interior are meteorites. They are generally believed to come from the core

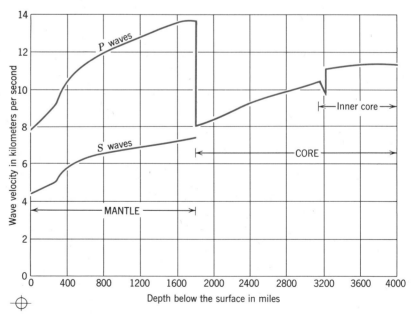

Figure 6.13 Graph showing the change in velocities of P and S waves at different depths below the surface. (Based on a diagram by Byerly, P., 1942, *Seismology*, Prentice-Hall, New York, 256 pp.)

THE DYNAMIC EARTH

of a fragmented planet or planetoid of the solar system. Meteorites, therefore, may bear some similarity to the core of the earth, based on the belief that members of the solar system have generally a common origin (Chapter 2). The physical and chemical properties of meteorites are in general agreement with what has been deduced about the earth's interior from other indirect evidence, the most important of which is derived from seismological studies.

Theoretical considerations confirmed by laboratory experimentation reveal that wave vibrations travel faster as the elasticity of the transmitting medium increases. It is also known that velocity decreases as density increases. Wave velocities increase with greater penetration within the earth, at least, to a depth of 1800 miles (Fig. 6.13). Hence, the deduction is that the elasticity of the earth increases more rapidly than density with depth. The question is, however, does the earth consist of a number of concentric shells, each of which permits waves to travel at a higher velocity than the shell immediately surrounding it, or is the increase in velocity gradual so that no sharp boundaries exist? Actually, the evidence points to both conditions; some definite boundaries (discontinuities) do exist, but between these the increase in velocity is gradual.

The Crust

The nature of the outer 10 to 30 miles of the earth can be deduced from several lines of seismic evidence. The Yugoslavian seismologist Mohorovičić, in examining seismic records from a great number of stations, observed that seismograms showed, at least, two pairs of *P* and *S* waves from a single earthquake. He deduced from this that the second pair must have been refracted from deep layers within the earth. For seismograms located at a greater distance from the epi-

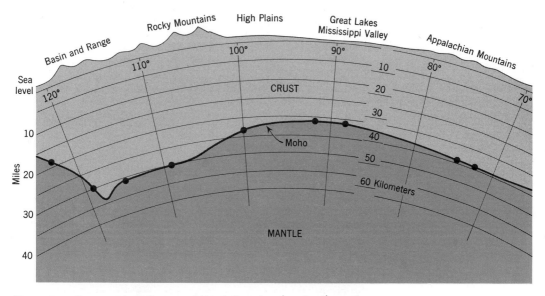

Figure 6.14 Cross section of a part of North America showing the variation in thickness of the continental crust as determined by seismic means. The black dots are points at which the base of the crust was actually established by seismic records. The vertical scale of the topography and crustal depths is exaggerated 20 times. The numbers along the top are degrees of longitude. (After Oliver, Jack, 1959, Long Earthquake Waves, *Scientific American*, v. 200, no. 3, p. 131-143. Copyright © 1959 by Scientific American Inc. All Rights reserved.)

center, however, he noticed that the pair of P and S waves which were "first" at the stations close to the epicenter were now "second." This, he reasoned, was because of the greater speed of travel possible at the greater depths, which allowed deep penetrating waves to speed up and "pass" the waves traveling a more direct route through the upper layers. Actually, Mohorovičić had noted three pairs of P and S waves from which he deduced the existence of the granitic crust (sial) underlain by a basaltic crust (sima) above the mantle. The base of the crust thus defined is known as the Mohorovičić discontinuity or Moho. At this boundary, 20 to 30 miles below the surface of continents, the velocities of both P and S waves increase abruptly. The speed of P waves increases from a velocity of about 6½ kilometers per second (4 miles per second) to about 8 kilometers per second (5 miles per second) as they cross the M-discontinuity at the base of the crust.

The depth to the Moho is greater beneath mountains than lowlands (Fig. 6.14) as deduced from reflected earthquake waves. It also lies deeper beneath the continents than beneath the ocean basins (Fig. 2.7). When scientists renew efforts to drill through the earth's crust to the Moho, they will undoubtedly select a drilling site in the ocean instead of on a continent.

That the crust beneath the continents differs from that beneath the oceans as indicated in Figure 2.7 is further suggested by two kinds of seismic data. The long-surface earthquake waves that travel through the rocks of the Pacific basin travel at speeds characteristic for basaltic material, suggesting that no sialic material is present there. In addition, for earthquake waves of equal epicentral distance, a seismic station on a continent records a greater intensity of reflected body wave energy than that of a Pacific station. The wave energy arriving at the continental station will have passed through the basaltic-granitic discontinuity, causing refraction; this refracted wave path strikes the surface at a higher angle, thereby producing larger reflected wave energy. Both these lines of evidence suggest a lighter granitic continent "floating" in a sea of basaltic material.

The Deep Interior

Earthquake waves that penetrate down to depths of 1800 miles are recorded at seismic stations up to about 103° away from the epicenter of the earthquake. Over this distance, stations record both P and S wave energy, indicating that the material they have passed through is solid. This zone of the earth, from the Moho to 1800 miles depth is the *mantle*. Passing through the mantle the P and S wave velocities increase gradually (Fig. 6.13). About 300 miles down there is an increase in the *rate* of velocity increase. There is no general agreement as to the significance of this flexure in the velocity-depth curve, but the

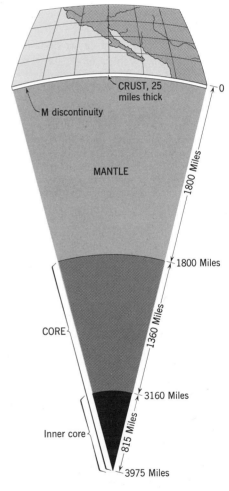

Figure 6.15 Diagram of a wedge of the earth showing the major zones and boundaries. For a summary of the physical properties of each see Table 6.c.

American geophysicist, Francis Birch, believes it indicates a boundary between materials of different composition rather than a boundary between two phases of the same material. About 500 miles down, the rate of velocity increase falls off somewhat.

From 500 miles to 1800 miles the mantle is apparently quite uniform in composition but increases in density. The most pronounced discontinuity below the Moho is the boundary between the mantle and the earth's *core*.

Earthquake waves that penetrate to depths of more than 1800 miles are drastically affected (Fig. 6.13). *P* waves penetrating to these great depths are next recorded on the surface at 143° away from the epicenter of the earthquake, but they arrive several minutes "late." *S* waves are recorded nowhere beyond 103° away from the epicenter. The 1800-mile-deep position must represent a major change in the earth's interior. The refraction of *P* waves and the damping out of *S* waves, so that a gap 40° in width on the earth's surface is produced in which no direct wave energy is recorded, is taken to mean a change from a solid mantle to a liquid *outer core* at a depth of 1800 miles. The gap of no record girdling the earth from 103° to 143° away from the epicenter of the earthquake is said to lie in the "shadow" of the earth's core, and is thus called the *shadow zone.*

The core material must be very dense, even if liquid, because the earth as a whole has a density of 5.5 whereas crustal rocks have densities of about 2.7 to 3.0; the difference must be made up in the deeper earth materials. The density of the core is probably 12 or more, but its exact composition is unknown. Some investigators favor an iron-nickel composition whereas others support the idea of an iron-silicate core.

Those *P* waves that penetrate directly through the core indicate an abrupt velocity increase at a "boundary" about 1360 miles below the mantle-core boundary (3160 miles below the earth's surface). A transition from an outer core made up of fluid nickel-iron to an *inner core* of solid nickel-iron would explain this change in velocity.

Table 6.c Summary of the Physical Properties of Various Parts of the Earth

Unit	Depth (Miles)	Description
Crust	0	Surface of the earth
		Sedimentary, igneous, and metamorphic rocks in the upper few miles. Granitic continents, basaltic ocean basins. Gradual increase in density with depth. Possible intermediate zone between base of granite and Moho.
Mantle	20–25	Moho (base of the crust)
		Rapid increase of velocity with depth between 100 and 500 miles. At about 300 miles there may be a compositional change. Rate of velocity increases slacks off between 500 and 1800 miles down.
Core	1800	Core boundary
		Material of the core lacks rigidity and is probably liquid. Density is 12 gm/cm³ or more. Inner core begins 3160 miles down where *P* waves increase in velocity, indicating a possible change back to the solid state. Both solid and liquid portions of the core may be of iron-nickel or iron-silicate composition.
	3975	Center of the earth

Modern views concerning the origin of the earth's magnetic field attribute it to electric currents generated by motions in the fluid core. A diagrammatic wedge of the earth is shown in Figure 6.15, and Table 6.c summarizes the physical properties of the different zones of the earth.

References

* Bullen, K. E., 1955, The interior of the earth, *Scientific American,* V. 193, No. 3 (September), pp. 56–61. (Offprint 804, W. H. Freeman and Co., San Francisco.)

* Elsasser, Walter M., 1958, The earth as a dynamo, *Scientific American,* V. 198, No. 5 (May), pp. 44–48. (Offprint 825, W. H. Freeman and Co., San Francisco.)

* Grantz, Arthur, George Plafker, and Reuben Kachadoorian, 1964, *Alaska's Good Friday earthquake, March 27, 1964,* U.S. Geological Survey Circular 491, 35 pp.

Hansen, Wallace R., 1965, *Effects of the Alaska earthquake of March 27, 1964 at Anchorage, Alaska,* U.S. Geological Survey Professional Paper 542-A, U.S. Government Printing Office, Washington, D.C., 68 pp.

Hart, P. J. (ed.), 1969, *The earth's crust and upper mantle,* Geophysical Monograph 13, American Geophysical Union, Washington, D.C., 735 pp.

* Hill, Mary R., (ed.), 1971, The San Fernando earthquake 1971; *California Geology,* V. 24, No. 4–5, pp. 59–88, Calif. Division of Mines and Geology, Sacramento, Calif.

Hodgson, J. H., 1964, *Earthquakes and earth structure,* Prentice-Hall Inc., Englewood Cliffs, New Jersey, 166 pp.

Iacopi, Robert, 1964, *Earthquake country,* Lane Book Co., Menlo Park, California, 191 pp.

* Leet, L. Don, 1948, *Causes of catastrophe,* McGraw-Hill Book Co., New York, 232 pp.

Oliver, Jack, 1959, Long earthquake waves, *Scientific American,* V. 200, No. 3 (March), pp. 131–143. (Offprint 827, W. H. Freeman and Co., San Francisco.)

Phillips, O. M., 1968, *The heart of the earth,* Freeman, Cooper and Co., San Francisco, 236 pp.

* Steinbrugge, K. V., 1968, *Earthquake hazard in the San Francisco Bay area: a continuing problem in public policy,* Institute of Governmental Studies, University of California, Berkeley, 80 pp.

Steinhart, J. S., and T. J. Smith (eds.), 1966, *The earth beneath the continents,* Geophysical Monograph 10, American Geophysical Union, Washington, D.C., 663 pp.

* Recommended for additional reading.

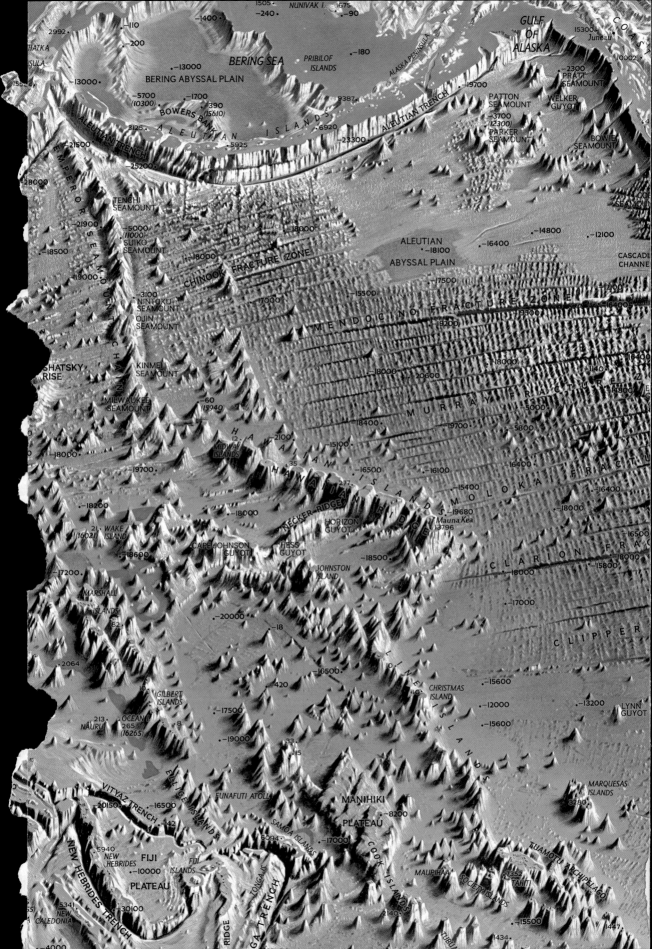

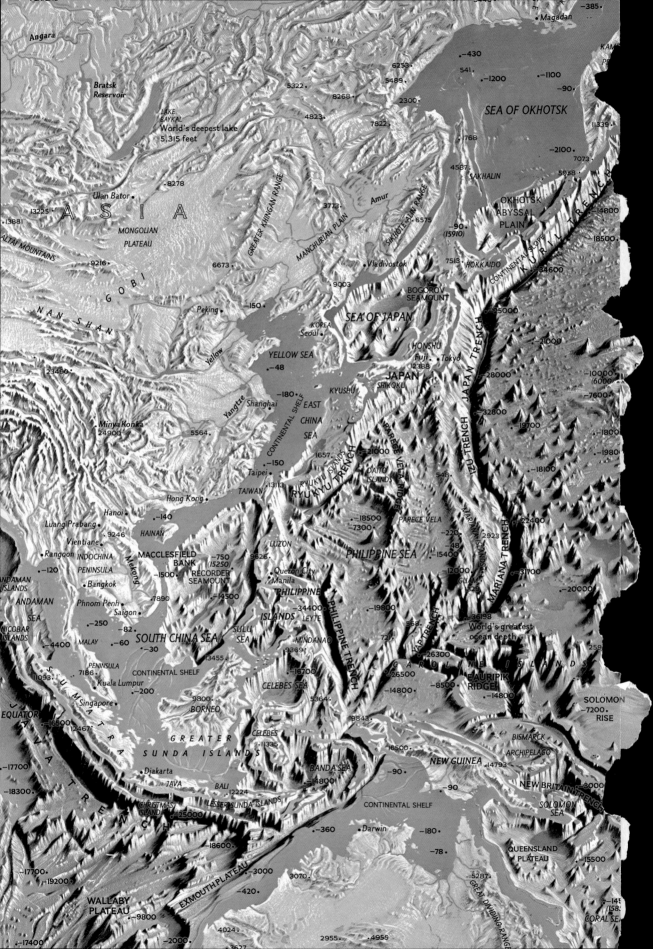

7 Global Tectonics

Tectonics is a word that refers to the major architectural features of the earth's crust. It includes a description of the broad geologic features of continents and ocean basins as well as an explanation of the forces responsible for their origin and distribution over the face of the earth. Prior to the space age, many geologists tended to view tectonics on a scale of continental or smaller dimensions, but beginning with the decade of the 1960's geologists and geophysicists began looking more closely at the grand tectonic framework of the entire earth's crust. *Global tectonics,* therefore, attempts to explain the present configuration of continents and ocean basins and the geologic phenomena associated with them in the light of dynamic geologic processes that have gone before as well as those that are operating now.

Chapter 2 introduces some basic concepts about global tectonics with particular atten-

Figure 7.1 Folded marine sedimentary strata in the Canadian Rocky Mountains of British Columbia. (Courtesy of E. L. Fitzgerald, Canadian Superior Oil, Ltd., Calgary, Alberta. This same photograph appeared as Figure 7 in Fitzgerald, E. L. and L. T. Brown, Disharmonic folds in Besa River Formation, Northeastern British Columbia, Canada, 1965, *Bulletin American Association of Petroleum Geologists*, v. 49, pp. 418-432.)

tion to the origin of continents and ocean basins, and Chapter 4 describes some of the geologic phenomena of the earth's crust. In this chapter the origin of mountainous regions and the evidence for continental drift are the main topics of discussion.

THE ORIGIN OF MOUNTAINS

Most mountain regions of the earth show evidence of crustal deformation on a large scale. Sedimentary strata have been con-

torted into intricate folds (Fig. 7.1), evidence for faulting on a large scale is commonplace, and igneous intrusions and volcanic eruptions are much in evidence. The geologist believes that if he can understand the complex relationships of these features as they reveal themselves in mountainous regions, he will be able to answer the question of how the mountains themselves came into being.

Not all the evidence bearing on the origin of mountains comes from existing mountain ranges, however. There are parts of the earth's surface that were occupied by ranges of lofty peaks in past geologic time. Those peaks are gone now because millions of years of erosion have finally removed them entirely or worn them down to mere hills. The sites of these ancient mountains still bear witness to the forces that produced them because their "roots" are still present, even though their topographic expression has vanished.

On the other hand, some earth features that are mountains in the topographic sense, as for example the peaks and pinnacles of the Grand Canyon, are of little value as a source of evidence to explain the intricately folded and faulted character of most other mountains. The "mountainous" aspects of the Grand Canyon have been produced by relentless erosion of horizontal strata that have not undergone much deformation since they were deposited as marine sediments long ago.

Types of Mountains

The chief characteristic of mountains from the tectonic point of view is the presence of certain geologic features that bear on their origin. These features include evidence of folding, faulting, igneous activity, or a combination of these elements. Although mountains usually exhibit all three of these components, it is possible to classify mountains on the basis of the most predominant geologic feature present. A simple classification, therefore, consists of five major types: fold mountains, dome mountains, fault block mountains, volcanic mountains, and complex mountains.

Fold Mountains. Examples of fold mountains of generally simple structure are the Appalachian Mountains of the eastern United States in which a series of parallel anticlinal and synclinal structures have produced a linear system extending from Alabama to New England. The simplicity of these structures is locally modified by extensive overthrust faults, a common associate of this style of mountain structure. Many of the Rocky Mountain ranges of the western United States exhibit broad fold structures modified by faults along their margins.

Dome Mountains. The Ozark Plateau of Missouri and Arkansas is a broad uplift in which strata dip away from a common center. Similar structures are also present in the Rocky Mountain region (Fig. 3.25).

Fault Block Mountains. The Basin and Range province (Fig. 19.5) of the southwestern United States is characterized by alternating mountain ranges of internally complex structures and parallel intervening valleys. The ranges were formed by uplift along normal faults on one or both their flanks. The east side of the Sierra Nevada in California, for example, has been produced by fault movement of this type. In contrast to the compressional forces indicated by many fold mountains, the fault block types may be due to horizontal elongation (tension) in the crust.

Volcanic Mountains. Some regions of the earth (for instance, the Hawaiian Islands and Iceland to name but two) are more-or-less dominated by volcanic processes. Others such as the volcanic cones of the Cascade Range of northwestern United States have volcanic accumulations atop geologic structures of great variety and complexity. (Volcanic mountains are treated more fully in Chapter 5.)

Complex Mountains. Mountain areas such as the Alpine chain of western Europe, the eastern part of New England, and the western ranges of North America, are systems in which a great variety of complex structures are found. Rock structures in those areas are incredibly complex, highly metamorphosed, and are nearly always accompanied by vast volumes of intrusive granitic rocks. It is these mountains, along with information about geologic processes operating on the ocean floors, that have provided the major clues to the understanding of global tectonics.

Characteristics of Mountain Systems

Any attempt to explain mountains and their origin must take into account those features, here greatly simplified, in which mountains differ from the low-lying areas of the earth. If we think of the geologic contrasts between the western one-fourth of the United States and the central interior as an example, we see that (1) the granitic crust (sial) is thicker

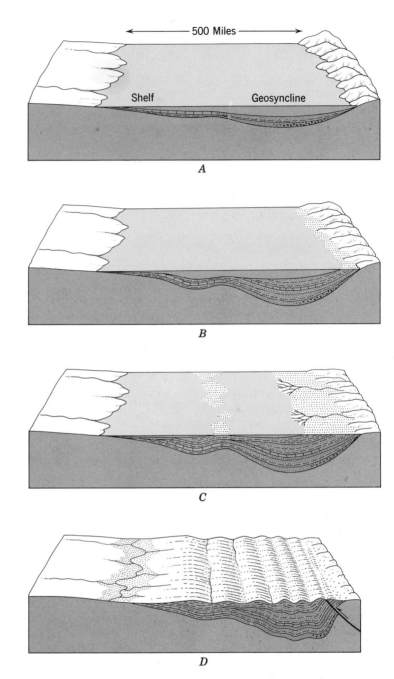

Figure 7.2 Series of block diagrams showing successive geologic episodes in the development of a folded mountain range from geosynclinal sediments.

beneath mountains than beneath plains (Fig. 6.14), (2) the stratified rocks in mountains are more deformed than those of plains areas and show great horizontal shortening, (3) stratified rocks of the mountains are generally ten times (or more) thicker than equivalent age rocks elsewhere, and (4) granitic rocks commonly occur at the cores of mountainous regions.

The Geosynclinal Concept

In the middle nineteenth century, a young New York geologist, James Hall, described many of the features listed above. He noticed that the rocks of the Appalachian Mountains of eastern New York were vastly thicker than equivalent rocks farther west, that they tended to be more coarsely detrital in the east, suggesting an eastward source, and that they were strongly deformed. Furthermore, he speculated that these sedimentary rocks, which contain many shallow water features such as mud cracks, ripple marks, coal and delta deposits, must have accumulated at the same rate as the surface on which they were deposited was sinking. Hall believed that the sinking of the basin was the result of the weight of the sediments being deposited. Physical considerations suggest that some more fundamental process is depressing the basin and that sedimentation merely keeps pace with subsidence. The observations made by Hall in New York were later duplicated all along the trend of the Appalachian Mountains. Hall is credited with the introduction of the idea of an elongate subsiding basin or geosyncline extending along the entire eastern edge of the North American continent long before the Appalachian Mountains came into being (Fig. 7.2).

Subsequent to Hall's work, it has been established that all mountain range systems possess characteristics more or less similar to those of the Appalachians. The geosynclinal concept is now universally accepted as an explanation of the accumulation of the thick layers of strata from which mountains are made. This is not to say that the geosynclinal concept explains the fundamental forces involved in mountain building, however, but rather, is the first in a series of events that leads eventually to the growth of a mountain range.

A further significant feature of geosynclinal areas are the sporadic episodes of crustal deformation separated, at least, at any given locality, by periods of quiescence. These episodes of deformation during which rocks were folded, metamorphosed, and intruded by molten magmas are termed orogenies, and they occur throughout the entire geologic record (Fig. 15.9). Their significance in piecing together the geologic history of any given area on earth is considered in more detail in Chapter 15.

Some geologists have pointed out a threefold history of geosynclinal development: first, a geosynclinal phase, characterized by extensive marine sedimentation; second, an orogenic phase during which the major deformation takes place as well as extensive marine volcanic activity and the intrusion of the earliest granitic bodies; and third, a plutonic phase, during which large portions of the geosynclinal sediments are invaded by massive granitic intrusives. To these could be added an uplift phase, as most mountain systems have experienced broad regional uplift late in their development.

This sequence of geosynclinal development cannot be applied rigidly to all parts of any given geosyncline, but it does describe the major processes at work in geosynclinal areas.

Theories of the Origin of Mountains

The Contraction Theory. The idea of a shrinking earth was postulated more than a hundred years ago. The idea still has a few adherents even though it has serious defects. The root of the contraction theory is the basic assumption that the earth was entirely molten at one stage in its very early history, an assumption accepted by most geologists, even by those who oppose the contraction theory. After the earth's crust solidified because of cooling, heat loss from the still molten interior caused the globe to contract.

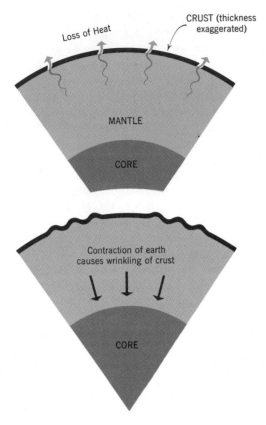

Figure 7.3 Cross sections of a part of the earth showing the contraction hypothesis as an explanation of the forces that cause crustal deformation. This hypothesis has been discarded.

Shrinking of the inner sphere set up compressional stresses in the crust, resulting in crumpling and folding of some of the crustal segments, much like the wrinkling of the skin of an apple when its interior shrinks while drying. But, unlike the skin of the apple which is uniformly wrinkled, the earth's "skin" contains only a few major wrinkles with vast areas of unwrinkled areas in between (Fig. 7.3). This, the contractionists argue, is because the original "skin" of the earth possessed certain zones of weakness marginal to the more rigid continental blocks. The weak zones represent the belts of folded mountains lying along the margins of the continents, as for example, the Rockies and Coast Ranges of western United States and the Appalachians of eastern North America. The theory was later modified to include periods of no compressional stress during which large blocks of crustal material collapsed or subsided, thereby explaining vertical crustal movements as well.

The major objection to the contraction theory is the very real possibility that the earth's crust is being heated by radioactivity just as fast or faster than it is losing heat. If this is true, the hypothesis is untenable because the earth would not be losing any heat and, therefore, could not be shrinking. In addition, the contraction hypothesis takes no account of the numerous normal faults of the earth that are the result of tension, a stress more likely produced by expansion rather than by shrinking.

The Convection Hypothesis. An ingenious explanation for the earth's mountain belts and other puzzling features of the globe was first introduced in the early 1930's by the Dutch geologist Vening Meinesz. This idea was based on the concept that heat is transmitted upward through the mantle to the base of the crust by convection. A heat source in the mantle, perhaps a center of intense radioactivity, was thought to generate movement in the form of giant convection cells. The mantle, being at a high temperature, would act like a very viscous material, and the convection current would, on reaching the base of the crust, spread laterally and exert a frictional drag on the crust. Velocities of rock flowage in the mantle would be of the order of a few centimeters per year according to reasonable estimates.

Where two laterally moving currents met, they would plunge downward (Fig. 7.4) and tend to drag the crust downward also. The deep oceanic trenches of the Pacific were considered a possible confirmation of this hypothesis. There, gravity measurements indicate a deficiency of normal gravity, termed a *negative gravity anomaly,* beneath the trenches. Such a deficiency may mean a greater than average depth to any particular density layer that could have been produced by a downward bending as envisioned in Figure 7.4.

If such convection cells can be maintained over a period of tens of millions of years or more, they can be invoked to explain all the pertinent features of mountain systems. The

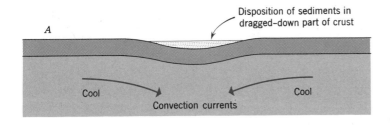

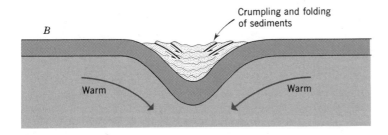

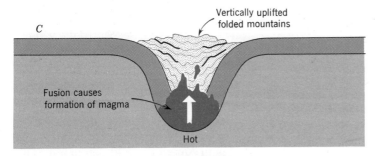

Figure 7.4 Sequence of diagrams showing the way in which convection currents could deform the earth's crust and cause folded mountains. (After Griggs, D., 1939, A theory of mountain-building, *American Journal of Science*, v. 237, pp. 611-650.)

early stage (Fig. 7.4*A*) provides the basin in which the geosynclinal sediments accumulate; the intermediate stage (Fig. 7.4*B*) accounts for the orogenic phase in which geosynclinal sediments are intensely deformed; and the final stage (Fig. 7.4*C*) explains the origin of granitic intrusive masses by down-bending of geosynclinal materials to depths where they are melted by high temperatures. The uplift phase of geosynclinal history (not represented in Figure 7.4) is accounted for by the diminution of the force that drives the convection movement. The light crustal materials that were pulled downward by convective force would rise isostatically after the convective force had died out.

Where convection currents rise to the base of the crust, they would bring excess heat to the surface. Such excess heat flow has been measured in many places along the mid-oceanic ridge systems to be discussed later. It is interesting to note also that the oceanic trenches mentioned above are sites of deficient heat flow, precisely what could be predicted from the convection hypothesis.

Convection currents provide a force capable of deforming the earth's crust without the necessity of having the force transmitted by the crust. The crust is considered to be too weak to transmit its own force of deformation.

Continental Drift. In 1912, Alfred Wegener, a German meteorologist, proposed his version of drifting continents. Earlier, others had speculated on this possibility because of

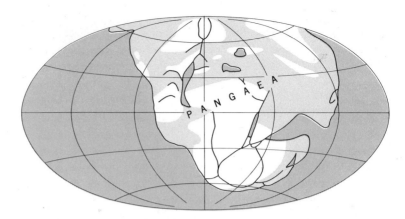

Figure 7.5 Pre-continental drift arrangement of continents according to Wegener. (After A. Wegener, from a drawing in DuToit, A. L., 1937, *Our wandering continents: An hypothesis of continental drift,* Oliver and Boyd, Edinburgh, 366 pp.)

the apparent geographic fit of the two sides of the South Atlantic Ocean. Wegener proposed, in addition, that not only were South America and Africa closer together than today, but that these continents, plus Australia, part of Asia, and Antarctica were part of a single protocontinent, which he called Pangaea (Fig. 7.5).

Wegener believed that the centrifugal force of the earth's rotation and tidal forces, acting on the sialic continental material, were sufficient to produce horizontal "gliding." Even though these forces are known to be far too weak to produce such motion, the idea was appealing because it appeared to afford an explanation of mountain structures. Wegener believed that the Cordilleran ranges of the Americas were produced as the crumpled leading edge of the North and South American continental plates "plowed" westward over a sea of subcrustal material. He explained the structures of the Alpine-Himalayan chain by the colliding of Africa and India against the Eurasian continent.

The theory was generally popular among European, South American, and South African geologists, but it fared poorly in North America at first. During the decade of the 1960's it experienced a great revival, largely on the basis of new evidence that will be discussed later in this chapter under the heading of *sea-floor spreading.*

Many geologists had accepted the validity of continental drift long before the sea-floor spreading concept developed, however. The reasons for acceptance then are still valid today. They are as follows:

1. *Map Fit.* The present geographic shorelines of western Africa and eastern South America would permit a reasonably good fit if the continents were moved together. The "fit" is greatly improved when the two continents are matched at the −3000-foot contour, a fit (Fig. 7.6) that is so remarkable it hardly can be the result of chance.

2. *Geologic Similarities.* Many of the rock strata of the same geologic age in South America, Africa, and Australia are remarkably similar. Red bed and volcanic rock sequences can be "matched" from one continent to another, a relationship that suggests they were formed in a single environmental basin that no longer exists because it has been fragmented by continental drifting.

An example of this technique of continental reconstruction is shown in Figure 7.7. Here, using a plausible predrift continental arrangement, the distribution of the Devonian Old Red Sandstone (350 m.y.) in the British Isles and its presumed equivalents in North America and elsewhere are explained as be-

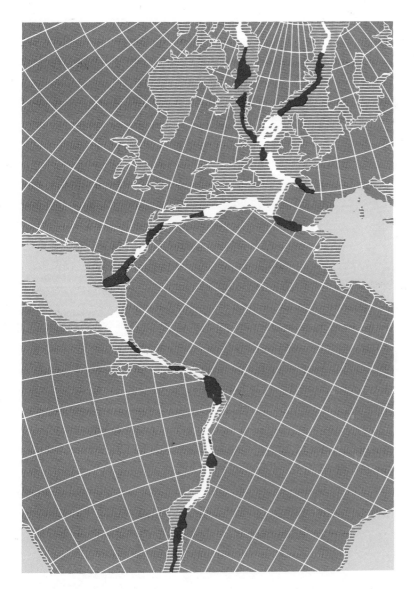

Figure 7.6 Possible arrangement of continents prior to the formation of the Atlantic Ocean. The continents are fitted together at the edges of their continental shelves (horizontally ruled areas) instead of the present shoreline. Gaps in the fit are shown in color and overlaps are shown in grey. (After Bullard, E. C., J. E. Everett, and A. G. Smith, 1965, The fit of the continents around the Atlantic, from Blackett, P. M. S., E. Bullard, and S. W. Runcorn (eds.) *A symposium on continental drift,* Phil. Trans. Royal Soc. London, v. 258, A1088, Fig. 8. Also in *The Origins of Oceans* by Edward Bullard. Copyright © 1969 by Scientific American, Inc. All Rights reserved.)

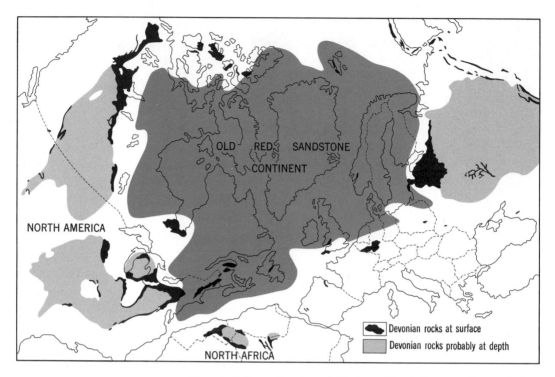

Figure 7.7 Reconstruction of Northern Hemisphere continents during the Devonian period (345 to 400 million years ago). The hypothetical Old Red Sandstone Continent is restored on the basis of the known distribution of sedimentary rocks of Devonian age at the earth's surface (black) and at depth beneath younger rocks (grey). This reconstruction assumes that the present Atlantic Ocean did not exist at this point in the geologic history of the earth, and that Europe and Africa were nestled closed to North and South America according to the hypothesis of continental drift. (From House, M. R., 1968, *Continental drift and the Devonian System*, University of Hull, Yorkshire, England, 24 pp.)

ing derived from a single landmass. This explanation of the origin of the similar red bed deposits of Great Britain and the New York-New England region is more credible than one that requires the continents to have remained in a fixed position since the beginning of earth history.

3. *Biologic Similarity.* Among the earliest evidences for the Gondwanaland continent are the close similarities in South America, Africa, India, and Antarctica of terrestrial plants, the Late Paleozoic *Glossopteris* fernlike flora, not found in the Northern Hemisphere, and the occurrence in both Brazilian and South African Permian rocks (250 m.y.) of the terrestrial aquatic reptile, *Mesosaurus.*

This creature could not have become dispersed across an intervening marine basin. Modern proof lies in the discovery of the fossil remains of *Lystrosaurus* in the Antarctic in 1969. This animal was a *land-dwelling* reptile of the Triassic age (200 m.y.), and lived on other continents in the Southern Hemisphere that now are separated geographically by wide expanses of the oceans. Had the continents been in their present relative positions during Triassic times, 200 million years ago, *Lystrosaurus* could never have become so widely dispersed.

4. *Climatic Evidence.* Many present geographic distribution patterns of desert

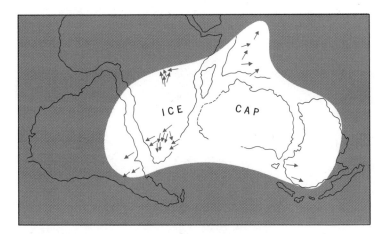

Figure 7.8 In parts of Africa, South America, Australia, Antarctica, and India, deposits of an ancient glacial ice cap were laid down during the Pennsylvanian and Permian geologic periods (225 to 310 million years ago). The distribution of these glacial deposits requires a closer grouping of the continents of the Southern Hemisphere such as is shown here. The arrows show the direction of movement of the ice cap as deduced from scratched and grooved rock surfaces on which the glacial deposits are lying. (After Du Toit, A. L., 1937, *Our wandering continents: an hypothesis of continental drift,* Oliver and Boyd, Edinburgh, 366 pp.)

sands, evaporite deposits, and coral reefs in ancient rocks are difficult to explain by modern geography. Most of their inexplicable distributions can be "solved" by one or more of the predrift land patterns that have been proposed.

The most convincing climatic arguments come, however, in consideration of ancient (Permo-Carboniferous) glacial deposits of the Southern Hemisphere. These rocks, clearly the result of the actions of continental glaciers on lowland areas, are now known from Australia, South America, Africa, and Antarctica. Those in Africa are now found at approximately 10° south latitude, far too close to the equator for continental glaciers to have existed. In addition, some of the glacial deposits in South America give evidence of having been deposited by ice moving onto the land from what is now the South Atlantic Ocean, a clear impossibility. These glacial anomalies can be satisfied if we assume that the continents were together and closer to

the south polar region before drifting began (Fig. 7.8).

5. *Structural Evidence.* Many of the mountain ranges and ancient mountain structures bordering the Atlantic Ocean run directly out to sea and appear to end abruptly. With the continental restoration shown in Figure 7.9, these structures fit quite closely. Mountain systems with similar structural histories are very compatible with the arrangement. Many of the rocks of the African and Brazilian shields can be closely matched in terms of rock type and orogenic episodes.

6. *Paleomagnetic Evidence.* One of the first modern studies that gave renewed impetus to continental drift was that of *paleomagnetism.* It refers to the history of the earth's magnetic field and is based on relict magnetism in certain rock types, particularly basic volcanic rocks.

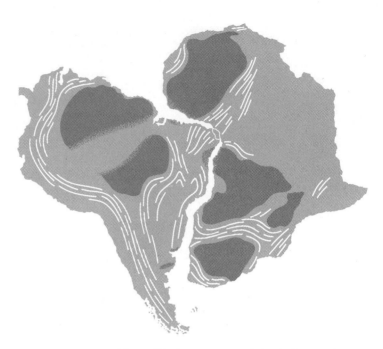

Figure 7.9 A reconstructed fit of Africa and South America. Colored areas indicate ancient stable areas on both continents, and the heavy lines show the trends of folded mountain systems of various geologic ages. The close geographic fit and the close match of major geologic features of both continents is very strong evidence in support of the continental drift hypothesis. (After Hurley, P. M., 1968, The confirmation of continental drift, *Scientific American*, v. 219, no. 6, December, pp. 52-64. Copyright © 1968 by Scientific American, Inc. All Rights reserved.)

The earth is essentially a large dipole magnet, and there is good reason to assume that throughout geological time the north and south poles of this magnet, the *geomagnetic poles,* have been more-or-less coincident with the poles of the earth's axis of rotation. It is known that the earth's rotational poles are not exactly coincident with the magnetic poles today, but the deviation is such that when considered over a period of 1000 to 10,000 years, the earth behaves as a uniformly magnetized sphere whose axis, called the *geomagnetic axis,* is so nearly coincident with the axis of rotation that variations can be ignored.

The positions of the geomagnetic poles can be determined from the relict magnetism in ancient rocks. Rocks owe their magnetic properties to the presence of small particles of magnetic minerals such as magnetite or hematite. In lavas, for example, these magnetic minerals become oriented in the earth's magnetic field that exists when the lava crystallizes. Providing no fundamental changes have been produced by subsequent geologic events such as remelting, the rocks retain a magnetism reflecting the pole positions at the time of formation. The relict magnetism of many rocks of different ages shows that the earth's magnetic poles have not always occupied the same positions in reference to the present geographic distribution of landmasses. A plot of the magnetic-pole positions of various times defines a path of *polar wan-*

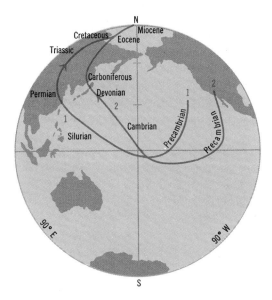

Figure 7.10 These curves trace the path of North Magnetic Pole from the Precambrian to the present as inferred from the paleomagnetic properties of rocks of different geologic ages. Line 1 is based on North American data, and line 2 is based on European data. (Cox, A. and R. R. Doell, 1960, Review of paleomagnetism, *Geol. Soc. America Bull.*, v. 71, figure 33.)

dering (Fig. 7.10). Its application to continental drift lies in the observation that polar wandering curves that are based on paleomagnetic data from different continents do not coincide. This discrepancy can be resolved only when it is assumed that the positions of the continents were changing ("drifting") relative to each other.

Sea-Floor Spreading. During the decade of the 1960's, the stimulating concept of sea-floor spreading took form. This hypothesis draws not only on the previous ideas of convection currents and continental drift but also embodies two new discoveries resulting from widespread oceanographic exploration. The first discovery was the recognition of a world-wide system of mid-oceanic ridges (Fig. 7.11) that everywhere are characterized by a rift valley along the axis of the ridges. The oceanic ridges are offset by numerous transverse fractures (Fig. 7.12) along which the distribution of earthquake epicenters suggests a movement of earth materials in a *transform* fashion. These ridges and transform offsets on the floors of parts of the Atlantic and Pacific oceans are vividly shown, although exaggerated, in Plates II and III.

The second major contribution to the hypothesis of sea-floor spreading was the discovery of symmetrical patterns of magnetic reversals on the sea floor (Fig. 7.13). These linear belts, determined by magnetic surveys, are displaced on both sides of oceanic ridges and represent rocks with normal polarity (as today) alternating with rocks of reverse magnetic polarity.[1] Terrestrial rocks also display reversals of the earth's magnetic field. These reversals have been recognized in rocks that can be dated with confidence back to about 3.5 million years ago, during which time there have been three reversals.

The present period of "normal" polarity began approximately 0.7 million years ago. Reversals have been recognized in rocks that are, at least, 76 million years old, since which time there have been 171 reversals of the magnetic poles.

If we return to the idea of convection currents, we will recall that mid-ocean ridges are zones of excessive heat flow. This suggests that volcanic and plutonic material rises beneath the ridges; on cooling on the sea floor as lava, this material will have impressed upon it the earth's magnetism of that time. As convection continues, however, the new material will force the earlier cooled material laterally, away from the ridge axis. If, during the passage of time, the earth's polarity has become reversed, a new magnetic polarity will be recorded in the rocks (Fig. 7.14). All this suggests a mechanism for drifting continents. It is visualized that the continental masses are carried along atop a layer of laterally migrating subcrustal material. Because a chronology of polarity reversals in terrestrial volcanic rocks has been established, it is possible to estimate that the rate of sea-floor spreading ranges from a few to several centimeters per year, sufficient to produce large-scale crustal migration over geologic time.

[1] Reversed polarity means that the north and south magnetic poles were reversed. The north *magnetic* pole was located at the south *geographic* pole. The cause of magnetic reversal is unknown.

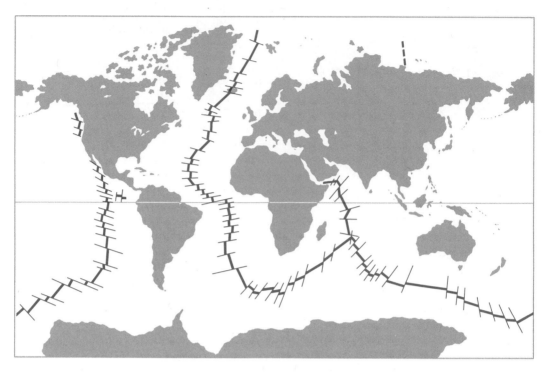

Figure 7.11 Worldwide system of submarine ridges and rises. Only a few of the numerous offsets are shown. See Plates II and III for greater detail of the transverse fractures along the ridge system. (After Drake, C. L., 1964, World rift system, *Transactions American Geophysical Union,* v. 45, no. 3, p. 436.)

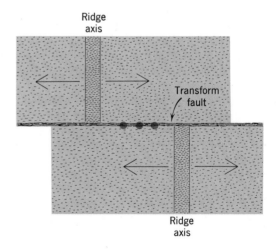

Figure 7.12 Simplified map of a part of the sea floor showing the offsetting of a segment of an ocean ridge along a transform fault. Earthquake epicenters (black dots) coincide with the trace of the fault along which the movment on each side is in the opposite direction. Arrows show direction of lava-spreading away from the ridge axes. (After Heirtzler, J. R., 1968, Sea-floor spreading, *Scientific American,* v. 219, no. 6, December, pp. 60-70. Copyright © 1968 by Scientific American, Inc. All rights reserved.)

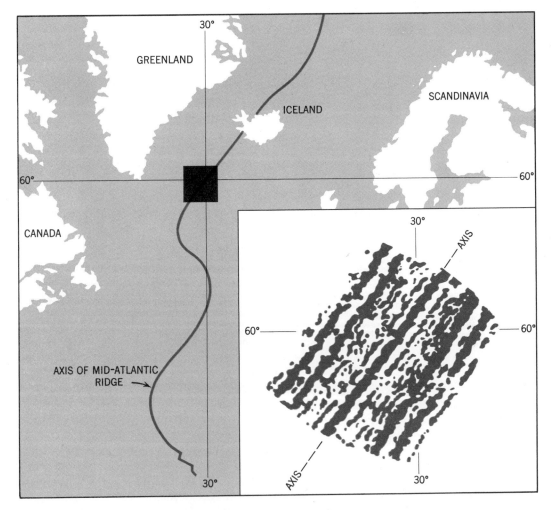

Figure 7.13 Map of sea-floor magnetic reversals on a part of the Mid-Atlantic ridge southwest of Iceland. Dark bands represent normally polarized belts, and light bands represent reversely polarized belts. The symmetry on each side of the central ridge is evident. (After Heirtzler, J. R., X. Le Pichon, and J. G. Baron, 1966, Magnetic anomalies over the Reykajanes Ridge, *Deep-sea Research,* v. 13, pp. 427-443.)

This type of spreading is confirmed by numerous observations. In some areas aligned volcanic cones within ocean basins are seen to be progressively older as distance from the ocean ridge increases. Drilling on the sea floor has shown that sediment thickness on the flanks of ocean ridges is small, but increases systematically farther from the ridge axis, suggesting that the areas of thicker sediment have been receiving sediments for longer periods as they slowly migrated away from the ridge area.

Drilling into the ocean floors has provided a further interesting clue as to the nature of sea-floor spreading. While the rocks of the continents are known to be more than 3 billion years old, no rocks on the ocean floors older than about 150 million years have been found. This indicates that the modern ocean basins are youthful earth features. Moreover, although new oceanic material is being produced along the mid-ocean ridges, it is destroyed elsewhere. This has suggested a process called *plate tectonics,* in which it is

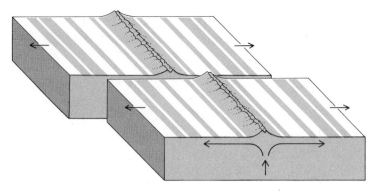

Figure 7.14 Block diagram of part of a mid-ocean ridge that has been offset along a transform fault. Upwelling of lava from beneath the central ridge supplies new material to the sea floor. The newly formed sea-floor lavas on each side of the ridge are magnetized during the cooling of the lava. This magnetization is "frozen" into the lava and will be normal or reversed, depending on whether the earth's magnetic field is normal or reversed at the time of cooling. Paired bands or stripes of lava on both sides of the ridge are shown in color if they have normal polarity and white if they are reversed. Each band, normal or reversed, is carried away from the central ridge by spreading of the sea floor as shown by the arrows. The magnetic lava bands farthest from the ridge are the oldest and their age has been determined by independent means, thereby permitting the establishment of a rate of sea-floor spreading on the order of a few to several centimeters per year.

visualized that several large plates or semi-rigid slabs of oceanic crust move away from the ocean ridges toward an ocean trench where they plunge downward owing to "collision" with another large plate. The continental crustal masses are carried along on the oceanic plates. The so-called American plate, comprising North and South America and the North and South Atlantic basins west of the mid-Atlantic ridge, is apparently moving westwardly. Along the west coast of South America the American plate apparently overrides the Pacific plate, forming a deep oceanic trench of large magnitude (Plate III). On the west coast of North America that part of the American plate east of the San Andreas fault is sliding horizontally past the Pacific plate, whose landward expression is the area west of the San Andreas fault. The East Pacific rise (Fig. 7.11) enters the Gulf of California, and its apparent extension appears again off the coast of northern California. The area between is occupied by the San Andreas fault

(Fig. 6.4). This major fault can thus be interpreted as a large transform fault offsetting a major oceanic ridge.

An interesting fact regarding sediment thickness off the Pacific coast may be explained by the collision of the American and Pacific plates. It is known that sediment volume off the west coast of the United States is one-sixth or less than off the east coast. It is believed that large volumes of sediment may be carried down, and, therefore, "lost" from view, along the steeply inclined (eastward) fracture zone separating the overriding plate and the overridden plate.

Figure 7.15 shows in schematic fashion a segment of the crust at two stages in the plate tectonics process. In stage A, a new rift system results from the breakup of a single continent. Stage B shows the creation of a new ocean basin by the further drifting apart of the two segments of the original continent. In addition, a new oceanic trench is produced by the impingement of crustal plate I against

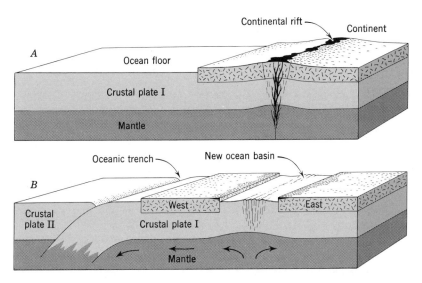

Figure 7.15 Schematic representation of plate tectonics. See text for explanation. (After Dietz, R. S. and J. C. Holden, 1970, Reconstruction of Pangaea: breakup and dispersion of continents, Permian to present, *Jour. Geophysical Research,* v. 75, no. 26, pp. 4940-4956.)

crustal plate II (not shown in stage A). Crustal plate I descends beneath crustal plate II and is absorbed in the mantle.

Figure 7.16 illustrates a set of restorations made by the American geologists, R. S. Dietz and J. C. Holden, in which an original landmass, Pangaea, is shown as it may have appeared in the Permian period, 225 million years ago (7.16*A*). This landmass was first disrupted (7.16*B*) into two major continental masses, Laurasia and Gondwana at the end of the Triassic period, 180 million years ago. Laurasia and Gondwana each experienced further breakup by the formation of new rift systems, with accompanied development of new oceanic crust, as well as various amounts of rotation. Two further stages are shown, in the late Jurassic, 135 million years ago (7.16*C*), and at the end of the Cretaceous period, 65 million years ago (7.16*D*). At this stage, Australia was attached to Antarctica, Canada was joined to Europe, Arabia was part of Africa, and India had not yet reached Asia.

Figure 7.16*E* shows the present continental distribution, and the amount of sea-floor spreading during the last 65 million years. On the assumption that present-day plate

movements will continue at their present directions and rates for the next 50 million years, Dietz and Holden have "predicted" that the Atlantic and Indian Oceans will grow at the expense of the Pacific Ocean; Australia will continue to drift northward against the Eurasian plate; Africa will drift northward, nearly closing off the Mediterranean; and the eastern portion of Africa will separate from the main part of Africa along the Rift-Valley system. They have further "predicted" that Baja California and the portion of California west of the San Andreas fault will become separated from North America and will drift to the northwest. In this event, Los Angeles will be opposite San Francisco in about 10 million years, and in 60 million years, Los Angeles will slide into the Aleutian trench!

The sea-floor spreading process, therefore, appears to provide for the zones of downbending that we recognize in the geologic record as geosynclines, for the deformation we see in mountainous regions, and the mechanism for drifting continents.

The Role of Gravity. Whatever the mechanism may be that brings portions of the crust to high elevations, gravity is a pervasive force

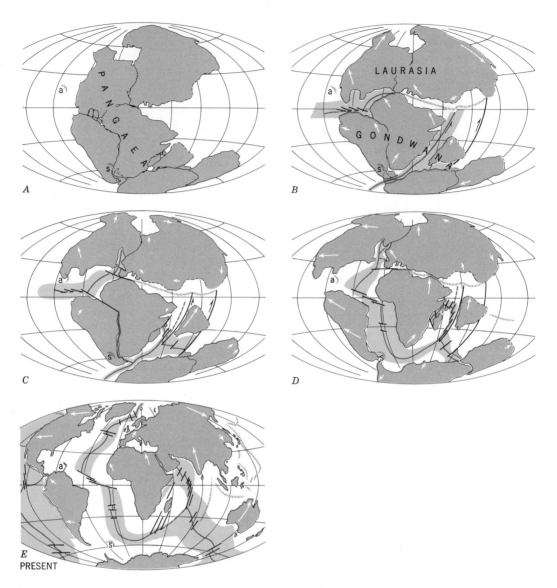

Figure 7.16 Restorations showing original continental mass Pangaea (A) as it appeared at the end of the Permian, 225 million years ago; (B) following its breakup into Laurasia and Gondwana at the end of the Triassic, 180 million years ago; (C) and (D) represent continental positions late in the Jurassic, 135 million years ago, and at the end of the Cretaceous, 65 million years ago, respectively; (E) present-day distribution of continents. a and s, representing the Antilles arc in the West Indies and the Scotia arc in the South Atlantic, appear on each diagram for a fixed geographic reference. Solid black lines and arrows indicate zones of slippage along plate boundaries; heavy colored lines are rifts; colored shading represents areas of new oceanic crust, black hatched lines are oceanic trench systems; white arrows indicate direction and amount of drifting. (After Dietz, R. S. and J. C. Holden, 1970, Reconstruction of Pangea; breakup and dispersion of continents, Permian to present, *Jour. Geophysical Research*, v. 75, no. 26, pp. 4939-4956.)

tending to bring materials to lower elevations. Some geologists have appealed to *gravitational sliding* to account for deformational structures that others explain by compressional forces. Gravitational sliding is not a rapid movement as the word, sliding, implies. Instead, it is a slow crumpling deformation due to the plastic yielding of rock layers tending to move from higher to lower elevations. Deformation of this type is analogous to the crumpling of a layer of snow as it slips down the slope of an inclined rooftop (Fig. 7.17).

Given sufficient time, sedimentary rock layers are capable of being deformed by gravity sliding, so long as an inclined surface is present to accommodate slippage along a plane. Many of the complex fold-and-fault structures of the Alpine and Himalayan chains are believed to be the products of large-scale movements caused by gravity (Fig. 7.18). Parts of the front (north) system of the Alps show indications of many miles of northward "sliding" from higher elevation to the south.

Figure 7.17 Snow layers sliding on a roof deform into folds, a process analogous to gravitational deformation of rock strata. Rock strata "slide" on gentler slopes over long periods of geologic time.

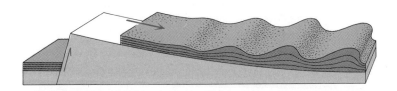

Figure 7.18 The hypothesis of gravitational sliding is based on the assumption that differentially uplifted sedimentary strata will deform by gravitational sliding. (Adapted from Holmes, A., 1965, *Principles of physical geology*, 2nd ed., Ronald Press, N. Y., 1285 pp.)

References

Bucher, W. H., 1956, Role of gravity in orogeneis, *Bull. Geol. Soc. America,* V. 67, pp. 1295–1318.

* Bullard, Sir Edward, 1969, The origin of the oceans, *Scientific American,* V. 221, No. 3 (September) pp. 66–75.

Carey, S. Warren, and others, 1958, *Continental drift, a symposium,* Geology Department, University of Tasmania, Hobart, 363 pp.

* Cox, Alan, G. B. Dalrymple, and Richard R. Doell, 1967, Reversal of the earth's magnetic field, *Scientific American,* V. 216, No. 2 (February), pp. 44–54.

Dickinson, William R., 1970, Global tectonics, *Science,* V. 168, No. 3936, pp. 1250–1259.

* Dietz, Robert S., and John C. Holden, 1970, The breakup of Pangaea, *Scientific American,* V. 223, No. 4 (October) pp. 30–41.

Gilluly, James, 1969, Oceanic sediment volumes and continental drift, *Science,* V. 166, pp. 992–994.

* Hammond, A. L., 1971, Plate tectonics: the geophysics of the earth's surface, *Science,* V. 173, pp. 40–41.

* Heirtzler, J. R., 1968, Sea-floor spreading, *Scientific American,* V. 219, No. 6 (December), pp. 60–70. (Offprint 875, W. H. Freeman and Co., San Francisco.)

* Hurley, Patrick J., 1968, The confirmation of continental drift, *Scientific American,* V. 219, No. 6 (December), pp. 52–64. (Offprint 874, W. H. Freeman and Co., San Francisco.)

Kay, Marshall, 1951, *North American Geosynclines,* Geol. Soc. America, Memoir 48, 143 pp.

King, P. B., 1969, *The tectonics of North America — A discussion to accompany the tectonic map of North America,* U.S. Geological Survey Prof. Paper 628, 94 pp.

* Menard, H. W., 1969, The deep ocean floor, *Scientific American,* V. 221, No. 3 (September), pp. 127–142.

Phinney, Robert A. (ed.), 1968, *The history of the earth's crust,* Princeton Univ. Press, Princeton, N. J., 244 pp.

Takeuchi, H., S. Uyeda, and H. Kanamori, 1970, *Debate about the earth,* rev. ed., Freeman, Cooper & Co., San Francisco, 281 pp.

Umbgrove, J. H. F., 1947, *The pulse of the earth,* 2nd ed., The Hague, Martimus Hijhoff, 358 pp.

* Vine, F. J., 1969, Sea-floor spreading — new evidence, *Jour. Geol. Education,* V. 17, pp. 6–16.

* Vine, F. J., 1970, Sea-floor spreading and continental drift, *Jour. Geol. Education,* V. 18, pp. 87–90.

* Recommended for additional reading.

8 Climate, Weathering, and Soils

. . . do you not see that stones even are conquered by time, that tall turrets do fall and rocks do crumble . . . ? Lucretius

The surface of the earth marks the juncture between the elements of the atmosphere and the materials of the lithosphere. The interaction of these components is the subject of this chapter. *Weathering* is the response of rocks to their physical and chemical environment at the interface of the atmosphere and the land surface. Two major weathering processes are generally recognized. *Physical weathering* is the mechanical breakdown or disintegration of rocks, whereas *chemical weathering* is the process of rock decomposition that results in chemical changes in the mineral constituents of the original rock mass.

Soil is one of the end products of rock weathering, but there are many places where soil is absent or poorly developed because the weathering processes have not had sufficient time to produce a soil, or the slope of the ground may be too steep to permit the weathered products to accumulate. Under these circumstances a rock outcrop acquires an appearance and configuration that reflects the composition and physical characteristics of the original rock and the nature and intensity of the forces of weathering (Fig. 8.1).

The Pinnacles in Crater Lake National Park, Oregon, are weathered remnants of volcanic material. (Courtesy of Oregon State Highway Department.)

Figure 8.1 The shape of this granite outcropping is determined by the more rapid rate of weathering along intersecting joint planes. Weathering is the combination of physical and chemical processes that reduces coherent bedrock to crumbling boulders and grains of sand. (Photograph by Tad Nichols.)

Before considering weathering processes and products in more detail, it will be necessary to examine some basic concepts of climatology because climatic factors, especially temperature and precipitation, are very important in determining the course of the weathering process. A knowledge of the basic climates that exist over different parts of the earth will therefore lead to a better understanding of weathering phenomena in general.

Climate

The climate of a geographic locality is determined by a number of factors that include temperature variations, precipitation pat-terns, surface elevations, prevailing winds, and the contrasting thermal properties of oceans and continents. In addition, the climate of a given area is strongly influenced by the latitudinal boundaries of that area because it is latitude that generally controls the gross temperature pattern of the world. Figure 8.2A shows a very rough relationship between mean annual temperature of a given point on the earth's surface and the latitude of that point (that is, distance north or south of the equator).

Temperature and latitude, however, do not provide a sufficient basis for classifying climates in detail. Figure 8.2A is merely a verification of a well-known fact that, *in general,* air temperatures are higher near the equator and lower near the poles.

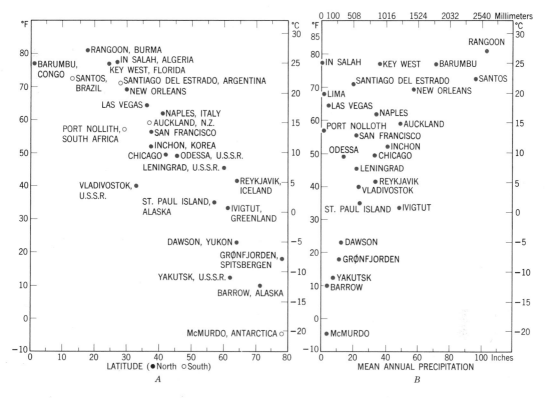

Figure 8.2 The mean annual temperatures of selected points plotted against latitude (A) and mean annual precipitation (B). (Data from Strahler, A., 1966 , *Physical Geography*, John Wiley and Sons, Inc. New York, pp. 643-658.)

By introducing a second factor, annual precipitation, a more precise characterization of the climate of a particular locality emerges. Figure 8.2B, for example, shows that even though two different geographical points have nearly equal mean annual temperatures, they can have quite different climates because of differences in precipitation. Compare, for example, the mean annual temperature and precipitation of In Salah in the Sahara Desert, Barumbu in the Congo Republic, and Key West, Florida. The mean annual temperature is the same for all three (within half a degree F.), but their different rainfall patterns put each in a different climatic zone.

A further refinement in characterizing a given climate involves a consideration of *seasonal* variation in temperature and precipitation. This is accomplished graphically on a *thermohyet diagram* on which the mean monthly temperature and precipitation are plotted for each month of the year. Several examples of thermohyet diagrams are given in Figures 8.3 and 8.4. These diagrams reveal a great deal about the climate of a particular place because they show at a glance whether the rainfall is equitably distributed throughout the year or whether there are wet and dry seasons. They also reveal whether there is a marked seasonal variation in temperature or if there are no winters and summers. The thermohyet diagram is thus a very useful technique for the presentation of the two most important climatological factors, temperature and precipitation.

Seasonal differences in temperature generally are characteristic of localities far removed from the ocean. A climate characterized by a high range between summer and winter temperatures is a *continental climate*.

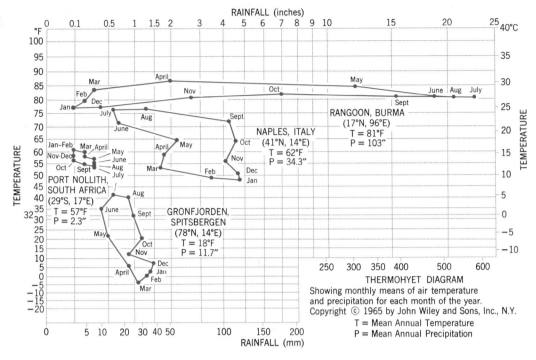

Figure 8.3 Thermohyet diagrams of selected geographical localities. (Data from same source as Figure 8.2.)

A lesser range in temperature between summer and winter is indicative of oceanic or *marine* (maritime) climates.

Classification of Climates. A number of climatic classification schemes have been devised by climatologists and geographers. The classification system adopted in this text (Table 8.a) consists of three broad climatic groups, each of which is more or less a reflection of the latitudinal control of the general temperature zones on earth. These groups are: I, Low Latitude Climates; II, Middle Latitude Climates; and III, High Latitude Climates. Each group is further subdivided into climatic categories in which factors such as precipitation, temperature, and seasonal variation of these parameters is taken into account. Thirteen major climatic types are included in this system, and theoretically the climate of any place on the earth's surface that is not in a mountain range can be accommodated in the system.

The mountainous regions of the earth do not fit into the system because altitude introduces a factor that is difficult to allow for in the scheme. Generally, increased elevation above sea level has the same effect as increase in latitude. The truth of this statement is borne out in the fact that the climates of high mountains lying in tropical areas are more akin to polar climates than to equatorial climates. Thus it is possible for Mt. Kenya on the equator in Africa to harbor glaciers at elevations around 16,000 feet above sea level. High mountains anywhere on earth, therefore, have weathering characteristics similar to those found in polar regions at much lower elevations.

Climate and Weathering. The great variability that exists between the climatic zones of the earth is a fundamental cause of the variation in the weathering process at different geographic localities. For example, rocks exposed at the surface in nonmountainous equatorial wet climates never experience freezing temperatures. Hence, the expansive force of ice formation in the physical disruption of rocks is not operative under

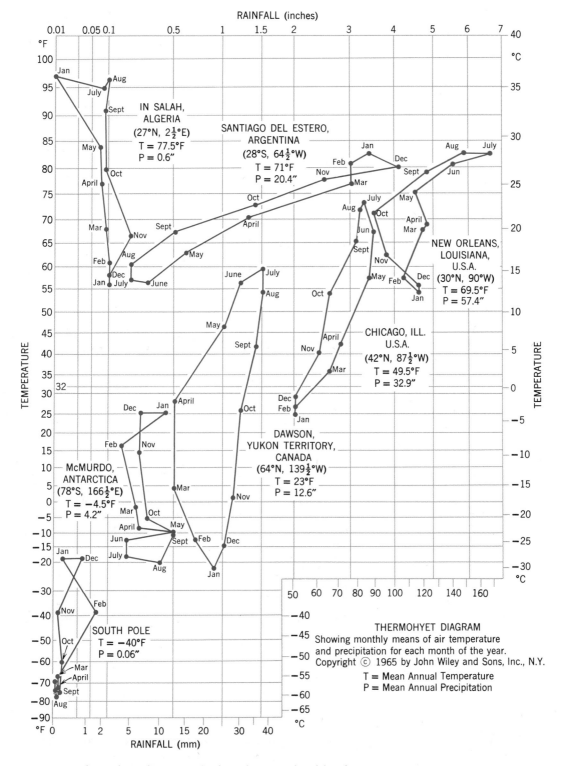

Figure 8.4 Thermohyet diagrams of selected geographical localities.
(Data from same source as Figure 8.2.)

Table 8.a Classification of Climates*

Climate	Distinguishing Characteristics
GROUP I. Low Latitude Climates	
1. Wet Equatorial Climate 10° N to 10° S latitude (Asia 10°–20° N)	Uniformly high temperatures prevail throughout the year. High annual rainfall may be distributed uniformly throughout the year or nonuniformly in a wet and dry season (monsoon type). Tropical rain forest is dominant vegetation.
2. Tropical Desert and Steppe Climate (15° –35° N and S latitudes)	Arid (desert) and semiarid (steppe) regions of low annual rainfall that falls sporadically throughout the year. Some months in a desert climate have no measurable rainfall. Moderate annual range in temperature with high maximum.
3. West Coast Desert Climate (15°–30° N and S latitudes)	Extremely dry but relatively cool with small annual temperature range. Frequent fog. Narrow belts along western coastlines.
4. Tropical Wet-Dry Climate (5°–25° N and S latitudes)	Hot, wet summers; dry winters. Generally high annual rainfall.
GROUP II. Middle Latitude Climates	
5. Humid Subtropical Climate (20°–35° N and S latitudes)	Temperate rainy climate with hot summers and copious rainfall. Cool winters. Characteristic of subtropical, eastern continental margins.
6. Marine West Coast Climate (40°–60° N and S latitudes)	Cloudiness and well-distributed precipitation with winter maximum. Annual temperature range is small. Summers warm to cool. Occurs along middle-latitude west coasts.
7. Mediterranean Climate (30°–45° N and S latitudes)	Wet-winter, dry-summer with moderate annual temperature range. Summers are warm to hot with extreme drought.
8. Middle Latitude Desert and Steppe Climate (35°–50° N and S latitudes)	Continental interior deserts and steppes with hot summers and cold winters. Low annual precipitation with uniform distribution.
9. Humid Continental Climate (35°–60° N latitude)	Strong seasonal contrast. Cold, snowy winters and warm to hot summers. Ample precipitation throughout the year, usually higher in summer.
GROUP III. High Latitude Climates	
10. Continental Subarctic Climate (50°–70° N latitude)	Very cold snowy winters and short, cool summers. Very high annual temperature range. Generally moist all year.
11. Marine Subarctic Climate (50°–60° N and 45°–60° S latitudes)	Relatively high precipitation. Moderate annual temperature range is less than more continental climate.
12. Tundra Climate (North of 55° N, South of 50° S)	Humid climate with severely cold winters and cool summers. Ground is perennially frozen. Low to moderate precipitation.
13. Ice cap Climate (Greenland, Antarctica)	Temperatures generally below freezing. Low to extremely low precipitation. Winters extremely cold; temperature of summer months averages below freezing. Some *days* above freezing.

* This classification system is a modified version of the one proposed by A. N. Strahler (1969) who followed a system developed by G. T. Trewartha (1954). (See references at end of this chapter)

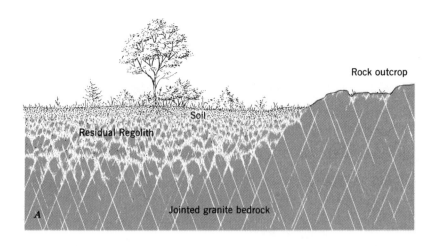

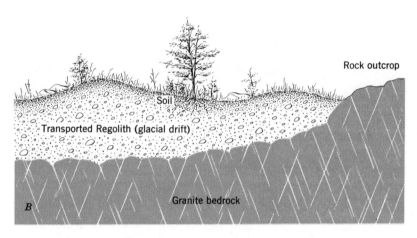

Figure 8.5 Diagram showing the relationship of soil, regolith, and bedrock. A residual regolith (A) is composed of the weathered products of the underlying bedrock. A transported regolith (B) contains fragments of rock types that are different from the local bedrock.

such conditions. On the other hand, the hot and wet climates of the tropics are conducive to strong chemical reactions between water and rocks. An increase in temperature of about 20° F. (10° C.) will increase the rate of chemical reactions by a factor of two or three. Conversely, a region that is dominated by a cold climate will reflect the effects of freezing and thawing on rock masses occurring there, but will exhibit little, if any, alteration due to chemical changes. In the middle latitudes, both chemical and physical weathering take place simultaneously

resulting in disintegrated and decomposed rock.

Bedrock, Regolith, and Soil

Geologic materials at the earth's surface can be categorized as bedrock, regolith, or soil (Fig. 8.5). Bedrock is a rock mass that lies in a natural position that is more-or-less undisturbed by surface agents. A block of sandstone that tumbles to the base of a cliff is no longer bedrock. Bedrock exposed at

Figure 8.6 Talus slopes on the steep flanks of the Garden Wall in Glacier National Park, Montana. The talus fragments are produced by frost action. (Photograph by H. S. Zumberge.)

the earth's surface is an *outcrop*. In many places, however, bedrock does not crop out at the surface because it is buried in its own weathered debris or it is covered by a regolith of some other uncemented material that has been emplaced by some geologic agent. *Regolith* is thus either *residual* or *transported* as shown in Figure 8.5. Both residual and transported regoliths contain soils in their upper parts.

Physical Weathering

Physical weathering is the mechanical breakdown of rocks by some means that does not involve a chemical reaction. The dominant force that produces rock disin-

tegration and physical disruption of soil and regolith is the growth of ice crystals. Minor physical weathering is accomplished by tree roots and the activity of some animals. Of unknown efficacy is the physical effect of temperature change on rock outcrops of coarse regolith, but circumstantial evidence indicates that rocks have been broken by alternate heating and cooling, especially in deserts.

Frost Action in Bedrock. The change of water from the liquid to solid state is accompanied by a volume increase of 9 percent. A daily *(diurnal)* temperature fluctuation across the freezing point of water is a common event in many regions of the earth. The repetition of the freeze-thaw cycle exerts great pressure on rocks when water

Figure 8.7 Altiplanation terrace above timberline near Hughes, Yukon. Frost action is the major process that produced the rock rubble seen in the lower right foreground. (Photograph by Troy Péwé.)

entrapped in small cracks and joints is crystallized into ice. Bedrock outcrops in mid to high latitudes and in mountainous regions are very susceptible to disruption by this process.

Individual joint blocks dislodged from a rock outcrop by frost action form *talus* if the outcrop is in the form of a cliff or steep slope (Fig. 8.6). If the rock crops out on a relatively flat surface, *block fields* are produced (Fig. 8.7). The process whereby extensive block fields at high elevations are produced is called *altiplanation* and usually involves the presence of a snowbank that supplies melt water by day which freezes at night. Although the exact mechanism of altiplanation is not understood, the climatic circumstances under which it operates leave little

doubt that the freeze-thaw cycle in the presence of water is the underlying cause.

Frost Action in Regolith. Water contained in regolith interstices expands when frozen, but even when a regolith is saturated with water, the expansion due to freezing may be barely noticeable. In silts and silty clays, however, water is drawn upward to the base of the frozen ground by capillary action during the freezing process. This *addition* of moisture during the freezing process causes discrete layers and lenses of ice to form in the layer of frozen ground as it thickens during the progression of the winter season. The growth of these ice masses causes the regolith to expand or heave in an upward direction; this process is called *frost heaving*.

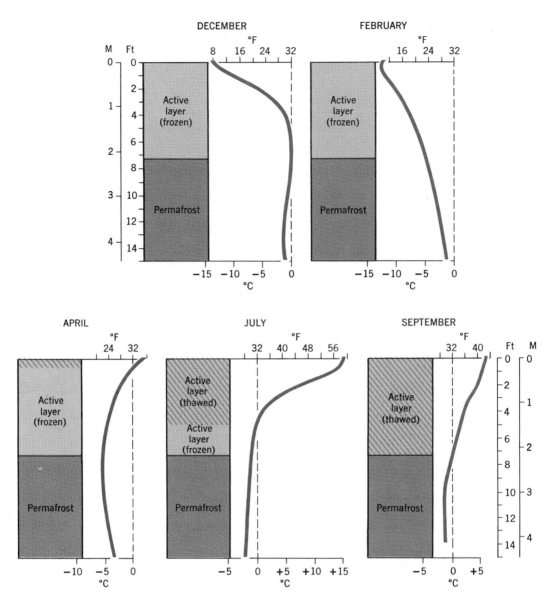

Figure 8.8 Diagram showing temperature-depth relationships in the permafrost of Fairbanks, Alaska at different months of the year. Notice the progressive thawing of the active layer from April (end of the freezing season) through September (start of freezing season). The top of the permafrost is defined as the depth at which the temperature remains at or below freezing continuously for at least two years. [Data from Stearns, S. R., 1966, *Cold Regions Science and Engineering, Part I, Sec. A2, Permafrost, (perennially frozen ground)*. U. S. Army Cold Regions Research and Engineering Laboratory, Hanover, N. H., 77 pp.]

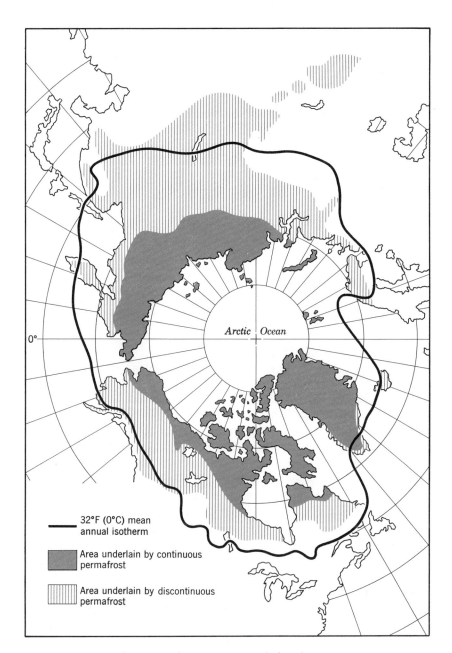

Figure 8.9 Map of the Arctic showing areas underlain by continuous and discontinuous permafrost. The heavy black line connects points at which the mean annual temperature is 32° F. (0° C.). (Based on Stearns, S. R., 1966, *Cold Regions Science and Engineering, Part I, Section A2, Permafrost (perennially frozen ground)*. U. S. Army Cold Regions Research and Engineering Laboratory, Hanover, N. H., p. 3.)

Frost heaving produces forces that are sufficient to crack or damage concrete footings and other structures that are in contact with ground. Such frost damage can be prevented if the structures are placed below the depth of maximum frost penetration. In North America east of the Rocky Mountains, the depth of maximum frost penetration ranges from a few inches in the southern United States to as much as five or six feet in southern Canada.

Highway engineers usually recommend the replacement of silty zones with gravel along proposed highway routes in cold climates to minimize damage from frost heave. The gravel will not permit the formation of ice lenses because the openings between the gravel particles are larger than capillary size. Hence, no water can be drawn upward during the freezing process.

Permafrost. In the nonglaciated land areas of the Arctic and Antarctic the temperature of rock or regolith is perennially below freezing to a depth ranging from a few feet to several thousand feet below the surface. This frozen ground is called *permafrost.* The top of the permafrost layer, the *permafrost table,* is defined by the maximum depth of thawing during the summer months. The surficial zone above the permafrost table in which thawing takes place each summer is called the *active zone* because it is there where many freeze-thaw cycles cause intense frost action. Figure 8.8 shows the temperature-depth regime in the active zone and upper part of the permafrost at Fairbanks, Alaska during the winter, spring, summer, and early fall. Notice that the temperature below the permafrost table remains below freezing throughout the year.

The thickness of the active zone is dependent on the amount of heat available for melting frozen ground. Mean annual temperature, the kind of vegetational cover, exposure to the sun during the summer months, and the thickness of the snow cover during the previous winter are the most important factors that affect the thickness of the active zone. The active layer is characterized by intense physical weathering due to frost action.

The permafrost region of the Northern Hemisphere is divided into two parts, the zone of *continuous permafrost* and the zone of *discontinuous permafrost* (Fig. 8.9). In the former, permafrost occurs everywhere beneath the active layer, and in the latter, unfrozen ground exists side by side with irregular-shaped masses of permafrost below the depth of summer thaw. The boundary between the continuous and discontinuous zones is roughly coincident with the 20° F. mean annual air temperature isotherm, and the southern boundary of discontinuous permafrost is defined roughly by the 32° F. mean annual air temperature isotherm.[1]

The active layer in the continuous permafrost ranges in thickness from less than one foot in the north to three to five feet at the boundary between continuous and discontinuous permafrost. The depth to the permafrost table in the zone of discontinuous permafrost ranges from a few feet to more than ten feet, depending on the local climatic and terrain conditions.

In general, the thickness of the permafrost increases toward the north. Near the southern boundary of the discontinuous permafrost, the permafrost ranges in thickness from a few inches to several feet. At the boundary of continuous and discontinuous permafrost, the thickness is about 200 to 300 feet, and in the high latitudes of the north polar regions, permafrost thicknesses are more than 1000 feet. On the Arctic slope of Alaska, it ranges from 700 to 1300 feet. In northern Canada the permafrost is 1500 feet thick, and in eastern Siberia, the thickness is reported to be 5000 feet, the thickest known permafrost in the Northern Hemisphere.

It has been estimated that about 20 percent of the earth's land surface is underlain by permafrost. Most of this lies in northern North America, Arctic Europe, and northern Asia. About one-half of Canada and the Soviet Union are underlain by perennially frozen ground. The development and exploitation of the Arctic regions cannot be

---

[1] An isotherm is a line connecting points of equal temperature.

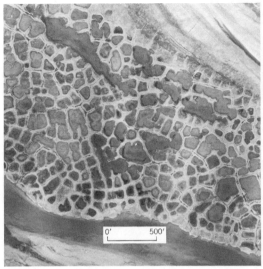

Figure 8.10 Left: Vertical aerial photograph of patterned ground underlain by permafrost, 60 miles southeast of Point Barrow, Alaska. (Photograph by R. F. Black.) Right: Oblique photograph of cultivated fields north of Harvington, Worcestershire, England. Relict patterned ground was produced by a colder climate that prevailed in this part of England during the last "Ice Age" or Pleistocene. No permafrost exists in this area at present. The scale in the center foreground is about the same as in the Alaskan photo. (Courtesy of F. W. Shotton and Arnold Becker.)

accomplished without an understanding of this widespread phenomenon. Man-made structures must be uniquely designed for permafrost conditions because excavations in permafrost tend to disrupt the thermal regime by allowing more heat to penetrate the ground.

The melting of clear ice masses and ice-cemented regolith is conducive to differential settling of man-made structures placed on or beneath a permafrost terrain. The planning and design of the 800-mile Trans Alaska Pipeline System to transport crude oil from Prudhoe Bay in Arctic Alaska to the year-round port of Valdez is an example of the difficult construction problems encountered in permafrost terrain.

Surface Phenomena in Permafrost Areas.
The ground surface of unconsolidated regolith in permafrost regions exhibits a peculiar phenomenon called *patterned ground* (Fig. 8.10A). Patterned ground takes on a variety

of forms depending on the ground slope and the kind of regolith. One widespread type of patterned ground consists of polygonal areas on the order of 30 to 300 feet across, each of which is outlined by a trough and low ridges. Each trough is underlain by an *ice-wedge* (Fig. 8.11) that forms in the following manner. During the very cold winters in permafrost regions the permafrost contracts because of thermal shrinking. The cracks thus formed divide the ground surface into roughly equidimensional cells. At the beginning of the melt season, water from melting snow or the active layer drains into the cracks and refreezes, forming vertical veins of ice. Repetition of this annual process over many centuries ultimately produces a network of ice-wedges (Fig. 8.12) which, when seen in plan view, show the unmistakable pattern of Figure 8.10A.

Vestiges of patterned ground that occur in areas not now underlain by permafrost

Figure 8.11 Ice wedge in organic-rich silt exposed during gold mining operations on Wilbur Creek near Livengood, Alaska. (Photograph by Troy L. Péwé.)

indicate that the southern boundary of permafrost was formerly at a much lower latitude (Fig. 8.10*B*). These conditions prevailed during the waning phases of the "Ice Age" when glacier ice covered much of North America, Siberia, and northern Europe as recently as 10,000 years ago (see Chapter 13).

Heating and Cooling of Rocks. In hot desert regions where daily maximum tem-peratures may reach 120° F. because of in-tense solar radiation, rock outcrops and surface stones are heated to very high temper-atures. Rapid cooling after sunset or quench-ing by rain causes thermal stresses in these rocks which supposedly can produce rup-ture. Rocks subject to hundreds of heating-cooling cycles in the laboratory have failed to verify this process, but close examination of pebbles and stones supposedly ruptured

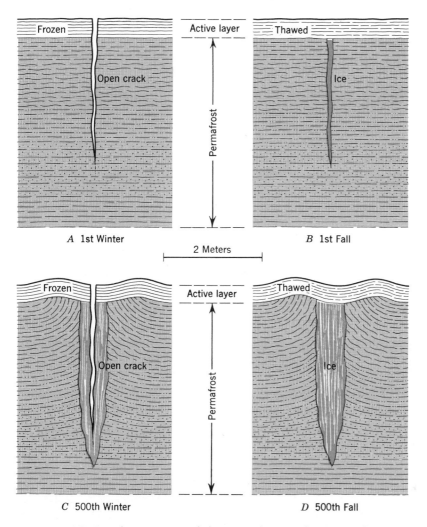

Figure 8.12 Series of cross-sectional diagrams showing the origin of an ice wedge in permafrost. (*A*) Severe winter temperatures causes surficial sediments to contract and crack. During the spring and summer, water from the thawed active layer seeps into the crack in the still-frozen sediments and freezes (*B*). This process is repeated over a period of several hundred years (*C* and *D*) until an ice wedge, such as shown in Figure 8.11, is formed. (After Lachenbruch, A. H., 1966, *Contraction theory of ice-wedge polygons: a qualitative discussion,* Proceedings: Permafrost International Conference, National Academy of Sciences-National Research Council, Publication 1287, Washington, D. C., pp. 63-71.)

by thermal stresses reveals that chemical weathering may have weakened the rock to a point where heating and cooling stresses could fracture it. In any event it must be pointed out that diurnal or seasonal air temperature changes do not penetrate very far beneath the ground surface, so if rocks are, in fact, ruptured by thermal forces, the process operates only at or very close to the ground surface.

Plant and Animal Activity. The physical intrusion of tree and plant roots into cracks in rocks and unconsolidated soils causes some disturbance of the regolith, but the effects of these forces in physical weathering are slight. Even where large trees are uprooted by the wind, the effect on mechanical stirring and mixing of soil particles is only locally important.

Earthworms cause mixing and loosening of soil particles in the upper part of the regolith (soil) in areas with temperate climates. The activities of these animals can be of some consequence. Actual measurements of the earthworm population in English pastures revealed a total weight of 1100 pounds of earthworms per acre. An added effect of these creatures is the mixing of organic material with soil particles in the upper mantle, thereby providing the raw material for the production of humic acids which add materially to chemical changes in the soil. Earthworms thus contribute to the chemical as well as physical changes in the upper mantle.

Burrowing rodents that live in steppe climatic zones cause physical disruption of the regolith, but their total effect is minor and only of local significance in comparison to other processes of physical weathering.

Chemical Weathering

Many rock types now exposed at the surface of the earth were formed under conditions different from those that now occur at the earth's surface. Igneous rocks such as granite came into being deep in the crust where both temperature and pressure are much higher than at the surface. Sedimentary rocks originated beneath the sea where the chemical and physical environment was far different from that which exists in places where they are now found at the earth's surface. Metamorphic rocks were formed under higher temperatures and pressures than those prevailing at or near the earth's surface.

When the mineral constituents of rocks are formed they tend to be in equilibrium with their chemical and physical environment, and so long as the environment remains unchanged, the minerals remain unchanged or stable. On exposure to a different environment, however, such as prevails in the various climatic regions of the earth, the constituent minerals of most rock types undergo chemical changes in the direction of a new state of equilibrium. Chemical weathering is the process whereby the mineral constituents of rocks are altered chemically under the conditions normally found at the earth's surface. In some cases, new minerals are formed that are more stable than the ones from which they were derived. In other cases, minerals are dissolved and removed in solution. Although chemical weathering is a process of mineral and rock *decomposition,* its effects are often manifested in physical changes as well.

Chemical weathering is most pronounced in wet equatorial regions where ample moisture and relatively high temperatures enhance the rate of chemical reactions. Under conditions of desert climates, chemical weathering is less intense, although most arid regions receive some rainfall during parts of the year. In polar regions, chemical weathering rates are at a minimum, but even there some chemical alteration takes place.

Oxidation. Minerals that contain iron in combination with sulfur are highly susceptible to chemical change through oxidation.[2] In the case of the oxidation of pyrite (FeS_2), for example, a new compound, limonite ($Fe_2O_3 \cdot nH_2O$), is formed. The sulfur replaced by the oxygen combines with water to form a weak solution of sulfuric acid (H_2SO_4). This acid further attacks other minerals in the rock, thereby accelerating chemical decomposition. Limonite is yellowish-brown in color and produces a brownish to yellowish stain on rocks containing sulfide minerals. A surface outcrop of rock heavily stained with limonite is often a clue to an important sulfide ore body at greater depth. Another oxidation product of iron-rich minerals is hematite (Fe_2O_3), which is red in color and responsible for red colors in rocks and soils.

[2] In strict chemical terms oxidation refers to an increase in positive valence or a decrease in negative valence (that is, the removal of one or more electrons from an ion or atom). As used here, oxidation refers simply to the chemical addition of oxygen to a mineral.

Figure 8.13 Leaching of calcium carbonate by surface water produced these channels along intersecting vertical joints in limestone that occurs in Hastings County, Ontario, (Photograph by M. E. Wilson, Geological Survey of Canada.)

Leaching. Although all minerals are soluble in water to some extent under certain conditions, some are more easily dissolved in water than others. *Leaching* is the removal of certain chemical components of a rock or mineral by solution, that is, mineral matter is actually dissolved just as table salt is dissolved in water. Limestone, dolomite, and rock gypsum are particularly susceptible to leaching.

Limestone is made up chiefly of the mineral, calcite, and the degree of solubility of calcite is related to the amount of carbon dioxide (CO_2) dissolved in natural waters. Water and carbon dioxide react to form carbonic acid (H_2CO_3), which attacks calcite as follows:

$$CaCO_3 + H_2CO_3 \rightarrow Ca(HCO_3)_2$$

calcite + carbonic acid → calcium bicarbonate

The product of this reaction, calcium bicarbonate, is soluble and is therefore removed in solution, a process that is prominently displayed on limestone outcrops in humid climates (Fig. 8.13).

Hydrolysis. This chemical reaction is very common in nature and involves the decomposition of a mineral and the reaction of some of its constituents with the water molecule itself. The weathering of feldspars is an example of this process. Orthoclase ($KAISi_3O_8$) is changed to kaolinite ($Al_2Si_2O_5$ $(OH)_4$) by hydrolysis. The transformation of feldspar to kaolinite is so commonplace that it is called *kaolinization.*

Hydrolysis causes expansion in rocks that contain abundant feldspar because the kaolinite occupies more space than the feldspar from which it was formed. This expansion causes the outer surface of a rock outcrop to spall or flake off in thin rinds or layers. This is a case where chemical weathering causes not only rock decomposition but rock disintegration as well. The process is called *exfoliation.* When exfoliation has acted on jointed rocks for a long time, the rock outcrop is reduced to a series of rounded knobs. Their spheroidal shape accounts for

Figure 8.14 The layered appearance of this granitic rock in Yosemite National Park, California is caused by exfoliation joints that roughly parallel the local ground surface.

the special name given to this weathering phenomenon, *spheroidal weathering.*

Exfoliation on a much larger scale occurs in granite rocks from which an overlying mass of rock has been removed by erosion. As the weight of the superincumbent rock is removed, the remaining rock masses expand slightly. This expansion produces concentric rock slabs more-or-less parallel to the surface of the outcrop (Fig. 8.14), and ranging in thickness from a few inches to 10 feet or more. The parallel joints separating the concentric slabs are invaded by moisture causing hydrolysis or, if the outcrop occurs in a cold climate, further disruption will be caused by frost action.

Organic Activity. Decaying plant material (humus) results in the production of humic acids that must also be considered agents of chemical weathering. Humus is the home of bacteria that are known to accelerate the process of chemical weathering. Some plants such as lichens can extract iron directly from the rock surfaces to which they are attached, thereby adding materially to chemical changes in the rock or regolith.

Weathering Products

The zone of weathering is the source of the material that is carried off the land surface to the oceans, either in solid form as clay, silt, and sand, or in solution. One would thus expect to find a relationship between the materials deposited in the ocean

Table 8.b   Products of the Weathering of a Granitic Type* Rock in a Humid Temperate Climate

| Major Mineral Constituents | (Percent) | Major Weathering Products | |
		Residual	Soluble
Feldspars			
Orthoclase ($KAlSi_3O_8$)	(35)	Kaolinite ($Al_2Si_2O_5$) $(OH)_4$	K, Si (slightly)
Plagioclase (Na, Ca, Al, silicate)	(35)	Kaolinite	Ca, Na
Quartz (SiO_2)	(25)	SiO_2 sand particles	
Biotite (K, Mg, Fe, Al, silicate)	(4)	Kaolinite	K, Mg, Fe
		Limonite ($Fe_2O_3 \cdot nH_2O$)	Si (slightly)
Pyrite (FeS_2)	(1)	Limonite	Fe, H_2SO_4

* An igneous rock having this composition is technically not a granite but a granodiorite because of the equal amounts of orthoclase and plagioclase feldspars.

and the weathered products on land. The relationship is not a simple one, however, because the weathered products are not always transported directly to the sea. During the time in transit from mountain top to sea bottom, weathered materials may undergo further change.

The matter is further complicated by the fact that some of the rocks undergoing weathering are sedimentary rocks formed in the sea at a much earlier period of geologic time. The constituent minerals of these rocks have already passed through one cycle of weathering—transportation—deposition, so it is not a simple matter to relate weathered products to ocean sediments.

Nonetheless, it is useful to consider the results of the weathering of a granite, a primary rock whose minerals were formed under conditions of temperature and pressure far different from the ones that obtain at the earth's surface. Granite or granitic rocks (rocks with a composition similar to a granite) crop out over hundreds of thousands of square miles of the continents, and their weathered products are not insignificant in terms of the material supplied in either the solid or dissolved state to the rivers that flow to the sea.

Weathering of Granitic Rocks. Consider the consequence of the decomposition and disintegration of an igneous rock containing four major minerals, orthoclase, plagioclase, quartz, and biotite, and one minor mineral, pyrite. Table 8.b summarizes the weathered products of this granitic rock whose mineralogic composition is given in the left-hand column. Assuming a constant climatic environment such as is found in the southeastern United States (humid continental), the weathered products produced by the chemical weathering of each mineral are shown in the two right-hand columns. Kaolinite is a residual from the weathering of the feldspars and biotite, and limonite is formed from the biotite and pyrite. The quartz undergoes no appreciable chemical change, although possibly a very slight amount may be dissolved. The residual material from the complete weathering of this rock would therefore consist of a mixture of loose quartz granules and kaolinite with lesser amounts of limonite which gives the residual regolith a yellowish-brown color. (Pure kaolinite is white.)

During the kaolinization of the feldspars and biotite, and the oxidation and hydration of the pyrite, certain elements have been freed, notably potassium (K), calcium (Ca), sodium (Na), and magnesium (Mg). Some iron (Fe) has also gone into solution, and the sulfur of the pyrite is now combined with water and oxygen as sulfuric acid. Note particularly that aluminum (Al) is not one of the elements liberated by weathering. It remains fixed in combination with silicon (Si) and oxygen as an insoluble residue in the kaolinite.

Stability of Minerals During Weathering.

All minerals do not succumb to chemical attack at the same rate. Some minerals are more resistant to decomposition than others, as has been demonstrated in the case of a granitic rock. The common rock-forming minerals can be arranged in a *stability* series as follows: quartz (most stable), muscovite, orthoclase, biotite, amphibole, pyroxene, plagioclase, and olivine (least stable). It can be deduced, therefore, that a granite with its characteristically high quartz and orthoclase content will inherently be more resistant to chemical weathering than a gabbro in which pyroxene and plagioclase predominate.

The stability series is the reverse order in which the same minerals crystallize from a magma (see page 35). This is to say that olivine and plagioclase are the least stable minerals and are among the first to crystallize from the melt, while quartz and muscovite, both of which are chemically stable insofar as weathering is concerned, are the last to crystallize.

Depth of Weathering.

If the ground surface of the rock exposed to weathering is flat or very gently sloping, the residual weathered products would accumulate to a considerable thickness over a period of time. The thickness of this residual regolith would depend not only on the surface slope but also on the rate of weathering and the length of time during which the weathering processes were operative.

Because the rate of chemical action doubles for an increase in temperature of about 20° F., the depth of weathering in the humid tropics is greater than in more temperate regions. In the New England states the residual regolith on weathered granite may be only 10 to 15 feet, but in the tropical climates of Africa and South America it may be 100 to 200 feet.

Caution must be exercised, however, in relating depth of weathering to *prevailing* climatic conditions because it is now well known that climates are not constant over long periods of geological time. Deep weathering profiles in places that are characterized by a modern temperate climate may, in fact, be relict from a previous climate of more tropical proportions.

Economic Deposits Produced by Weathering

The chemical weathering of certain rock types may produce residual materials of economic value. These deposits are called *residual ores.* The mineral content of these ores is determined by the composition of the fresh unweathered rock, the intensity and duration of chemical decomposition, and a favorable topographic setting to permit the accumulation (rather than erosion) of the weathered residue.

Laterite.

This name comes from the Latin word *later,* meaning brick, and it is applied to any reddish-brown or yellowish residual mantle in tropical or subtropical regions. Laterites are characteristically high in the hydrous iron and aluminum oxides ($Fe_2O_3 \cdot nH_2O$ and $Al_2O_3 \cdot nH_2O$) and low in silica (SiO_2). Silica is more readily leached under tropical climates than under any other climatic zone.

The vast majority of laterites that occur in South America, Africa, India, Southeast Asia, and the Caribbean have no commercial value except as a building material and highway brick. These uses of laterite were more extensive in prehistoric times than now, as in the case of the ancient walled city of Angkor Thom in Cambodia, where a temple and other abandoned edifices of the ancient Khmer civilization were constructed from laterite.

A laterite must contain about 50 percent iron in the form of hydrous iron oxide or an equal amount of aluminum as hydrous aluminum oxide in order to qualify as an ore deposit. These ore deposits are known as *lateritic iron ore,* or in the case of aluminum, *bauxite.* The tropical wet-dry climates are especially conducive to the process of laterization.

Care must be exercised also in distinguishing laterite from *latosol.* The latter is a soil term and refers only to the upper part of a lateritic regolith. Latosols are considered in more detail under the section on soils.

An example of the chemical changes that take place during laterization can be seen in Table 8.c, which shows a comparison of the chemical compositions of the bedrock and the residual regolith. Notice that the original rock is made up of one-third silica (mainly in the form of silicate minerals rather than quartz) whereas the weathered concentrate has only a little more than one percent silica. Even more pronounced is the loss of magnesium (Mg); only a trace is found in the laterite. On the other hand, both iron and aluminum oxides showed large percentage increases in the weathered residue, not because these substances were *added* during the weathering process but because others were lost.

Table 8.c Chemical Composition of a Basic (Mafic) Igneous Rock (Dunite) and Its Residual Regolith (Laterite) from Guinea in Equatorial West Africa*

Constituent†		Fresh Rock (Percent)	Weathered Regolith (Percent)
SiO_2		33.9	1.3
Al_2O_3		1.6	8.2
Fe_2O_3		8.6	76.8
FeO		7.4	—
CaO		0.5	Trace
MgO		36.7	Trace
H_2O		10.6	12.9
Other		0.2	0.3
	Totals‡	99.5	99.5

* After Bonifas, M., 1959, *Contribution à l'étude géochimique de l'alteration latérique*, Mem. Carte Alsace-Lorraine, No. 17, Strasbourg, 159 pp.

† These are chemical constituents, *not* mineral constituents.

‡ Percentage totals do not always equal 100 percent because of slight errors in the making of a chemical analysis.

SOILS

The upper part of the regolith is occupied by soil. It is in the soil where the maximum effect of the interaction between the atmosphere and the surface of the earth is seen. One important factor present in soil to a much higher degree than the rest of the regolith is the biological activity of plants and microorganisms. This activity includes not only the living plants and animals but also the chemical reactions involving dead organic material. The degree to which soils can support the growth of natural vegetation or agriculture crops is a measure of its fertility, and this fertility derives not only from decayed organic matter and atmospheric components such as nitrogen (N), but also from available inorganic nutrients such as K, Ca, and others.

The science that deals with soils from the agricultural point of view is called *pedology*. Because of his interest in soil productivity, the *pedologist* defines soil as a ". . . collection of natural bodies on the earth's surface, containing living matter, and supporting or capable of supporting plants."[3]

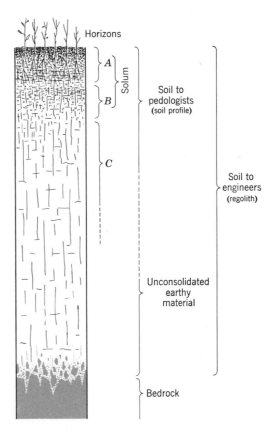

Figure 8.15 Generalized soil profile showing the various morphological units and soil terminology as used by pedologists and engineers.

[3] *Seventh Approximation,* Soil Conservation Service, U.S. Department of Agriculture, Washington, D.C., 1960, p. 1.

The term, soil, is used in a much broader sense by engineers who are concerned with the utilization of the materials on or near the earth's surface for constructural purposes. To the engineer, soil is an unconsolidated or uncemented earthy material that cannot be classified as bedrock (Fig. 8.15). In the engineering sense, then, soil is synonymous with regolith as previously defined. In the paragraphs that follow, the pedological rather than the engineering usage will be employed because of the close connection between soil genesis and weathering processes.

Prior to the late nineteenth century soil science was geologically orientated. Not until the work of the Russian geologist, Dokuchaev, did pedology become a science in its own right. Dokuchaev observed that five factors are responsible for soil development: (1) parent material, (2) climate, (3) organisms, (4) topography, and (5) time. Collectively, these five factors determine what a soil will look like or, as a scientist would say, what its *morphology* is. When one considers that all of the five soil-forming factors except the action of organisms are geological factors, it is hardly surprising that pedology evolved from the science of geology. It also explains why, in an elementary text on geology, it is appropriate to consider this subject at all.

The main concern of the geologist with soil is the relationship between soil formation and weathering. Inasmuch as both weathering and soil development are time-dependent, the geologist can learn much from soil science about the rates of weathering under certain geological and climatic conditions. When Dokuchaev and his successors were studying soils in the last century, they were unaware of the relatively recent climatic change that had affected large geographic zones of the earth. Hence they assumed that the soils that now exist in the world were formed under the conditions of climate that now prevail. Yet it is known beyond question that constancy of climate cannot be presumed anywhere on earth. Indeed, there is ample evidence from independent sources to show that arid lands were wetter in the past, or that temperate zones were rather recently cloaked in an arctic climate.

Ultimately, then, geology can learn much from pedology that will be helpful in working out the details of recent geological history as it relates to climatic change. Past climatic conditions are reflected in ancient soils that are buried beneath younger sediments. Such soils are called *paleosols,* and the science dealing with them is *paleopedology.*

Before considering the classification and distribution of soils, however, it is necessary at this point to introduce some aspects of soil morphology and nomenclature.

Soil Morphology

The depth to which the soil-forming processes are operative in the regolith defines the thickness of the soil layer. This layer of soil is generally about 3 feet thick and is usually not homogeneous from top to bottom. The soil layer is characterized by different zones or *horizons* that are roughly parallel to the ground surface and may be distinctly visible to the naked eye because of color, texture, presence or absence of calcium carbonate, or other visible properties.

Most soils have three horizons (Fig. 8.15), but in some soils the horizons may be so poorly developed that no horizonation is visible. In other cases a horizon may be absent. From the surface downward, three horizons, *A, B,* and *C,* are identifiable in a maturely developed soil. These make up the *soil profile*. The *A* horizon ("top soil") and the *B* horizon ("subsoil") together constitute the *solum*. It is in the solum where the effect of the five soil-forming factors reach their greatest manifestation, and it is the nature of the *A* and *B* horizons of the solum that differentiates one soil from another.

Perhaps the most important factor that causes the formation of the *A* and *B* horizons is the downward percolation of water through the soil profile. In humid regions there is sufficient water to allow certain materials to be removed from the *A* horizon and redeposited either in the *B* horizon or at some depth lower in the soil profile. Removal of soluble material and exceedingly fine clay particles from the *A* horizon is called *eluviation,* and the redeposition of these same substances lower in the profile is called *illuviation*. For

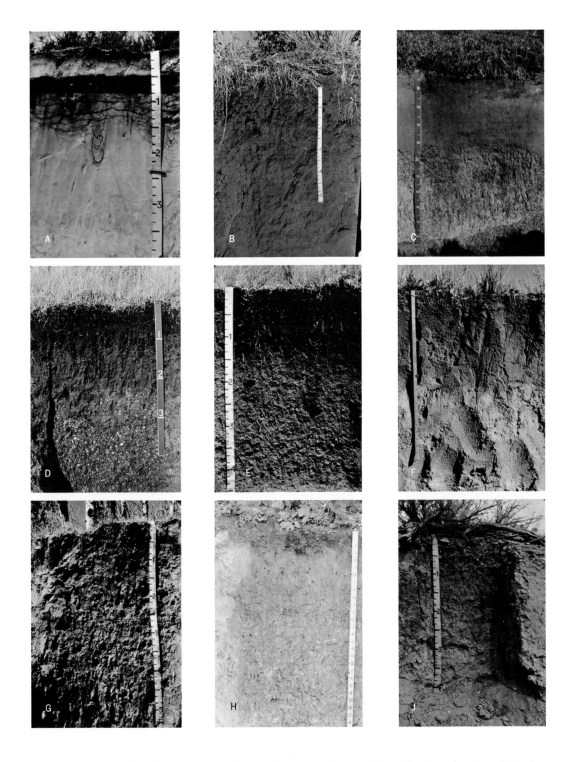

A Spodosol (Podzol) Holland; **B** Oxisol (Latosol) Sao Paulo, Brazil; **C** Ultisol (Red-Yellow Podzolic) Queensland, Australia; **D** Mollisol (Chernozem) South Dakota, U.S.A.; **E** Mollisol (Brunizem) Iowa, U.S.A.; **F** Mollisol (Chestnut) Alberta, Canada; **G** Alfisol (gray-wooded) Wyoming, U.S.A.; **H** Alfisol (gr.-br. Podzolic) Iowa, U.S.A.; **J** Aridisol (gray-desert) Nevada, U.S.A.

Plate I Representative soil profiles. Equivalent names of older classification are given in parentheses. (Courtesy of Roy W. Simonson, United States Department of Agriculture. Scales for C, F, and H are metric. All others are in feet.)

example, calcium carbonate is thoroughly leached from the A and B horizons of soils in regions of ample rainfall. Also, clay particles of extremely small size are mechanically washed from the A horizon by downward percolating rainwater and redeposited in the B horizon. Generally, the A horizon is called the eluvial horizon and the B horizon is the illuvial horizon.

In arid regions, on the other hand, water may actually move *upward* through the solum because of intensive evaporation, thus causing a lesser degree of differentiation between the A and B horizons. Concentrations of calcium carbonate are found at much shallower depths in soils of arid lands than soils of the more humid regions.

The C horizon of the soil profile is roughly equivalent to the regolith and is aptly called the parent material of the soil. Some soils that have had insufficient time for full development contain only an A and a C horizon.

Soil scientists study soil in much greater detail than has been given here. The three horizons are subdivided further, which in some cases results in as many as eight horizons, each designated by the letter of the main horizon to which it belongs plus a subscript (for example, A_0, A_1, A_2, A_3, B_1, B_2, etc.). These refinements are important in pedology but do not add materially to the basic principles of soil science as outlined here.

Soil Classification

A soil is produced over a period of time by the interaction of climate and organisms on a certain parent material that occurs in a particular topographic setting. Early classification schemes by the Russians in the nineteenth century and later modifications by Americans in the twentieth century generally reflected the point of view that the geographical distribution of different kinds of soils was largely controlled by the climatic conditions that prevailed in the regions where those soils were found.

Because considerable time is required for the development of a soil profile, the actual genesis of a soil (that is, development of the solum) cannot be observed in nature nor can it be reproduced in the laboratory. By and large, then, soil genesis must be inferred from soil morphology. Up until the mid-twentieth century, soil scientists believed that they could differentiate the combined effects of the five soil-forming factors as they were manifested in the soil profile. And, to a large extent they succeeded. Laboratory studies on the physical and chemical properties of soils verified some of the theories based on field observations and also provided additional means whereby subtle differences in soils could be explained.

As more and more soils were studied, however, and as more and more complexities were revealed, it became obvious to some of the leading pedologists that the intricacies of soil morphology were not so easily explained by the accepted ideas of soil genesis. Furthermore, it was no longer a simple matter to fit all the different kinds of soils into a tidy classification in which climate was the major controlling factor of soil genesis, even when differences due to topography, vegetation, and time were taken into account. Put in other terms, it was no longer possible to classify soils on a genetic basis when the only clue to genesis was the morphology of the soil.

By 1960, American pedologists in the United States Department of Agriculture had discarded the older climatic classification and had replaced it with one that is now the official system in the United States and a number of other countries. Table 8.d gives the major categories (soil orders) in this system, and Plate I illustrates the visible characteristics of some of these soils.

The United States Department of Agriculture Soil Classification System. This system divides the soils of the earth into ten major categories called *orders.* Each order is further subdivided on the basis of more refined morphologic characteristics, but for the purposes of this book, they will be ignored. Generally speaking, the soils of most of the ten orders are climatically controlled in their gross properties as can be seen in Figure 8.16. Below the rank of order, however, climatic influence is of less importance in dis-

Table 8.d Distinguishing Characteristics, Climatic Range, and Vegetation of the Ten Soil Orders According to the United States Department of Agriculture Soil Classification System*

Soil Order	Climatic Range	Natural Vegetation	Distinguishing Characteristics
Entisol	All climates, from polar regions to subtropics, arid to humid. No climatic significance.	Highly variable, from forests and grass, to desert shrubs.	Little or no horizonation with no more than A and C horizons. Common in recently transported materials such as river floodplain deposits, in regoliths highly resistant to weathering such as quartz sands, and on steep slopes subject to rapid removal of weathering products. Occur on flat to steep slopes.
Vertisol	Humid to arid. Hot dry seasons followed by season of higher precipitation.	Mostly grass and woody shrubs.	Soils of high clay content that form cracks in dry season and expand during wet season. B horizon commonly absent. Occur on flat surfaces as well as sloping uplands. Parent material widely variable but always characterized by high content of clays that expand when wet.
Inceptisol	Humid and subhumid areas from Arctic to the tropics. Do not occur in desert regions.	Mostly forests but some grass.	Weak horizon differentiation with no significant eluviation or illuviation. Parent material not weathered. Appreciable accumulation of organic material in A horizon. Occur on land surfaces ranging from flat to moderately steep slopes.
Aridisol	Arid, semiarid, steppe.	Sparse grasses, shrubs, cactus, and other desert plants.	Calcium carbonate (lime), calcium sulfate (gypsum), or sodium chloride concentrated in various parts of the profile. Lime may cement soil particles to form *caliche*. Horizons not well differentiated, generally. Colors range from gray in steppe climates to red in true deserts. Little organic matter in A horizon.
Mollisol	Humid, subhumid, semiarid. Mostly cold winters, hot summers.	Dense grass.	Dark soils due to heavy concentration of organic matter in the A horizon well mixed with mineral constituents. Lower B or upper C horizon may have secondary lime accumulation.

Table 8-d, continued

Soil Order	Climatic Range	Natural Vegetation	Distinguishing Characteristics
Spodosol	Humid continental to subtropical.	Coniferous forest, heather, some deciduous forest.	An A_2 horizon of organic matter over a mineral A_2 horizon with pale colors due to lack of iron oxides and organic matter. B horizon illuviated with organic matter, iron oxides, and aluminum oxides; generally brown or black in color. Soils do not develop in parent material high in clay.
Alfisol	Cool humid to subhumid.	Generally deciduous forest.	Thin A_1 horizon rich in humus and thicker light brownish-gray A_2 horizons over B horizon illuviated with very fine clay particles and with small amounts of organic matter and iron oxides. Parent materials relatively young (geologically speaking) and usually calcareous.
Ultisol	Humid, sub-tropical, and tropical.	Mixed forest to rain forest.	Thin A_1 horizon with some humus and thicker A_2 horizon of pale to moderate color over B horizon illuviated with very fine clay particles and some iron and aluminum oxides. More strongly weathered and leached than Alfisols. Parent materials generally old and strongly weathered but some un-weathered feldspars and micas remain.
Oxisol	Tropical and subtropical.	Rain forest.	Somewhart darkened and fairly thick A_1 horizon over B and C horizons of great thickness. Aluminum and iron oxides are residual concentrates through loss of other components, including silica. High in clay but porous, permitting ready downward movement of water. Parent material is deeply weathered with no original minerals remaining.
Histosol	Moist to wet.	Swamp, marsh, and bog vegetation.	Solum made up chiefly of organic materials under conditions of very poor drainage. No visible horizons. Commonly called bog soils, muck, or peat.

* Prepared with the help of Roy W. Simonsen, Soil Conservation Service, U.S. Dept. of Agriculture.

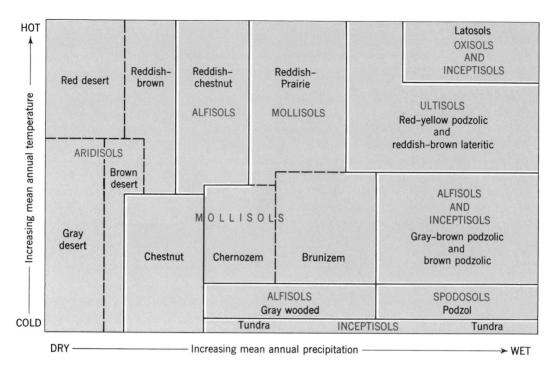

Figure 8.16 Diagram showing the relationship of the soil classification system of the United States Department of Agriculture (colored lettering) compared to the soil names of the previous classification system (black lettering). See Table 8.d for a description of the different kinds of soils shown above.

tinguishing the submembers of the orders. Table 8.d summarizes the climatic range, natural vegetation, and distinguishing characteristics of each of the ten orders. The names of the orders are coined words in which Latin and Greek roots are used. The names of some comparable soils in the older classification are also given in Figure 8.16, and the world distribution of soils is shown in Figure 8.17.

United States soil scientists of the Department of Agriculture are constantly refining this system, which was adopted officially in the 1960's. They are of the opinion that this system is better than the old for a number of reasons. To begin with, the United States system is based on observable soil characteristics instead of on inferred genetic factors. The new system is one based solely on the properties of the objects to be classified rather than on the inferred origin of those properties. Because of this approach to clas-

sification, soils that have had a complex history of origin are more amenable in the new system as compared to the old. For example, soils that have developed under more than one climatic regime during their life histories retain profile characteristics produced by each climatic episode. These *polygenetic soils* were difficult to accommodate in the older system but are less so in the new one. Another example of "problem" soils in the older classification schemes are those that have been cultivated by man. Plowing disturbs the upper part of the soil profile and destroys the original characteristics of, at least, part of the solum. In the old system, these soils were classified on the basis of properties that they were *presumed* to have had in their natural or virgin state. The new system accommodates these soils more readily.

Although it is inaccurate to claim that *all* pedologists in the United States have adopted

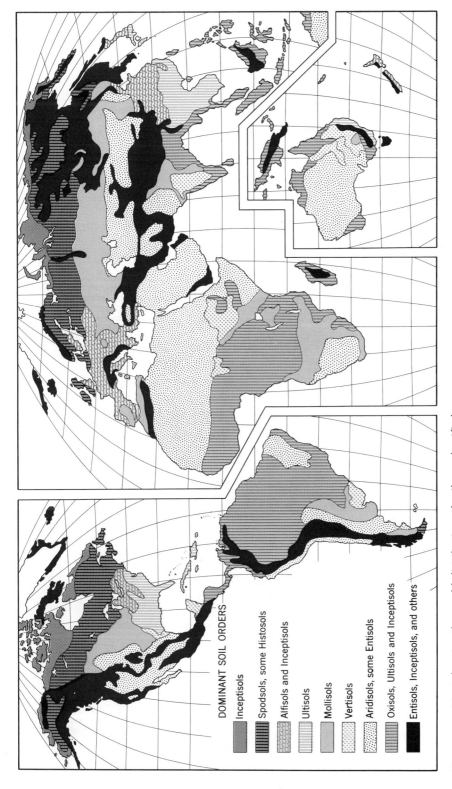

Figure 8.17 Map showing the world distribution of soils as classified by the United States Department of Agriculture. See Table 8.d for description of the soil orders. (Courtesy of Roy Simonsen, U. S. Soil Conservation Service, Department of Agriculture.)

DOMINANT SOIL ORDERS

Inceptisols

Spodsols, some Histosols

Alfisols and Inceptisols

Ultisols

Mollisols

Vertisols

Aridisols, some Entisols

Oxisols, Ultisols and Inceptisols

Entisols, Inceptisols, and others

the new scheme of the United States Department of Agriculture, the fact that the system has been adopted as the *official* system of the United States Soil Conservation Service and other Department of Agriculture units will be a powerful influence in keeping the system in vogue until some future group of pedologists in the United States or elsewhere devises a better one.

References

* Baldwin, M., C. E. Kellogg, J. Thorp, 1938, Soil classification: *Soils and Men, Yearbook of Agriculture 1938,* U.S. Government Printing Office, Washington, D.C., pp. 993–995.

Brown, R. J. E., 1970, *Permafrost in Canada,* University of Toronto Press, Toronto, 234 pp.

Goldich, S. S., 1938, A study in rock weathering, *Jour. Geology,* V. 46, pp. 17–58.

Keller, W. D., 1955, *The principles of chemical weathering,* Lucas Brothers, Columbia, Missouri, 88 pp.

Lachenbruch, A. H., 1963, Contraction theory of ice-wedge polygons: a qualitative discussion, *Proceedings, Permafrost International Conference,* National Acad. Sciences—National Research Council, Pub. 1287, Washington, D.C., pp. 63–71.

Loughnan, F. C., 1969, *Chemical weathering of silicate minerals,* Oliver and Boyd, London, 150 pp.

Maignien, R., 1966, *Review of research on laterites,* UNESCO Natural Resources Research IV, Paris, 148 pp.

* McNeil, Mary, 1964, Lateritic soils, *Scientific American,* V. 211, No. 5 (November) pp. 97–102. (Offprint 870, W. H. Freeman and Co., San Francisco.)

* Millar, C. E., L. M. Turk, and H. D. Foth, 1965, *Fundamentals of soil science,* Chapter 10, Origin and classification of soils; and Chapter 11, Great soil groups, John Wiley and Sons, Inc., New York, 491 pp.

Papadakis, J., 1969, *Soils of the world,* Elsevier Publishing Co., New York, 208 pp.

* Péwé, Troy L., 1966, *Permafrost and its effect on life in the north,* Oregon State University Press, Corvallis, Oregon, 40 pp.

Reiche, P., 1950, *A survey of weathering processes and products,* University of New Mexico Press, Albuquerque, 95 pp.

Simonson, R. W., 1963, Soil correlation and the new classification system, *Soil Science,* V. 96, pp. 23–30.

Stearns, S. Russell, 1966, *Cold regions science and engineering part I, section A2, Permafrost (perennially frozen ground),* Cold Regions Research and Engineering Laboratory, Hanover, N.H., 77 pp.

* Strahler, A. N., 1969, *Physical geography,* 3rd ed., Chapter 13, Climate Classification and Climatic regimes, John Wiley and Sons, Inc., New York, 733 pp.

Taber, Stephen, 1929, Frost heaving, *Jour. Geology,* V. 37, pp. 428–461.

Thorp, J., and G. D. Smith, 1949, Higher categories of soil classification: order, suborder, and great soil groups, *Soil Science,* V. 67, pp. 117–126.

* Trewartha, G. T., 1954, *An introduction to climate, Part II, The world pattern of climates,* McGraw-Hill Inc., New York, 395 pp.

Washburn, A. L., 1956, Classification of patterned ground and review of suggested origins, *Geological Society of America Bulletin,* V. 67, pp. 823–866.

* Recommended for further reading.

9 Gravitational Movement of Earth Materials

Decaying rock pinnacles in Arizona. (Photograph by Tad Nichols.)

Thus we perceive how motion may be produced by the combined action of the decomposition and gravitation of large masses of rock.   Playfair

The earth's gravitational field is controlled by a force that tends to draw all materials toward its center. The gravitational force is omnipresent and causes rocks and regolith to move in a downslope direction when the forces that tend to keep them at rest or in a fixed position are reduced.

This chapter is concerned with gravity-induced movements of earth materials in which a transporting medium such as ice, water, or air is not necessarily involved. The movement of rock and debris under the influence of gravity is also known as *mass movement* or *mass wasting*. Such movement may be imperceptibly slow or extremely rapid, and may involve a single rock fragment, a few grains of sand, or half a mountainside. The movement may have both a vertical and horizontal component, or it may be vertical only as in the free fall of rocks from a cliff face.

The settling or subsidence of parts of the earth's surface is also a form of mass movement. The chief cause of this subsidence is compaction related to excessive withdrawal of fluids (water or crude oil) from underground reservoirs, and is therefore discussed in the next chapter on groundwater.

Mass wasting is closely allied with weathering because materials subject to gravity-induced movements usually have been weakened by physical or chemical weathering beforehand, but many uncemented sediments that have been only slightly weathered also succumb to mass wasting. As a geological process, mass wasting is directly responsible for the configuration of much of the earth's landscape, but many of the characteristic landforms resulting from earth movements go unnoticed by the casual observer.

Spectacular and, not infrequently, catastrophic effects of gravity-induced earth movements occur as landslides and mudflows. Some of these destructive dislocations of large earth masses are triggered by the activities of man, but the great majority of them occur as the result of natural processes. In either case, large-scale earth movements can create havoc in populated areas and wreak great destruction on man-made installations. Mass wasting is therefore an important aspect of man's natural environment, particularly in urban areas where the pressure for more building space has led to the practice of locating new dwellings on sites that are highly susceptible to mass movement.

Factors Causing Slope Instability and Slope Failure

The surface of the land is dominated by slopes that range from sheer cliffs and precipices in mountain regions, river gorges, and coastal bluffs to the gentler terrain of the foothills and lowlands. Large expanses of uniformly horizontal surfaces are rare in nature, and many of the surfaces that appear to be completely flat to the naked eye have a slight inclination.

All earth materials in the solid state, bedrock or regolith, possess some strength. That is to say, they have a greater or lesser capacity to retain their shape and position under natural conditions. A natural slope that retains a constant angle of inclination over a period of time is said to be stable or in a condition of equilibrium. This is simply a way of saying that the strength of the rock or regolith on which the slope has been developed is strong enough to counteract the force of gravity that tends to pull rock and soil in a downward direction.

The most important factor in maintaining slope stability with respect to gravity is the *shear strength* of the materials on which the slope is developed. The shear strength of natural materials is that property which tends to keep rock or regolith from gravitational dislocation either by slippage along some potential plane of failure or by internal deformation.

Rocks vary greatly in their shear strength. Some rock masses are strong in parts but have weak zones in the form of joints, faults, fractures, or bedding planes. Weathering produces further weakening along these zones, thereby setting the stage for future failure of the rock. The inclined contact between regolith and bedrock is another example of a slope that may be intrinsically unstable with respect to gravity.

Materials with high clay content are capable of absorbing much water under which condition their shear strength is low throughout the entire mass rather than just along preexisting planes of weakness.

Slope failure is a response of the slope material to a change in the conditions that control its stability. For any given material there are two kinds of changes that tend to decrease slope stability. One is a reduction in the shear strength, already defined, and the other is an increase in the *shear stress*. A shear stress is a force tending to cause two adjacent parts of a solid to slide past one another along a plane or a number of parallel planes.

Reduction in Shear Strength. Most of the weathering processes produce a loss of shear strength in rock outcrops and regolith. Chemical weathering loosens the bonds between adjacent mineral grains in igneous and metamorphic rocks, dissolves cementing material in detrital sediments, and produces clay minerals that are intrinsically of low shear strength. Some rocks or regoliths are initially weak insofar as their shear strength is concerned, and weathering merely aggravates an already unstable situation. Clays and shales are particularly low in their resistance to shear and are more susceptible to mass wasting than other rock types.

Water plays an important role in the shear strength of earth materials. In small amounts, water contained in the pores between grains of sand or silt may actually increase the shear strength of those materials but, in larger amounts, water may cause swelling of clay minerals and, in very large quantities, water can change an otherwise coherent mass of shale or clay to a "soupy" mixture that acts more like a fluid, with practically no shear strength, than a solid. Many rapid movements of slope materials occur during or following heavy rains or when thawing snow or frozen ground releases copious moisture to underlying materials.

Increase in Shear Stress. Gravity is the main force that produces a shearing stress in earth materials, but since gravity is a constant with respect to mass wasting, it cannot be responsible for an increase or decrease in shear stress. A shearing stress may be increased by (1) increase in weight of the material, (2) removal of support, or (3) earth vibrations.

The increase in weight of material on a slope may be brought about by the addition of new material on top or by saturation with water. Removal of support is accomplished by natural or artificial means and involves the increasing of the inclination of all or part of the slope. The undercutting at the base of a river bank by running water or the attack of waves against a sloping shore are examples of natural processes that remove support. Excavations during construction projects accomplish the same thing. Earth vibrations result from earthquakes, the nearby passage of heavy trucks or trains, blasting, and the like. In many cases, vibrations or tremors act as a mechanism for triggering slope failures in situations where slope instability has already been created by other means.

Types of Slope Movements

Slope movements are divided into four general categories: (1) free fall, (2) slippage or sliding along one or more discrete planes, (3) flowage, and (4) creep. While this classification is a systematic means of differentiating one from the other, it must be borne in mind that many slope failures involve more than one kind of movement during the time in which the dislocation of slope material is in progress.

Table 9.a is a simplified presentation of some of the more commonplace examples of slope movement. The first two kinds of movement, free fall and sliding, result from shear failure along a plane or a series of parallel planes. The third form of movement, flowage, represents deformation of the whole mass rather than failure along discrete planes or zones of weakness. Creep is the slowest of all movements and may involve either slippage or flow, depending on the material involved, the slope angle, and other factors. No attempt is made to include every possible combination of movements because the emphasis is on the principles involved

Table 9.a Examples of Earth Movements on Slopes According to the Kind of Movement*

Kind of Movement	Example
Free Fall	Rock fall, debris fall
Sliding	Rock slide, debris slide (landslides), slump
Flowage	Debris flow, mudflow, solifluction
Creep	Normal creep, rock glacier

* Simplified and modified from Sharpe, C. F. S., 1938, *Landslides and Related Phenomena*, Columbia University Press, New York, 137 pp.

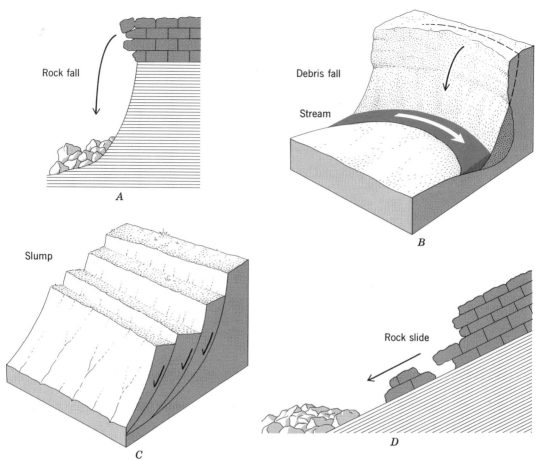

Figure 9.1 Some common types of gravitational movements of earth materials: *A* and *B*, free fall; *C* and *D*, failure by slippage along one or more discrete planes.

rather than an exhaustive classification. Diagrammatic sketches of some of the main types are shown in Figures 9.1 and 9.2, and photographs of real examples are shown where appropriate in the pages that follow.

Movement by Slippage Along Planes of Weakness

Free Fall. As the term clearly implies, movements in this category result when material at the base of an escarpment is removed, thereby leaving the overlying rock layer or regolith unsupported. The two examples, *rock fall* and *debris fall,* are illustrated in Figure 9.*A* and *B*. In mountainous

regions, diurnal freezing and thawing of water along joints is an additional process that abets the production of rock or debris falls.

Sliding. Three kinds of sliding movement are listed in Table 9.a. *Rock slides* occur where planes of weakness such as bedding or joint planes are more-or-less parallel to the surface of the rock outcrop (Fig. 9.1*D*). *Slump* can develop in either bedrock or unconsolidated material, and is characterized by a rotational movement of the slumped block as it slips along a curved plane of failure (Fig. 9.1*C*).

Debris slides are among the most spectacular kinds of failures and occur in a number of different ways. Debris slides and rock slides are known by the more familiar term,

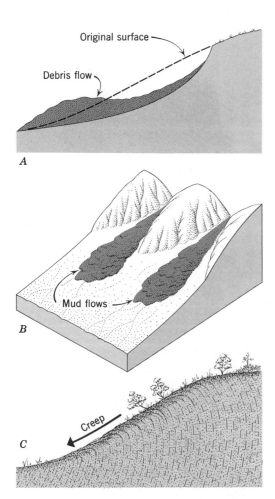

Figure 9.2 Diagrams of different kinds of gravitational movement. Flowage produces debris flows (A) and mud flows (B), both of which move rapidly in comparison to creep (C).

landslides. Some geologists also use the term, *rock* or *debris avalanche* to describe such phenomena, but generally avalanche is used more appropriately in reference to snow or ice slides in mountainous regions of the world. In that sense, an avalanche is a debris slide in which the debris is mostly snow, or snow and ice, with some rock debris mixed in.

One of the most spectacular landslides (debris slides) of modern times is the one that blanketed a large segment of the lower Sherman Glacier in the rugged Chugach Mountains of southeastern Alaska (Fig. 9.3). The Sherman landslide was triggered by the Alaska earthquake of March 27, 1964. The earthquake, whose epicenter was 80 miles to the east, released a rock slide on the face of a peak near the south flank of the glacier. The landslide started as a rock slide composed of a massive sandstone and metamorphic rock that was shaken loose along a bedding plane from the upper 160 feet of the peak. The rock mass, roughly 1500 feet long, 1000 feet wide, and 500 feet thick, broke instantaneously into fragments that plunged 2000 feet down a 40-degree slope and slid from that point as far as 3 miles from its point of origin across the gentle slope of the lower Sherman Glacier. On the opposite side of the valley the slide climbed 80 feet up the valley wall. Debris from the slide covered an area of about 3.4 square miles with an average thickness of 10 to 20 feet.

The slide is believed to have reached a maximum velocity of 115 miles per hour before it settled over the glacier as a rather uniform blanket of debris. The most interesting aspect of the Sherman slide is the explanation of how it could move over such a broad aerial expanse. Detailed studies by a landslide expert, Ronald L. Shreve, revealed that the slide actually traveled on a cushion of compressed air that was entrapped beneath the debris as it reached the base of the peak from which it broke loose. This compressed air acted as a lubricating layer that allowed the entire mass to spread with uniform thickness over a broad expanse at an extremely high velocity. This explanation also accounts for certain morphological characteristics of the debris surface that can be explained in no other way.

One of the most devastating avalanches in recent history was the one triggered by the Peruvian earthquake of May 31, 1970. This catastrophic event killed many thousands of people and destroyed entire villages. The earthquake loosened ice, rock, and debris from the north peak of Nevados Huascarán. These materials moved about 12,000 feet vertically and 9 miles horizontally in less than four minutes. The havoc created by this landslide of rock and ice moving with a velocity of well over 100 miles per hour is almost beyond description. The volume of ma-

Figure 9.3 Debris slide (also rock avalanche or landslide) on the Sherman Glacier, produced by the 1964 Alaska earthquake. The linear features are flow lines that indicate the direction of movement of the mass. (Photograph by Austin Post, U. S. Geological Survey.)

terial moved by gravity during these few minutes is more than ordinary erosive agents could move in a century or more.

Movement by Internal Deformation

A solid that deforms by changing shape is said to deform internally. That is, movement is not restricted to a single or a few planes of failure but is accomplished by the differential gliding or shifting of constituent particles with respect to one another. Movement of this kind is akin to the flow of highly viscous fluids if copious water is mixed with the material. Certain clays are transformed almost instantaneously into jellylike masses when vibrated, and other materials with high clay content move as a plastic rather than as a liquid. Plastic flow is flow that results in the deformation of the solid without any fracturing or rupture of the mass. Plastic materials have some shear strength that must be exceeded by a shearing stress before movement can begin, whereas a true fluid will continue to flow until its surface reaches a horizontal attitude.

The velocity of movement involving flowage ranges from a few feet per month to several feet per minute and depends on the degree of fluidity. Material moving by flowage is generally, but not universally, confined to some preexisting channel or valley.

Debris Flows. These flows are highly variable in water content. The flow of dry sand has been observed in situations where supporting material of the sand body has been removed. In most cases, however, debris flows occur on slopes made up of unconsolidated regolith containing a high water content and a wide range of particles, from clay size through large boulders (Fig. 9.2A). Some debris flows start out as debris slides but change to debris flows once movement has begun.

Debris flows with high water content have a consistency not unlike wet concrete, and while in motion, they have the capacity to transport or dislodge objects as large as an automobile. Highly fluid debris flows are very destructive of the terrain over which they pass, and those that occur in populous areas disrupt or destroy many of the cultural features along the courses they follow. Debris flows of high fluidity have been known to flow into river valleys thereby forming a natural dam behind which a lake forms. Other flows spread out over valley floors as lobate masses that remain for centuries (Fig. 9.4).

Mudflows. Mudflows are debris flows in which, at least, half the materials involved are in the clay, silt, or sand size range (Table 3.d). Mudflows invariably move in well-defined channels that are occupied by streams at other times, and they may carry larger fragments or blocks of rock that are picked up along the way. A mudflow progresses down-valley in a series of surges in

Figure 9.4 Debris flow west of the Pahsimeroi River in south-central Idaho. (Photograph by John S. Shelton.)

Figure 9.5 Lobate tongue of a mudflow in Kern County, California. The pattern of cracks was produced by desiccation of the flow some time after it ceased to flow. (Photograph by William B. Bull, U. S. Geological Survey.)

intervals ranging from a few seconds to several hours. Some mudflows are highly fluid and form digitate blanket deposits where they emerge onto the floor of a gentler sloping surface (Fig. 9.2*B* and 9.5).

Mudflows occur most frequently in environments where a thick, fine-grained material lies on a steep slope and where copious quantities of water are intermittently available from heavy cloudbursts or melting snow. Volcanic ash and other volcanic debris ejected during a volcanic eruption becomes a debris flow when melting snow on the volcano's flank provides prodigious quantities of water (see Chapter 5).

Solifluction. A phenomenon commonly but not exclusively associated with permafrost regions of the world is the downslope *flowage* of water-saturated regolith lying on gently sloping terrain. *Solifluction* (from *solum*, "soil," and *fluere*, "to flow") is the name of this process. In a sense, solifluction slope movements are miniature wet debris flows of low velocity and seasonal duration. Abundant water is derived from snow melt or

the thawed active layer in permafrost. Vegetational mats consisting of grasses, sedges, mosses, and other tundra plants are moved downslope by the underlying semifluid mass. Tundra slopes in Arctic and sub-Arctic climates exhibit evidence of solifluction in the form of small turf terraces, small hummocks, and other small landforms (Fig. 9.6).

Creep. The slow, imperceptible downhill movement of rock and soil debris is *creep*. Gravity is the force causing such movement, and manifestations of it can be observed in the weathered zone of steeply dipping sedimentary rock layers and the dislocation of cultural features situated on them (Fig. 9.2*C*). Creep is not dependent on water content, although the freezing and thawing of water in slope debris will accelerate the process. During a freeze-thaw cycle small stones and clumps of soil particles are thrust outward at right angles from the slope and lowered in a vertical direction when the small ice pinnacles beneath them are melted, as sketched in Figure 9.7. This process is more precisely defined as frost creep. *Frost creep* is a

Figure 9.6 Oblique aerial photograph of solifluction lobes near Tolovana, Yukon Territory, Canada. (Photograph by Troy L. Péwé.)

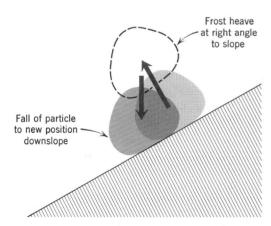

Figure 9.7 Diagrammatic drawing showing mechanics of downslope movement of a single soil particle due to frost creep. The particle is first thrust orthogonally away from the slope by the freezing of water in the soil beneath. When thawing removes the ice support, the particle falls vertically to a new position (shown in color) downslope from its original position.

"ratchetlike" movement not to be confused with solifluction which is a true flow phenomenon.

Rock Glaciers. High mountainous regions provide the topographic conditions and climatic environment conducive to the production of frost-derived rock rubble at the bases of cliffs and steep slopes. This angular rock debris may accumulate as talus (Chapter 8) or, under certain climatic conditions, it may move down-slope as a *rock glacier*. Excavations into rock glaciers reveal that the interstitial space between rock fragments contains ice, which accounts for the glacial-like movement and appearance of these forms (Fig. 9.8). Ice moves under the influence of gravity by plastic deformation with velocities on the order of a few feet per year. This rate of flow is comparable with other creep phenomena in that it is imperceptible to the casual observer. However, actual flow rates

Figure 9.8 Well-developed rock glacier south of Carbondale, Colorado. (Photograph by John S. Shelton.)

can be measured by precise surveying techniques carried out over a period of years.

A rock glacier that no longer moves is called inactive. Some inactive rock glaciers in Alaska are identified by a covering of lichens and other vegetation on their down-valley ends. Inactive rock glaciers are usually less than 100 feet thick whereas active rock glaciers are on the order of 150 feet thick.

References

* Embleton, C., and C. A. M. King, 1968, *Glacial and periglacial geomorphology* (Chapter 23, Periglacial mass movements and slope deposits), St. Martin's Press, New York, 608 pp.

Erickson, G. E., G. Plafker, and J. F. Concha, 1970, Preliminary report on the geologic events associated with the May 21, 1970, Peru earthquake, *Geological Survey Circular 639*, Washington, D. C., 25 pp.

* Hutchinson, J. N., 1968, Mass movement, in *The Encyclopedia of Geomorphology*, (R. W. Fairbridge, ed.), Encyclopedia of Earth Sciences Series, V. III, pp. 688–695.

Krynine, D. P., and W. R. Judd, 1957, *Principles of engineering geology and geotechnics, Chap. 17, Landslides and other crustal displacements*, McGraw-Hill Book Co., New York, 730 pp.

Marangunic, C., and C. Bull, 1968, The land-slide on Sherman glacier, *The great Alaska earthquake of 1964, V. 3, "Hydrology,"* National Academy of Sciences, Washington, D. C., pp. 383–394.

* Sharpe, C. F. S., 1938, *Landslides and related phenomena,* Columbia Univ. Press, New York, 137 pp.

* Shreve, R. L., 1966, Sherman landslide, Alaska, *Science,* V. 154, pp. 1639–1643.

Terzaghi, K., 1950, Mechanism of land-slides, *Geological Society of America, Berkey Volume,* pp. 83–123.

* Varnes, D. J., 1958, Landslide types and processes, Chapter 3 of *Landslides and Engineering Practice* (E. B. Eckel, ed.),

National Academy of Sciences-National Research Council, Highway Research Board Special Publication 29, Washington, D. C., 232 pp.

Wahrhaftig, C., and A. Cox, 1959, Rock glaciers in the Alaska Range, *Geological Society of America Bulletin,* V. 70, pp. 383–436.

Washburn, A. L., 1967, Instrumental observations of mass-wasting in the Mesters Vig district, northeast Greenland, *Meddeleser om Grønland,* V. 66, No. 4, C. A. Reitzels Forlag, Copenhagen, 296 pp.

* Recommended for further reading.

10 Groundwater

A newly drilled water well near Westfield, Massachusetts, being pumped to test its yield or discharge. (Courtesy of the Johnson Division, Universal Oil Products.)

. . . he digged the hard rock with iron, and made wells for water. Ecclesiastes

Groundwater is water that occurs in a saturated zone of variable thickness and depth below the earth's surface. Cracks and pores in rocks and unconsolidated deposits make up a large underground reservoir where part of the water in the hydrologic cycle is stored. The volume of water in the groundwater reservoir of the land areas contains the largest fraction of liquid water on the land areas (Table 10.a). An estimated 1,000,000 cubic miles of water is contained in all the earth's underground reservoirs to a depth of one-half mile. The volume of water beneath North America is so large that if it were distributed on a per capita basis, each person's share would be 750 million gallons.

Groundwater not only functions as a geologic agent in the role of solvent, carrier, and depositer of minerals, but it also serves as a source of water for use by man.

This chapter will deal with the basic principles of the origin, occurrence, and movement of groundwater, and its extraction by man through wells. Problems of supply will also be examined in the light of ever-

Table 10.a Estimated World Water Supply*

Water Item	Approximate Volume Cubic Miles (1000's)	Percent of Total Water
Water in land areas:		
Freshwater lakes	30	.009
Saline lakes and inland seas	25	.008
Rivers (average instantaneous volume)	.3	.0001
Soil moisture and vadose water	16	.005
Groundwater to depth of about 13,000 feet	2,000	.61
Icecaps and glaciers	7,000	2.14
Total, land areas (rounded)	9,100	2.8
Atmosphere	3.1	.001
Ocean basins	317,000	97.3
Total, all items (rounded)	326,000	100.00

* From U. S. Geological Survey, 1967, as reported in Todd, D. K. (ed.),
1970, *The Water Encyclopedia*, Water Information Center, Port Washington,
N.Y., p. 62.

increasing demands for more water in a world of irrigated agriculture, industrial expansion, and swelling metropolises. Finally, natural phenomena related to the geologic activity of ground water such as limestone caverns, hot springs, geysers, and the like will be considered.

The Origin of Groundwater

The philosophers or "scientists" of the ancient world knew of the existence of underground water because water could be observed emerging from the ground as springs. Moreover, water has been withdrawn from wells since early biblical times. Many myths and fantasies were fabricated by the ancients to account for the origin of underground water, but it was not until the late seventeenth century that two Frenchmen attacked the question in a scientific manner. Pierre Perrault (1608–1680) and Edmé Mariotté (1620–1684) made some hydrologic meas-

urements over a three-year period on part of the land drained by the River Seine. They measured the rainfall on this area and the amount of water discharged by the Seine at a point in Burgundy. These measurements showed that only one-sixth of the precipitation that fell on the land ended up as flow of the River Seine. The other five-sixths was "lost" somewhere between the point where the rain struck the earth and the point where the flow of the Seine was measured.

This simple experiment was the first conclusive proof that there was more than enough precipitation falling on the land to account not only for the flow of rivers but also for the source of underground water.

The scientific perceptiveness of Perrault and Mariotté paved the way for establishing the elements of the *hydrologic* equation, which can be expressed in a general form as follows:

Precipitation = Runoff + Infiltration + Transpiration + Evaporation

The last three elements on the right-hand side

of this equation represent the five-sixths of the precipitation on the River Seine drainage that did not appear as runoff in the river. *Infiltration* is the seepage of rainwater or snowmelt into the ground, *transpiration* is the use of water by plants, and *evaporation* is the change from the liquid to vapor state. Hydrologic measuring techniques have been vastly improved since the days of Perrault and Mariotté, and they all have substantiated the fundamental principle, that groundwater originates as precipitation which infiltrates the ground and fills the voids in rocks and regoliths.

Groundwater that originates from precipitation is called *meteoric water.* Two very minor sources of groundwater are *connate water,* salt water entrapped in sediments that accumulated in shallow seas of the geologic past, and *juvenile water,* the water that comes chiefly from volcanic emanations in the form of water vapor. Neither one contributes significantly to the total volume of subsurface water.

Occurrence of Groundwater

The outer portion of the earth's crust is made up of material ranging from dense granite with almost no pores, to loose, unconsolidated gravel with many voids between mineral grains. The volume of the pores in a rock or sediment is expressed as a percent of the total volume of the material, and is known technically as *porosity.* Porosity depends on the shape and packing of the grains plus the degree of sorting (uniformity of size) (Fig. 10.1).

To illustrate, a 5-gallon pail is filled with coarse, dry sand. To this is added one gallon of water which just saturates the sand. The porosity of the sand is established by the relationship,

$$\text{Porosity} = \frac{\text{Volume of voids}}{\text{Total volume}} =$$

$$\frac{1 \text{ gallon}}{5 \text{ gallons}} \times 100 = 20 \text{ percent}$$

All the materials of variable porosity near the upper part of the earth's crust can be considered as a potential storage place for groundwater and, hence, are called the *groundwater reservoir.* The total volume of water contained in the groundwater reservoir in any particular area is dependent on (1) the porosity of the rock, (2) the rate at which water is added to it by infiltration, and (3) the rate at which water is lost from it by evaporation, transpiration, seepage to surface water courses, and withdrawal by man.

To understand the conditions of occurrence of groundwater, consider the zones penetrated while drilling a hole through a homogeneous and isotropic (that is, the same in

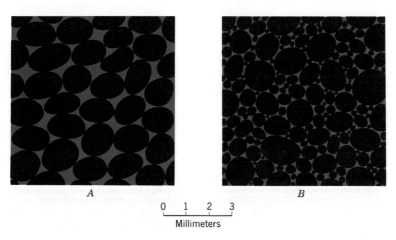

Figure 10.1 Diagrammatic sketch of well-sorted (*A*) and poorly sorted (*B*) sediments. Porosity of *A* is greater than *B* because the grains of *A* are all about equal size. In *B,* the smaller grains fill the spaces between the larger grains, thus reducing the volume of void space.

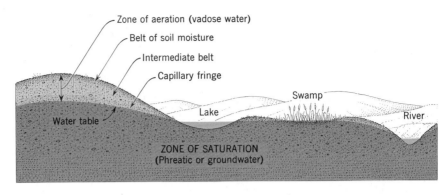

Figure 10.2 Cross-sectional diagram showing the zones of sub-surface moisture and the relationships of groundwater to surface water. (After Meinzer, O. E., 1923, *Outline of groundwater hydrology*, U. S. Geological Survey, Water Supply Paper 494, p. 23.)

all directions) material such as sand. Within a few feet of the surface the soil is slightly damp, depending on the recency of the last rainfall. Below this *belt of soil moisture* a zone of increasing moisture content, the *intermediate belt,* will be encountered, and still deeper the sand will be very wet where moisture is held by molecular attraction in the *capillary fringe.* These three belts comprise what is known as the *zone of aeration* or *vadose zone* as shown diagrammatically in Figure 10.2.

Eventually, the hole will penetrate sand in which all the voids were filled or saturated

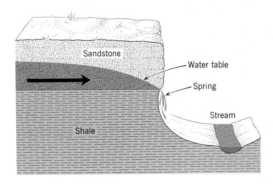

Figure 10.3 This spring is formed by groundwater seepage along the contact of the permeable sandstone and the relatively impermeable shale. The water table is said to be "perched." The arrow shows the direction of groundwater movement.

with water. This is the *zone of saturation* or *phreatic zone,* and the undulating plane separating the vadoze zone from the phreatic zone is the *water table.* In humid climatic zones, the water table is a subdued replica of the land surface. Where the water table lies at or very near the ground surface, swampy conditions exist, and a lake is merely a closed surface depression that has a bottom below the water table.

Springs are points at the earth's surface where water escapes from the groundwater reservoir and becomes incorporated in the surface drainage system. Springs usually occur along valley walls (Fig. 10.3), but wherever the water table intersects the ground surface, a spring occurs. Seasonal fluctuations of the water table also affect the discharge from springs to the extent that many of them cease to flow during periods of drought.

The depth to the water table fluctuates as the amount of infiltration changes. It is nearer the surface during wet seasons and deeper during dry seasons. If water ceased to be added by infiltration, the water table would eventually flatten out, because the water in the zone of saturation is constantly moving toward lower points on the water table, although such movement may amount to only a small fraction of a foot per day in silts and other fine-grained materials.

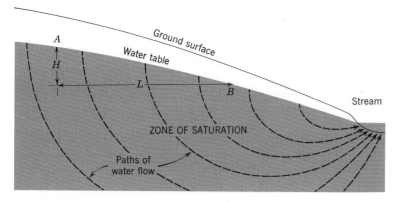

Figure 10.4 Cross-sectional diagram showing the flow of ground water in a uniformly permeable material. The difference in elevation between points *A* and *B* on the water table is *H,* and the distance between them is *L.* The hydraulic gradient is *H/L.* (After Hubbert, M. King, 1940, The theory of ground water motion, *Jour. Geol.,* v. 48, p. 930.)

Movement of Groundwater

Because the water table has high and low points on it, it is not in equilibrium. In order that equilibrium may be approached, however, water moves from the high points on the water table to points lower down. The rate at which this movement occurs is dependent on two factors: (1) the ability of the porous medium to transmit the water, and (2) the *hydraulic gradient,* usually expressed as the ratio between the difference in elevation of two points on the water table (in the direction of flow) and the distance between them (Fig. 10.4).

The ability of a rock or unconsolidated sediment to transmit groundwater is *permeability.* It is not to be confused with porosity, since the latter is only an expression of how much void space exists in the rock. The *coefficient of permeability,* on the other hand, is the quantity of water passing through a certain cross-sectional area of the water-bearing material in a definite time under a hydraulic gradient of 1 (that is, $H/L = 1.00 = 100$ percent). The United States Geological Survey defines the permeability coefficient, *P,* as the number of gallons of water transmitted through one square foot of an aquifer in one day under a hydraulic gradient of 100 percent. Laboratory and field tests have

yielded values of *P* that range from less than one gallon a day per square foot of shale or clay to 10,000 or more gallons a day per square foot of coarse gravel.

In a more general sense, clays and shales are considered to be relatively impermeable whereas sandstones, sands, and gravels are classified as permeable materials. Permeable materials in the zone of saturation will yield appreciable quantities of water to a well whereas relatively impermeable strata will yield little if any water to a well. Even though a shale may have a porosity equal to that of a sandstone (about 20 percent), the shale has a low permeability because the size of the openings between the individual mineral grains of the shale are too small to permit the easy passage of water through them.

The materials of the earth's crust vary greatly in their coefficients of permeability, and it is this variation that complicates the pattern of groundwater movement as well as the occurrence of water available for man's extraction by wells.

Aquifers and Wells

A well is a man-made hole or pit in the ground from which water can be withdrawn,

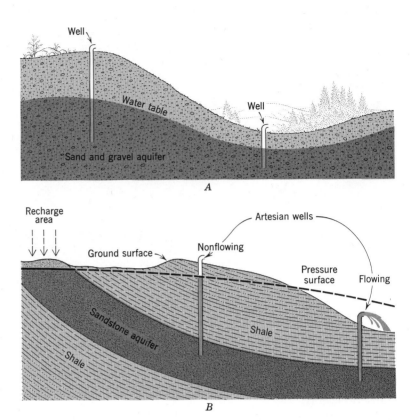

Figure 10.5 The two basic types of aquifers, unconfined (*A*), and confined (*B*). The water level in unconfined aquifers rises only to the water table, whereas the water level in artesian wells is governed by the pressure surface. A flowing artesian well is one in which the pressure surface lies above the ground surface.

and the geologic material that yields the water to the well is an *aquifer*. The amount of water yielded by a well is dependent on many factors, some of which, such as the well diameter, are inherent in the well itself. But all other things being equal, the permeability and thickness of the aquifer are the most important.

Aquifers vary in depth, lateral extent, and thickness, but in general all aquifers fall into one of two categories, *unconfined* and *confined*. An unconfined aquifer is one in which water-table conditions prevail because of the absence of a layer of relatively impermeable material on top (Fig. 10.5*A*). Confined aquifers are those which are capped with a relatively impermeable stratum that restricts the movement of the water (Fig. 10.5*B*); the water is thus under pressure and will rise in a

well that penetrates it. Such wells are called *artesian*.

The level to which the water will rise in an artesian well is determined by the highest point on the aquifer. The water in an artesian well cannot rise to this full height, however, because the friction of the water moving through the aquifer uses up some energy. The flow of water through a confined or artesian aquifer may be likened to the flow of water through an inclined pipe filled with sand (Fig. 10.6). If no flow existed, and no leakage occurred from the aquifer, then the levels in all artesian wells tapping the same aquifer would form a horizontal surface because of the principle that water contained in a vessel of any shape will seek a level surface.

The fact that the water level in an artesian well may rise all the way to the surface of the

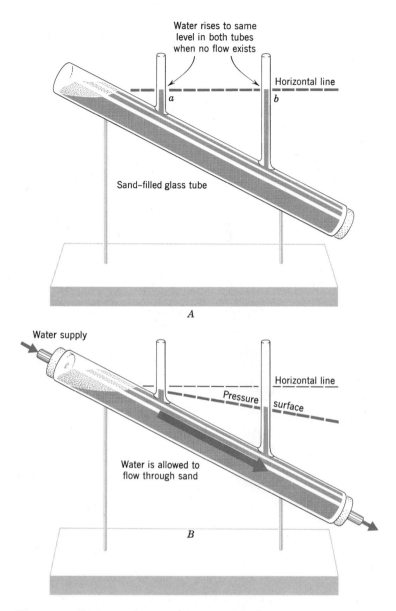

Figure 10.6 Diagram of an experiment in which
the principles of artesian wells are demonstrated.
A: a glass tube filled with permeable sand is saturated with water. The stopper in the lower end of
the tube prevents the movement of water through
the sand, and water in the two vertical "wells"
stands at the same elevation. B: when water is
allowed to flow through the sand, the water level
in the two vertical tubes defines the pressure surface. The difference in elevation of the water in
the two vertical tubes is H, and the distance between them is L. The hydraulic gradient is H/L.

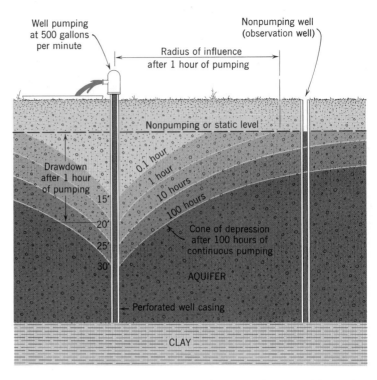

Figure 10.7 Diagram showing the development of a cone of depression around a pumping well drawing water from a homogeneous aquifer. Note the relationship of drawdown to time since pumping began, and the growth of the radius of influence to time. See text for further details.

ground is a matter of topographic circumstance rather than an inherent peculiarity of the artesian aquifer beneath that particular well. If the *pressure surface* (Fig. 10.5*B*) lies above the ground surface, the wells will be *flowing artesian wells,* but if the pressure surface is below ground level, the wells will be artesian but nonflowing, and will require a pump to bring the water to the surface.

Pumping of Wells

When a well pump is activated, the water level in the well immediately declines so that a hydraulic gradient is established toward the well forming a *cone of depression* if the aquifer is unconfined, and a *cone of pressure relief* if the aquifer is confined (Fig. 10.7). The difference between the *static* or nonpumping level of the water in the well, and the pumping level is the *drawdown;* and the maximum distance from the pumped well at which the effects of pumping are felt is the *radius of influence.* The drawdown increases and the radius of influence expands until the flow toward the well is balanced by recharge from the aquifer. In all aquifers, the drawdown *rate* decreases with time. The example in Figure 10.7 illustrates the changes in drawdown and radius of influence with time: Notice that the drawdown increased from 20 to 25 feet in 9 hours (drawdown after 10 hours of pumping minus drawdown after one hour of pumping). But, to increase the drawdown by another 5 feet, 90 additional hours of pumping is required.

If several closely spaced wells all draw water from the same aquifer, their respective cones of depression may overlap. This results in greater drawdown in each well, and can produce a general lowering of the water table in a heavily pumped area.

GROUNDWATER PROBLEMS

Problems of Supply

The water requirements for the United States are increasing at a phenomenal rate. In 1965 the 50 States and Puerto Rico used an average of about 311 bgd (billion gallons per day). This represents an increase of 15 percent over 1960 when about 270 bgd were used. During the same 5-year period the population increased only about 7.8 percent, which means that the daily per capita use has increased. In 1960 the per capita use was 1500 gpd (gallons per day), whereas in 1965 it was 1600 gpd.[1] Estimates for 1980 indicate a total use somewhere between 400 and 410 bgd, or 1740 to 1780 gallons per capita per day.

About 19 percent of all water used annually in the United States comes from groundwater sources. If the percentage of groundwater supplies is to remain constant or to increase slightly in the years ahead, the problem of locating new supplies and of making maximum use of existing aquifers will pose a continuing challenge to geologists and groundwater engineers.

Water Use Defined. Water use refers to water withdrawn from a surface or underground source for one of four categories, public supply, rural (domestic and livestock), irrigation, and industrial (including thermoelectric power generation and air conditioning). These uses are commonly referred to as *withdrawal uses* and are included in the figure of 311 bgd for 1965 in the United States. Table 10.b shows the amount of surface and subsurface water withdrawn for each of these categories in 1965. These figures show that the largest user of surface water is industry, and the largest user of groundwater is irrigated farm land.

Consumptive Use. A further refinement of the term water use or withdrawal use is necessary. Part of the water withdrawn for use in one of the four categories already defined is returned to the hydrologic cycle in liquid form after being used, and part of the water withdrawn is consumed in the process of

---

[1] These figures refer to total use for all purposes (that is, public supply, irrigation, industrial, etc.). *Individual* or domestic use is about 150 gpd per capita. Annual increases in this category are about 2 percent.

Table 10.b Estimated Water Use in the 50 United States and Puerto Rico for 1965*

Type of Use	Billions of gallons per day			
	Surface Water	Ground-water	Totals	Percent
Rural	1	3	4	(1)
Public Supply†	15	8	23	(7)
Irrigation	74	42	116	(38)
Industrial	160	8	168	(54)
Totals:	250	61	311	100

* Based on Murray, C. R., 1965, *Estimated use of water in the United States,* U.S. Geological Survey Circular 556, U.S. Government Printing Office, Washington, D.C., 53 pp.

† Water supplied by municipal and other public waterworks to 152,500,000 urban dwellers in 1965 amounted to 155 gallons per person per day. This amount is expected to increase by about 2.8 percent annually until 1980 when the rate of increase will drop to about 2 percent per year.

being used. For example, a town brewery withdraws water from a well for use in making beer. Some of the water is used to keep the vats clean, wash the floors, and other similar housekeeping chores. The volume of water thus used leaves the plant via the sanitary sewers and ends up in the municipal sewer plant where it is purified (hopefully) and discharged into the river flowing through the town. This water is available for reuse downstream and, hence, as far as the brewery is concerned, this water was used but not *consumed*. It is thus a *nonconsumptive use*. However, the water that was incorporated in the beer itself was consumed in the brewing process and is thus called a *consumptive use*.

Any number of similar examples could be cited, but in every one of the four categories of withdrawal use, consumptive and nonconsumptive uses must be separated in order to get a clear picture of water supply and demand. Of the 311 bgd of water withdrawn in the United States during 1965, 78 bgd, or about 25 percent, was in the form of con-

sumptive use. The largest consumptive use is in water used for irrigation (Fig. 10.8). The consumed portion of the water withdrawn for irrigation is lost by evaporation and by plant use, a combined process known as *evapotranspiration*.

The consumptive use of water is particularly acute in the states of Texas, Arizona, and California where much of the irrigation water is derived from groundwater reservoirs.

Consequences of "Overuse" of Groundwater Resources. Water withdrawn from groundwater sources is replaced by natural recharge. When the withdrawal rate exceeds the recharge rate, a corresponding decline in the water table or pressure surface results. Substantial lowering of the water table in West Texas between Lubbock and Plainview (Fig. 10.9) reflects a net loss in the water content of the major aquifer underlying this region (the Ogallala sandstone). In the San Joaquin Valley of California where groundwater is profusely used for irrigation, the water table has declined more than 100 feet in some places. Along the Santa Cruz River in southern Arizona near Tucson the overdraft of groundwater resulted in a decline of the water table by more than 60 feet in some places between 1962 and 1966.

Water withdrawal from aquifers in excess of recharge can lead to serious changes in both the aquifer and the ground surface above the aquifer. When an aquifer is dewatered it compacts into a smaller volume, so that its porosity is reduced and the surface above the aquifer collapses or subsides. Subsidence due to aquifer dewatering is irreversible. *Groundwater subsidence* has resulted from excessive pumping in many parts of California (Fig. 10.10) and in Las Vegas, Nevada. Large sections of Mexico City have subsided several feet because of the removal of large volumes of water from the underlying groundwater reservoir. Visitors to that city can observe the effects of subsidence in cracked buildings and dislocated streets and sidewalks. Entire buildings have visibly settled and tilted since they were originally built.

Depletion of groundwater supplies in coastal regions results in *seawater intrusion* into the aquifer. This condition along the

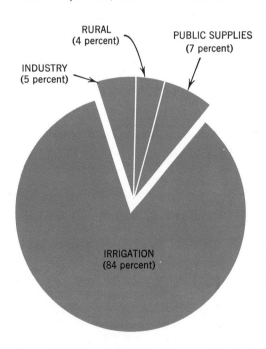

Figure 10.8 The total consumptive use of water in the United States is 78,000 million gallons per day (1965 estimate). Most of this loss is sustained by irrigation. (From Murray, C. R., 1968, *Estimated use of water in the United States,* U. S. Geological Survey, Circular 556, Washington, D. C., p. 8.)

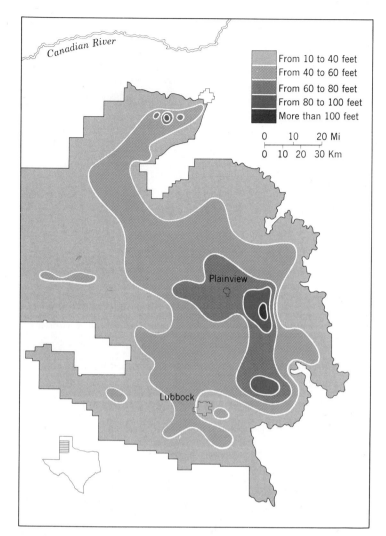

From 10 to 40 feet
From 40 to 60 feet
From 60 to 80 feet
From 80 to 100 feet
More than 100 feet

Figure 10.9 Map showing decline of the water table between 1938 and 1962 in a part of west Texas. The overdraft is due mainly to the pumping of groundwater in excess of natural recharge. (After a map in *The Cross Section*, v. 8, no. 12, High Plains Underground Water Conservation District No. 1, Lubbock, Texas, 1962.)

coast of California in the Los Angeles area prompted the introduction of surface water into the aquifer by means of injection wells as a means of driving out the saltwater and impeding its further landward encroachment into the aquifer (Fig. 10.11).

Increasing the Supply. Obviously the withdrawal of groundwater in excess of natural recharge cannot continue indefinitely.

Since it is unlikely that a rapidly expanding community will reduce its withdrawal rate and thereby curb its potential for more expansion, one of the solutions to declining water tables is *artificial recharge*. This technique requires putting water back into the ground reservoir either through wells or surface infiltration ponds. The state of Texas has a grandiose plan whereby water from the

Figure 10.10 These cracks in the ground are the result of subsidence over an area in Fresno County, California where groundwater has been withdrawn at a rate in excess of natural recharge. The aquifer beneath this area has collapsed, thus reducing the porosity irrevocably. Once an aquifer has reached this state, its original water-holding capacity cannot be restored by any known means. (Photograph by William B. Bull, U. S. Geological Survey.)

Mississippi could be diverted to the water shortage areas of the High Plains and used, in part, to recharge groundwater aquifers depleted by irrigation withdrawals. Certain industrial plants in communities in the eastern United States are recharging the groundwater reservoir with river water during the winter months, and then pumping water from underground for air conditioning and other uses in the summer when the demand is high. Underground storage of water is more efficient than surface storage because the consumptive loss through evaporation of groundwater is nil.

Besides importation of water from areas where there is a surplus to areas where there is a shortage, new supplies from unexploited aquifers will be sought. In burgeoning metropolitan areas, however, these sources are likely to be some distance from the points of use because the nearby sources have all been discovered.

Nevertheless, exploration will continue and geological information will be the basis on which new groundwater supplies are located. While a great many people still believe that underground lakes, rivers, and "veins" can be found by some sort of "hocus pocus" in which bent wire rods or forked branches are employed in the hands of "dowsers," "diviners," or "water witches" (Fig. 10.12), there is no scientific evidence to support the claims of these medieval practitioners. Anyone who is inclined toward the view that this is an unfair statement can explore the art of dowsing at greater length in the book by Vogt and Hyman which is listed in the references at the end of this chapter.

Groundwater Pollution. The fact that groundwater lies beneath the earth's surface, does not protect it from pollutants. Undesirable dissolved materials from both solid and liquid waste sources may contaminate a groundwater supply to the extent that water pumped from it is no longer fit for use by man.

Solid wastes generated by the urban environment are "disposed of" in a number of ways. In an appallingly large number of cases, the disposal method still consists of the ancient practice of depositing refuse in open dumps, many of which are located indiscriminately wherever land is available. A more modern practice of solid waste disposal is the use of the sanitary landfill which

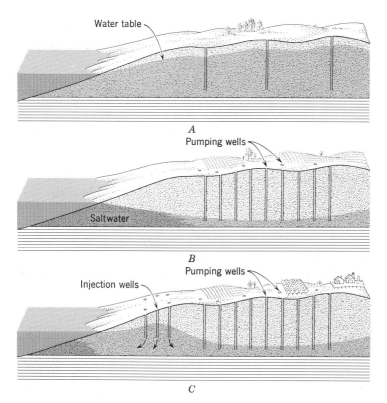

A

B

C

Figure 10.11 Hypothetical situation similar to aquifer conditions in coastal regions. *A:* water table slopes toward the coast with only a few pumping wells. *B:* excessive pumping from many wells lowers the water table so that its gradient is reversed, thereby allowing saltwater intrusion. *C:* freshwater, returned to the coastal zone through injection wells, forms a groundwater ridge between the ocean and the zone of fresh groundwater. The zone of saltwater encroachment is reduced and further intrusion is halted.

Figure 10.12 A "dowser" according to the cartoonist, Johnny Hart. (By permission of Johnny Hart and Field Enterprizes, Inc.)

consists of alternating layers of compacted refuse and regolith. Some landfills are developed on essentially flat land sites, but others are formed in natural depressions, gullys, and ravines, or in man-made pits.

When rain falls on a waste disposal site, some of the waste materials are dissolved or leached and may be carried by percolating waters into the underlying groundwater reservoir. Open dumps are more suceptible to the leaching process than landfills, but even the latter may allow the introduction of organic and inorganic constituents into the groundwater reservoir. These substances may destroy the use of an aquifer as a source of potable water.

The contamination of groundwater from solid waste disposal sites may spread many miles from the pollution source in a decade or less. The extent to which a pollutant will spread and the time required for its travel depends on the permeability of the materials in a landfill and beneath a dump, as well as the hydraulic gradient and permeability of the underlying aquifer.

From the foregoing it is obvious that the selection of solid-waste disposal sites should be made with due regard to the protection of surface and subsurface water supplies. This can be accomplished only by the study of the hydrologic conditions of the proposed disposal site before it is approved for use. Many states require that such studies be made, and regulate the location of sanitary landfills by law. This practice must be expanded if the public health is to be protected as the per capita volume of solid waste continues to mount each year.

Liquid wastes, many of which contain toxicant chemicals or are radioactive, are often "disposed" by injection into deep underground rock formations. For example, oilfield brines, which are unwanted by-products resulting from the production of crude oil, have been injected into porous rock formations for many years. Texas, with about 20,000 brine injection wells, along with the other oil-producing states in the United States, inject an estimated 3 billion gallons of brine underground every year. If the brine is returned to the same rock layer from which it was extracted with the crude oil, the poten-

tial for contaminating the overlying freshwater aquifers is slight. Not all brine is returned to its host formation, however. In these cases, the brine may come into contact with fresh groundwater, thereby significantly increasing the salt content of the aquifer.

The experience of oil-field brine injection cannot be related directly to the underground disposal of other industrial liquid wastes. The pore space of all rocks below the water table is occupied by some fluid, and if additional fluids are injected under pressure, the original fluids will be displaced. This means that under certain conditions a liquid waste disposed by underground injection methods could contaminate potable groundwater and render it unfit for human use.

In some cases, the effects of liquid waste injection are wholly unpredictable. An example is the case of the Rocky Mountain Arsenal 10 miles northeast of Denver, where liquid waste was disposed through an injection well. Between March 1962 and September 1963, and from September 1964 to February 1966, about 165 million gallons were injected into a fractured gneiss about 12,000 feet below the surface. The number of low intensity earthquakes in the immediate vicinity of the Arsenal increased substantially after the injection began, and decreased significantly during the period of no injection, from October 1963 to August 1964. The consensus is that the increase in the number of earthquakes was due to the injection of wastes. The increased liquid pressure apparently reduced the internal friction so that slippage along fractures was more frequent during injection periods than before or after. The "pollution" in this case was an increase in the number of earthquakes, the ultimate result of which is not predictable.

Although the underground disposal of liquid wastes may solve the problem of surface pollution, it may also introduce the possibility of groundwater pollution. Even though great care is exercised in studying the geological and hydrologic setting of the disposal site, and even though the best engineering skills are used in the design of the injection wells, the possibility of the waste liquids becoming intermingled with fresh groundwater cannot be dismissed. Considerable research on this problem is necessary in order that future laws regulating the practice of underground liquid waste disposal can be based on scientific fact rather than human guesswork.

GEOLOGIC WORK
OF GROUNDWATER

Groundwater accomplishes geologic work on a large scale. Underground caverns or caves represent the solution action of groundwater on soluble sedimentary rocks such as limestone, dolomite, gypsum, and rock salt. In the sense that rock materials are removed during cavern formation, groundwater is an erosive agent.

Groundwater acts also as an agent of transportation because dissolved mineral matter is carried in solution by the percolating underground water. Also, under certain conditions, the dissolved material carried in groundwater solution is redeposited, thereby making groundwater an agent of deposition. Groundwater erosion, transportation, and deposition thus constitute the three roles played by subsurface water as a geologic agent.

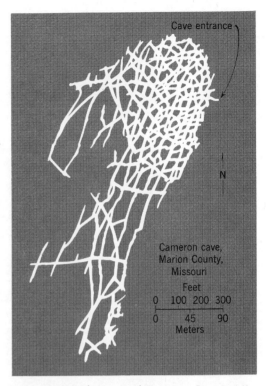

Figure 10.13 This map of Cameron Cave in Missouri shows an intersecting network of cave passages that are controlled by a joint pattern in the limestone bedrock. (After Bretz, J Harlen, 1956, *Caves of Missouri*, Missouri Geological Survey and Water Resources Division, Rolla, Missouri, p. 56.)

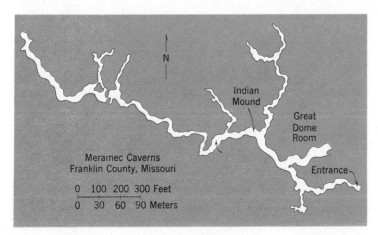

Figure 10.14 Map of Meramec Caverns in Missouri. The branching pattern is not as obviously controlled by joints as in the case of Cameron Cave shown in Figure 10.13. (After Bretz, J Harlen, 1956, *Caves of Missouri*, Missouri Geological Survey and Water Resources Division, Rolla, Missouri, p. 167.)

Figure 10.15 Stalactites, stalagmites, and columns in Floyd Collins Crystal Cave, Kentucky. Notice the water drops on tips of stalactites. (Photograph by William Austin.)

Limestone Caverns. Underground caves are attractive natural features that provide scenic pleasure for the vacationing public and a place of scientific studies for the geologist and speleologist.[2] Caverns are subterranean openings in soluble rock strata. They usually form a three-dimensional network or system of chambers and passageways frequently controlled by the joint system inherent in the original rock (Fig. 10.13). The growth of some caves is thus influenced by the joints and bedding planes in carbonate rocks, but other cave systems are less obviously joint-controlled in that they exhibit a branching or linear pattern (Fig. 10.14).

The question of cave origin is an interesting problem. Are caves formed above or below the water table? Although it is true that most of the known caves lie in the zone

[2] One who studies caves; not necessarily a geologist.

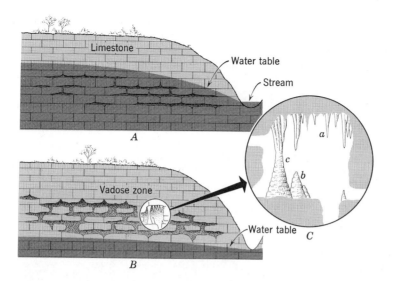

Figure 10.16 Diagram showing the two-cycle hypothesis of cave origin. *A:* the initial solution of the limestone is produced along joints and bedding planes beneath the water table. *B:* lowering of the water table by downcutting of the stream leaves the caverns in the vadose zone. Vadose deposits include stalactites (*a*), stalagmites (*b*), and columns (*c*), all of which are composed of calcium carbonate.

of aeration (vadose zone), it does not necessarily follow that this has been the case throughout the life history of the cave. In fact, the most obvious process taking place in known caverns is the *deposition* of calcium carbonate in the form of *stalactites, stalagmites,* and other *dripstone* features (Fig. 10.15).

From a geologic point of view, considerable evidence suggests that caves undergo a two-cycle history. The first cycle involves the dissolving of the rock along joints and bedding planes by circulating groundwater in the zone of saturation (Fig. 10.16*A*). The second cycle begins after the caverns emerge above the water table either by uplift of the land or by lowering of the water table. Once above the water table, the cavern is modified by the action of underground streams that can cause further enlargement by erosion, while calcium carbonate or other soluble compounds accumulate on the walls, ceiling, and floor of the cave when vadose water seeps into it (Fig. 10.16*B*). Although the

two cycle theory is not accepted by all geologists, it hardly can be denied that both phreatic and vadose features are common phenomena of limestone caves.

Karst Topography. The name karst was first applied to a limestone region along the Dalmatian coast of Yugoslavia. The name, karst topography, is now applied to countless other regions of similar geologic nature where the solution action of groundwater has dissolved portions of the underlying rock, resulting in caverns. A surface depression, called a *sink,* is formed when the roof of a cave collapses. As time progresses, the sinks enlarge and surface streams flow into the underground network of caverns. Karst topography is therefore characterized by the presence of many sinks and a system of underground streams flowing through solution passages in the rock. In the United States, karst topography is well displayed in Florida, southern Indiana, and in Kentucky especially in the vicinity of Mammoth Cave.

Figure 10.17 Eruption of Old Faithful Geyser in Yellowstone National Park, Wyoming is caused by the heating of groundwater in contact with hot underlying volcanic rock. (Photograph by George Grant, National Park Service.)

Geysers and Hot Springs. About once every hour, Old Faithful Geyser in Yellowstone National Park shoots out a column of scalding water nearly 150 feet in the air (Fig. 10.17). Since 1872 when Yellowstone became the first national park in the United States, millions of people have witnessed this spectacular display.

The water spouted from geysers is groundwater that has come in contact with hot (but not molten) igneous rock. Groundwater collects in irregular tubelike openings and although its temperature at depth may rise about 100° C., boiling does not occur. This is because the weight of the overlying water raises the pressure which, in turn, raises the boiling point. However, the superheated water expands and spills some of the water column over the lip of the surface orifice, causing a reduction of the pressure at depth and, hence, a lowering of the boiling point. At the moment the pressure is lowered, the

superheated water flashes into stream and violently drives out the remaining water. After the eruption is over, the process begins anew as more water seeps into the tube and becomes heated.

Travertine. Around the orifices of geysers and especially where hot springs discharge at the surface, calcium carbonate is released from solution and accumulates as a porous or solid mass known as travertine (Fig. 10.18). Impurities of iron and other substances impart variegated colors of red, brown, yellow, and black to some travertines. The carbonate deposits in caves are also known collectively as travertine.

Concretions and Geodes. Minor features associated with groundwater deposition are nodular masses in sedimentary rocks known as *concretions*. They range in size from a fraction of an inch to several feet and are commonly ellipsoidal or discoidal in shape. Some are highly irregular but they all rep-

Figure 10.18 Travertine terraces at Mammoth Hot Springs, Yellowstone National Park, Wyoming as they appeared in 1921. (Photograph by W. T. Lee, U. S. Geological Survey.)

Figure 10.19 These ancient logs in the Petrified Forest National Monument of Arizona were petrified while entombed in sediments below the water table. Subsequent erosion by surface water has removed the enclosing sediments and exposed the more resistant fossil logs.

resent the accumulation of calcium carbonate or other materials around a nucleus in the host rock, and literally grow in place.

Geodes are small cavities lined with crystals of quartz, calcite, or other compounds deposited by groundwater containing mineral matter in solution. In reality, a geode is a miniature cave in the cycle of filling by groundwater deposition.

Petrifaction. Wood or other organic material becomes petrified when its cells are replaced with mineral matter such as silica. Shells, bones, leaves, or entire logs may be completely replaced by mineral matter carried by groundwater. The Petrified Forest National Monument in Arizona contains hundreds of petrified logs scattered over many square miles (Fig. 10.19). Growth rings and other minute structural details of the cellular structure are commonly preserved during the process, and various elements in small amounts impart a variety of colors to these ancient logs.

References

Bretz, J Harlan, 1956, *Caves of Missouri,* Missouri Geological Survey and Water Resources Division, Rolla, Missouri, 490 pp.

* Leopold, L. B., and W. B. Langbein, 1960, *A primer on water,* U. S. Department of Interior, Geological Survey, U. S. Government Printing Office, Washington, D. C., 50 pp.

Meinzer, O. E., 1923, *The occurrence of ground water in the United States,* U. S. Geological Survey, Water Supply Paper 489, 321 pp.

Meinzer, O. E., 1942, *Hydrology, physics of the earth,* V., IX, McGraw-Hill Book Co., New York, 712 pp.

Piper, A. M., 1965, *Has the United States enough water?,* U. S. Geological Survey Water-Supply Paper 1797, U. S. Government Printing Office, Washington, D. C., 27 pp.

Piper, A. M., 1969, *Disposal of liquid wastes by injection underground—neither myth nor millennium,* Geological Survey Circular 631, U.S. Geological Survey, Washington, D.C., 15 pp.

Schneider, W. J., 1970, *Hydrologic implications of solid-waste disposal,* Geological Survey Circular 601-F, U. S. Geological Survey, Washington, D. C., 10 pp.

* Thomas, H. E., and L. B. Leopold, 1964, Ground water in North America, *Science,* V. 143, pp. 1001-1006.

Todd, David K., 1959, *Ground water hydrology,* John Wiley and Sons, New York, 336 pp.

U. S. Department of Agriculture, 1955, *Water yearbook of agriculture,* U. S. Government Printing Office, Washington, D. C., 751 pp.

Vogt, Evon Z., and Ray Hyman, 1959, *Water witching USA,* University of Chicago Press, Chicago, 248 pp.

* Recommended for further reading.

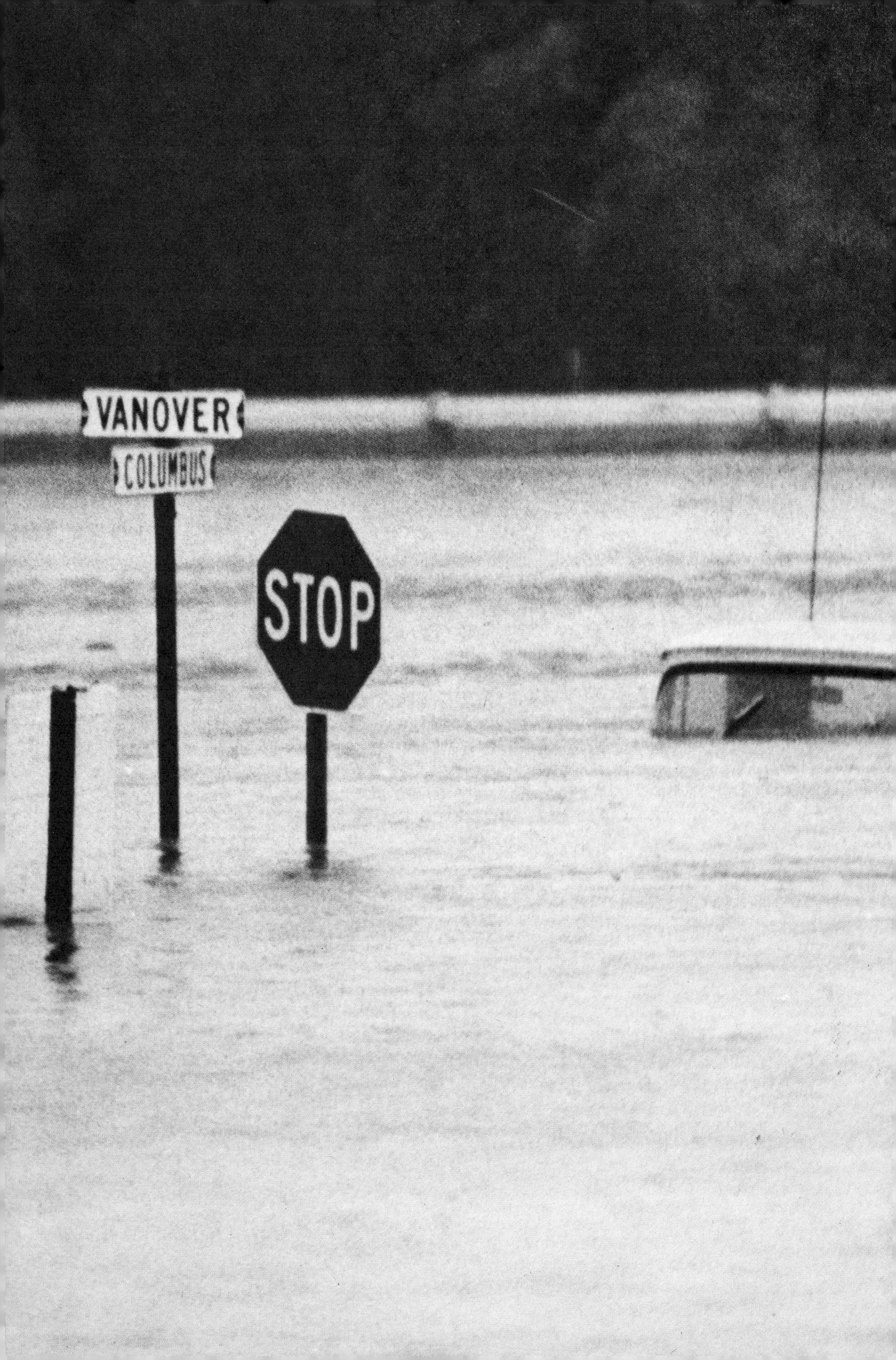
VANOVER
COLUMBUS
STOP

11 Rivers

One of the most important elements of the hydrologic cycle is the return of precipitation that falls on land to the ocean basins via the network of river channels that lace the continents. Although these rivers and their tributaries contain only 300 cubic miles of water at any one time compared with the more than 326,000,000 cubic miles of water on the entire earth (Table 10.a), they account for a vast amount of geologic work in eroding the land and transporting weathering products to the sea. The Mississippi River alone carries an average of 830,000 tons per day in suspension at Baton Rouge, Louisiana, and the Colorado River moves about 410,000 tons per day through the Grand Canyon. The inescapable conclusion to be drawn from the observations on these and other rivers is that running water is a potent agent of erosion.

In this chapter some of the major characteristics of rivers and drainage systems will be investigated, the role of running water in

Floodwaters of the Ohio River, July 1969. (Photograph by Don Roese, courtesy of the Akron Beacon Journal.)

sculpturing the surface of the earth will be examined, and the relationship of rivers to man and his environment will be considered.

Running Water on the Land Surface

Precipitation that falls on the land surface as rain or snow is recorded as inches or millimeters of water per year for a given gauging station. In the United States more than 10 thousand gauging stations reveal that the average annual precipitation for the 48 conterminous states is about 30 inches (760 mm).

Precipitation is the primary source of water that runs off the land surface. *Runoff* includes *channel flow,* the water which flows in established channels, and *overland flow,* the flow of rainwater or snowmelt over the land surface before reaching a stream. Some overland flow may never reach an established drainageway because it evaporates or infiltrates the ground en route.

Rivers and Climate

Rivers in Humid Climates. A river that carries runoff in a humid region of the world usually flows all year long, although the volume of water flowing past a certain cross section of river may show considerable variation from month to month or even day to day. So long as the river channel contains some flowing water the year around it is said to be a *perennial stream.* During a time of high flow, or high stage, a perennial stream is fed by groundwater seepage, precipitation, and overland flow; but during periods of no rainfall or lack of snowmelt, the flow in a perennial stream is sustained by contributions from the groundwater reservoir alone. In humid regions the water table is intersected by the stream channels as shown in Figure 11.1*A* so that groundwater moves toward the stream channel. A stream that is fed by groundwater seepage is an *effluent stream.* Most perennial streams are effluent, and the volume of water contained in them when no overland flow is available to feed them is called the *base flow.*

Rivers in Arid Climates. Aridity is caused by low precipitation and high evaporation. Rivers in these climatic zones are of two types. *Ephemeral streams* flow only during and immediately after a rain. *Intermittent streams* carry water only during wet seasons when frequent rains and some base flow sustain them. They cease to flow during dry seasons, however, because the water table is too low to provide any base flow. Ephemeral streams are *influent streams* (Fig. 11.1*B*), and actually lose water to the groundwater reservoir. The water is not "lost" insofar as man is concerned, however, because such water is still available from wells that can be drilled to tap these groundwater supplies. In terms of runoff, influent streams show dramatic reductions in flow in a downstream direction because the river water is progressively "lost" to the groundwater reservoir.

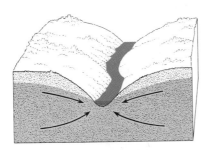

A Effluent stream

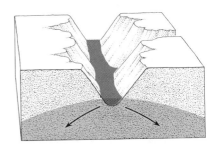

B Influent stream

Figure 11.1 An *effluent* stream is supplied with water from the groundwater reservoir in humid regions. An *influent* stream supplies water to the groundwater reservoir in arid regions.

THE DYNAMIC EARTH

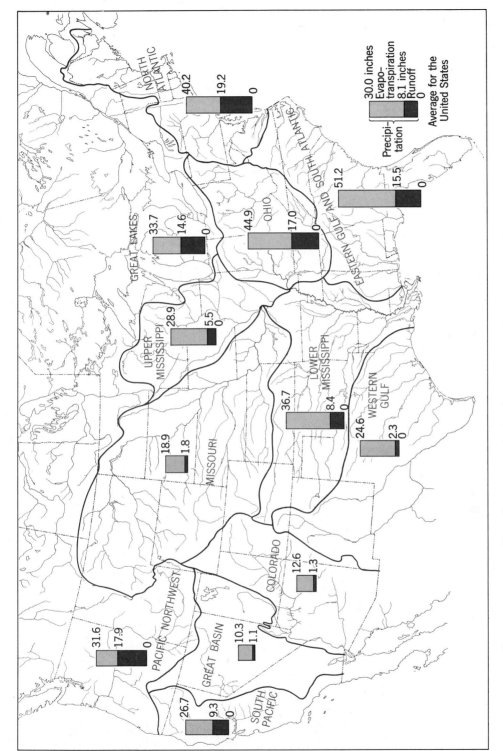

Figure 11.2 Annual precipitation, runoff, and evapotranspiration by major drainage areas in the 48 conterminous United States. (From Report of The Interior and Insular Affairs Committee, House of Representatives, U. S. Congress, 1952, *The physical and economic foundation of natural resources*, v. II, *The physical basis of water supply and its principal uses.*)

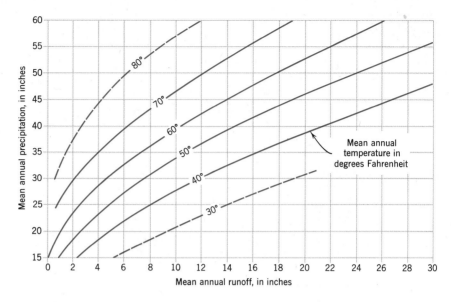

Figure 11.3 A series of graphs showing the relationships of precipitation and runoff to mean annual temperature. (From Langbein, W. B. and others, 1949, *Annual runoff in the United States,* U. S. Geological Survey Circular 52, 14 pp.)

Runoff

Runoff is a measure of all the water exported by a river through a particular channel cross section from the drainage basin upstream. The flow of water at a particular gauging station is expressed as a volume per unit time such as cubic feet or cubic meters per second, gallons or liters per day, or acre-feet per year.[1] Because precipitation is generally stated in inches or millimeters per year, it is convenient to convert stream-gauge measurements to the same units, thereby permitting a direct comparison between precipitation and runoff in any given drainage area.

Figure 11.2 reveals a wide variation in the ratio of runoff to precipitation for the major drainage areas of the conterminous United States. For the Columbia River Basin, runoff amounts to 56 percent of the annual precipitation, and for the West Gulf Coast Region, the annual runoff is only a little over 9 percent of precipitation. In the Mississippi River drainage basin, the ratio of runoff to precipitation is different for each of the tributary drainage basins.

Reasons for these differences in runoff-precipitation ratios are many in number and interrelated, but it will suffice to examine some of the more obvious ones.

Factors Controlling Runoff. Four broad factors influence the runoff from any given drainage basin. The larger the area of the watershed,[2] the more complex is the interrelationship of these factors. They are (1) climate, (2) geology, (3) topography, and (4) vegetational cover.

The classification of climates presented in Chapter 8 is based primarily on temperature and precipitation and the seasonal range of these two parameters. Temperature influences the runoff-precipitation ratio as shown in Figure 11.3. The higher the mean annual temperature for a given drainage basin, the greater the evaporation, hence, less water is

[1] An acre-foot is the volume of water on one acre of land covered to a depth of one foot. One acre-foot equals about 326,000 gallons.

[2] Watershed is synonymous with drainage basin.

available for runoff. The runoff-precipitation values in Figure 11.2 show, in general, that the lower percentage runoff values correlate with watersheds in warmer climates, and higher values are found in regions with cooler climates. Exact correlation between runoff and climate is not to be expected because of the other three factors that influence runoff.

Another climatic factor that exerts a control on runoff is the intensity and duration of rainfall. A slow drizzle lasting several hours will produce less runoff than a heavy downpour lasting 10 or 15 minutes, even though the total rainfall for each event is the same.

A geologic factor that exerts a considerable influence on runoff is the kind of soil, regolith, or bedrock that constitutes the surface material in the drainage basin. Loose-textured soils and permeable bedrock absorb more rainfall than heavy clay soils and relatively impervious bedrock. Permeable geologic materials have high water storage capacities and hold large volumes of water in

the groundwater reservoir, some of which will be returned later to the surface drainage courses as base flow.

As an example of how geologic materials affect stream flow consider the drainage basins of the Tippecanoe River and Wildcat Creek, both of which drain comparable areas and both of which lie in the same climatic zone of Indiana, U.S.A. If the discharge of each stream is plotted for the water year, which begins on October first, a *hydrograph* is the result (Fig. 11.4). The Wildcat Creek hydrograph shows several prominent peaks, each of which represents the response in stream discharge to a heavy rainfall on the watershed. The Tippecanoe River, however, shows no sharp peaks on its hydrograph, even though the intensity and duration of rainfalls are no different there than on the Wildcat Creek watershed. The difference in flow characteristics between these two streams reflects the fact that the Wildcat Creek drainage basin is underlain by relatively impermeable clay whereas the Tippe-

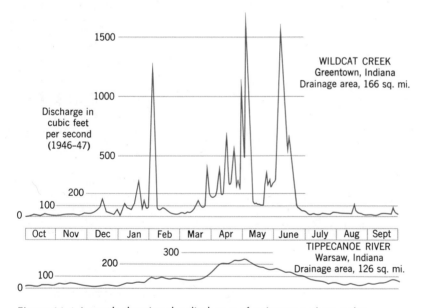

Figure 11.4 A graph showing the discharge of a river at a given point during the water year (October 1 — September 30) is called a *hydrograph*. These two hydrographs show a great difference in two rivers draining areas of similar size in the same climatic region. The difference in runoff characteristics of the two is ascribed to the different geologic materials of the drainage basins. (From Langbein, W. B. and J. V. B. Wells, 1955, *Water, yearbook of agriculture*, U. S. Department of Agriculture, Washington, D. C., p. 57.)

canoe basin is underlain by permeable sand. Rainfall is shed rapidly from the one but is partially absorbed and stored by the other.

The topography of a drainage basin influences runoff rates. Steep slopes permit more rapid runoff than gentle slopes. Watersheds containing lake basins tend to retard runoff because they act as surface storage areas for rainwater and snowmelt. Some of the runoff that collects in a lake basin is short-circuited back into the atmosphere by evaporation.

Vegetation influences runoff by intercepting rain or snow before it reaches the ground. Rainfall of low intensity and short duration may not produce any runoff because all the precipitation has been trapped in vegetational foliage from whence it evaporates shortly after the rain has subsided. Heavy jungle foliage and dense forest cover also tend to soften the impact of falling rain, thus protecting the surface from severe erosion by overland flow and allowing more time for infiltration of water into the soil and into the groundwater reservoir. A lush grass cover also reduces the velocity of overland flow, but sparse grasses and shrubs offer little protection from the frequent heavy downpours in the arid climatic zones where they occur.

Some vegetational types such as willow trees, cottonwood trees, and salt cedars have deep root systems that draw considerably more water from the ground than is needed to sustain their life processes. These *phreatophytes* flourish along the margins of intermittent or ephemeral river channels in arid and semiarid regions. Dense stands of salt cedars in parts of Arizona and New Mexico transpire thousands of gallons of water daily at the expense of stream flow and groundwater storage.

Drainage Patterns

The land drained by a river and its tributaries constitutes a *drainage basin,* and the boundary between adjacent drainage basins is the *divide.* Major drainage basins can be subdivided into their component drainage basins as defined by major tributaries. For example, the Mississippi River drainage basin and the Colorado drainage basin (Fig. 11.2) are major basins that are separated by the Continental Divide. The Mississippi River drainage basin is subdivided into separate basins for its major tributaries such as the Missouri and Ohio rivers, and each of these can be further subdivided on the basis of tributaries flowing into them.

Where several adjoining but independent drainage basins are grouped together for broad hydrologic analysis, for instance, those comprising the West Gulf Coast area in Figure 11.2, the area is designated as a *drainage area* instead of a drainage basin.

The map pattern of a trunk stream and its tributaries is called a drainage pattern. Drainage patterns reflect the details of geologic structure of the materials on which the drainage pattern is developed, a generalization that holds true for perennial, ephemeral, and intermittent streams. Where the geology is complex and consists of many different rock types in juxtaposition, the drainage pattern will reflect this complexity. Some of the more common drainage patterns and their geological significance are given in the following paragraphs and illustrated in Figure 11.5. All gradations between these basic patterns can be found in nature.

Dendritic. This pattern (Fig. 11.5A) develops on horizontal sedimentary rocks or homogeneous regolith and is so named because of its similarity to the branching system of an oak or chestnut tree.

Trellis. Dipping or folded strata of alternating resistant and nonresistant beds give rise to this pattern (Fig. 11.5B). Characteristically, the tributaries enter the main stream at right angles.

Radial. Most commonly the radial pattern (Fig. 11.5C) is diagnostic of a volcanic cone or an intruded rock mass that has domed the land surface.

Parallel. This pattern (Fig. 11.5D) is characteristic of the configuration of drainage lines on broad inclined surfaces fringing the base of mountainous areas in arid climates. The streams flowing in these channels are commonly intermittent or ephemeral.

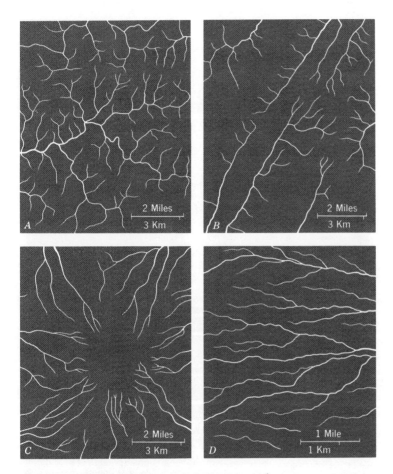

Figure 11.5 Maps of some representative types of drainage patterns. *A:* dendritic, Marietta Quadrangle, Ohio-West Virginia; *B:* trellis, Clearville Quadrange, Pennsylvania; *C:* radial, Mt. Hood Quadrangle, Oregon; *D:* parallel, Casa Grande Quadrangle, Arizona.

River Profiles and Channel Patterns

Base Level and the Stream Profile. The lowest level to which a river can erode the valley in which it flows is the *base level of erosion.* The ultimate base level for rivers entering the ocean is sea level, but temporary base levels along the stream course exist in the form of resistant rock layers, lakes, or artificial dams. No part of a drainage basin can be eroded by running water to an elevaion below its base level.

A *stream profile* is a graph on which the elevation of points on the stream bed are plotted against the distance along the stream channel from its mouth (Fig. 11.6). All profiles show a concave form upward. Over a period of geologic time a stream profile becomes smoother and adjusts its slope and channel cross section to the discharge so that just the velocity needed for transporation of the sediment delivered to it is maintained. Such a river is said to be *graded.* Normally, the graded condition of a stream is reached first in its lower reaches and progresses upstream, although in some rivers different reaches may become graded regardless of their position with respect to mouth or headwaters.

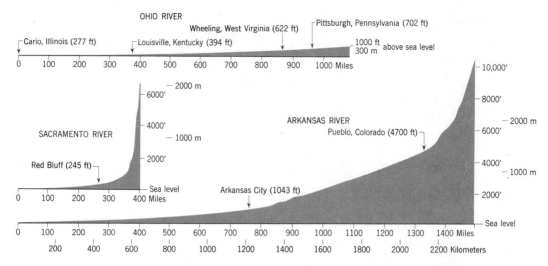

Figure 11.6 Examples of some river profiles. Each river has a concave upward profile, although the degree of concavity is different for each of the examples. Elevations of certain points on the profiles are in feet above sea level, and the distance along the bottom of each profile is in miles from the river's mouth. (After Gannett, H., 1901, *Profiles of rivers in the United States,* U. S. Geological Survey Water Supply Paper 44, Washington, D. C., 100 pp.)

Figure 11.7 The braided channel of the Waimakariri River, South Island, New Zealand. (Photograph by Tad Nichols.)

Channel Patterns. When viewed from the air, rivers tend toward three basic configurations, braided, meandering, or straight.

Braided channels are those which consist of a number of smaller channels that divide and reconnect downstream (Fig. 11.7). The material that separates the channels is a sandbar or island, some of which become stabilized with vegetation.

Meandering channels have a sinuous course characterized by a series of arcuate or S-shaped bends called *meanders,* a term derived from the Menderes River in Turkey. Meandering is a channel form toward which many rivers tend to adjust themselves because it is the form in which the river can best distribute its expenditure of energy.

Straight river courses are relatively rare in nature and observations have shown that even straight channels show a tendency to flow in a sinuous path within the confines of their straight channels. Rivers that flow in relatively straight channels for long distances are usually controlled by some linear characteristic of the bedrock such as a fault or contact between two steeply dipping formations.

River Discharge and Channel Geometry

A hydrograph (Fig. 11.4) shows the change in discharge of a river at a given point. Discharge at that point is the product of the mean velocity of the river and the width and depth of the water-filled channel. If the velocity is measured in feet per second and the width and depth are measured in feet, the discharge is expressed in cubic feet per second (cfs).

At a given point on a river flowing in its own deposits, velocity, width, and depth change as the discharge changes. For example, note the relationship of discharge to the width, depth, and velocity on the East Branch of Brandywine Creek at Seven Springs, Pennsylvania (Fig. 11.8). A systematic change in channel cross section and velocity takes place as discharge increases. This relationship holds true not only for a relatively small stream such as the East Branch of the Brandywine but also for larger rivers like the Mississippi.

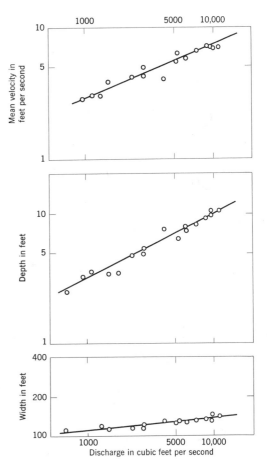

Figure 11.8 Graph showing the relationships between discharge, channel width, channel depth, and velocity, East Branch of Brandywine Creek at Seven Springs, Pennsylvania. (After Wolman, W. G., 1955, *The natural channel of Brandywine Creek, Pennsylvania,* U. S. Geological Survey Professional Paper 271, Washington, D. C., p. 11.)

In a downstream direction on the same river, channel depth and width both increase to accommodate larger discharges caused by tributaries entering the main stream. In some trunk streams, velocities increase in a downstream direction while in others velocities remain constant or decrease slightly.

Erosion and Transportation by Running Water

The material moved by a river is its *load.* This load is derived from the river channel

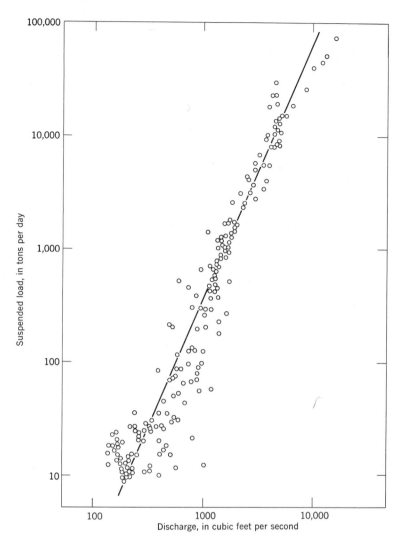

Figure 11.9 Graph showing the relationship between discharge and suspended sediment in Brandywine Creek at Wilmington, Delaware. (From Wolman, M. G., 1955, *The natural channel of Brandywine Creek, Pennsylvania,* U. S. Geological Survey Professional Paper 271, Washington, D. C., p. 20.)

and from debris delivered to the channel by overland flow, mass movement, or ground-water seepage. A river's total load consists of solid particles and dissolved compounds. The solid portion of the load is made up of the *bed load* and *suspended load.* The maximum load that a stream can move for a given dis-charge is its *capacity.*

Suspended Load. The flow of water in a river is *turbulent,* that is, the water particles do not follow parallel paths but move in swirls and eddies in a general downstream direction. This turbulent action is what keeps the clay, silt, and sand grains in suspension. The suspended load increases with discharge as shown in Figure 11.9. The distribution of suspended load at various water depths in the Missouri River at Kansas City, Missouri is shown in Figure 11.10.

Bed Load. As the discharge of a river de-clines some of the larger particles are dropped from the suspended load to the channel bed. They may still be moved downstream as bed load, however, by a pushing, rolling, or

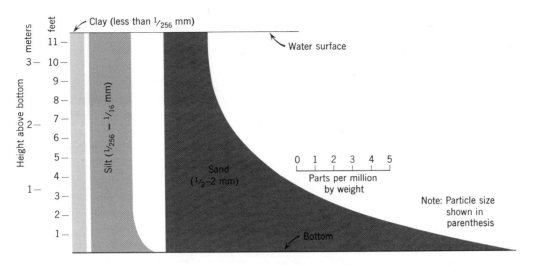

Figure 11.10 This graph shows the distribution of sand, silt, and clay versus water depth in the Missouri River at Kansas City, Missouri on January 3, 1930. Notice the heavy concentration of sand near the river bottom and the even distribution of silt and clay at all water depths. (Adapted from L. Straub *in* Meinzer, O. E., (ed.), 1942, *Hydrology*, McGraw-Hill Book Co., New York, p. 625.)

jumping motion. The size of the largest particle that can be moved by a stream under a given discharge is a measure of the stream's *competence*. A stream's capacity and competence both increase with increasing discharge.

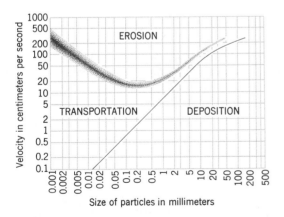

Figure 11.11 A graph showing the water velocities required to move particles of various size on a stream bed. See text for further explanation. (After Hjulström, F., 1935, Studies of the morphological activity of rivers as illustrated by the River Fyris, *Upsalla Geological Institute, Bulletin* 25, p. 298.)

Figure 11.11 is a diagram that is worthy of close examination because it reveals much about the interrelationship between bed load, suspended load, and velocity with respect to clay, silt, and sand particles. Consider, for example, a hypothetical situation in which a river is flowing in a channel composed of sand grains, all of which are one millimeter in diameter. Suppose further that the river is at a very low stage (low discharge) with a velocity less than one centimeter per second, so that all of the sand grains are lying at rest; the stream has neither suspended load nor bed load, although it may be carrying some dissolved load.

If the discharge begins to increase gradually in this hypothetical situation, Figure 11.11 reveals that the sand grains will not be picked up by the water until a velocity greater than about 20 centimeters per second is attained. If the discharge declines, however, the sand grains will remain in suspension and will continue to be part of the stream's load even after the velocity has subsided to a level below that which was initially required to lift the sand grains off the bed. In this particular case, the sand grains would not come to rest

Table 11.a Chemical Composition of the Dissolved Material in the Waters from the Mississippi, Pecos, and Mahanadi Rivers* (Values Are in Percent)

Constituent	Mississippi River (New Orleans, La., U.S.)	Pecos River (New Mexico, U.S.)	Mahanadi River (Cuttack, India)
CO_3	34.98	1.54	27.06
SO_4	15.37	43.73	1.08
Cl	6.21	22.56	2.04
NO_3	1.60	n.d.	7.44
PO_4	n.d.†	n.d.	0.72
Ca	20.50	13.43	15.78
Mg	5.38	3.62	4.62
Na	8.33	14.02	5.92
K		0.77	1.64
SiO_2	7.05	—	33.45
Al_2O_3	0.45	0.33	n.d.
Fe_2O_3	0.13	—	0.25
Totals	100.00	100.00	100.00

* From Rankama, K. R. and Th. G. Sahama, 1950, *Geochemistry*, University of Chicago Press, p. 272 (with permission).
† Not determined.

on the stream bed until the velocity had dropped to 10 centimeters per second. Less energy is required to keep a sand grain in suspension than is needed to lift it into suspension.

Of additional interest in Figure 11.11 is the fact that silt and clay particles require a higher velocity to be picked up from the stream bed than medium or coarse sand, but once they are picked up from the stream bed, they remain in suspension at much lower velocities than the larger particles. The higher pickup velocities for clay and silt results from the fact that they do not project up into the stream bed high enough to receive the full impact of the flowing water. The small particles also have a certain cohesiveness that must be overcome before they can be moved by running water.

Dissolved Load. Chemical weathering produces materials that are soluble in water, and these dissolved substances ultimately find their way into the stream systems. Table 11.a shows the percentage of the various chemical constituents contained in three rivers that drain basins lying in different climatic zones. Dissolved loads are derived from the chemical weathering of rocks and regolith so some correlation between the chemical composition of river water and the climactic environment is to be expected. In areas of abundant rainfall, such as the drainage basins of the Mississippi and Mahanadi rivers, the chief dissolved component is carbonate (CO_3). In arid regions, like the Pecos River drainage basin of New Mexico, sulfate (SO_4) and chloride (Cl) are dominant. Silica (SiO_2) is relatively high in tropical rivers because it is dissolved during the weathering processes of laterization (Chapter 8). On the other hand, the aluminum and iron oxides (Al_2O_3 and Fe_2O_3) are relatively low in all rivers because they are relatively insoluble.

The dissolved load of most rivers eventually reaches the oceans, but some rivers empty into lakes with no outlets to the sea. The oceans are thus supplied with dissolved minerals from the land, and lakes with no drainage to the sea become highly charged with dissolved substances.

GEOLOGICAL WORK OF RIVERS

Denudation of Land Areas

The fact that rivers carry a load of disintegrated and decomposed rock waste to the oceans means that the continents gradually are being worn down. By measuring the loads carried by rivers to the oceans, it is possible to establish rates at which the process of downwasting or *denudation* is taking place for a given drainage basin. A denudation rate is an expression of the time required for the removal of a hypothetical layer of rock or regolith uniformly distributed over the entire drainage basin upstream from the point at which the load was measured. A denudation rate is an average rate of downwasting for a drainage basin and does not mean that the land is being worn down at that rate in any particular place. Some regions of steep slopes are eroded more rapidly than others in the same watershed. Denudation rates provide a convenient means whereby comparisons can be made between different areas with respect to the exogenous processes acting on them.

Both suspended and dissolved loads need to be measured if any meaningful results are to be obtained in the calculation of denudation rates. Bed load should also be taken into account, but because no reliable technique has yet been devised for measuring bed-load transport, and because on a continental scale bed load probably represents only a small fraction of the total load delivered to the oceans, it has been neglected in computations involving denudational rates.

A denudational rate for a drainage basin is determined by measuring the load carried out of the drainage basin by the trunk stream of the system. It is assumed that the entire stream load is contributed from the land surface of the watershed and thus is a measure of the lowering of the land surface. It is further assumed that 165 pounds of stream load is equivalent to one cubic foot of surface rock. Using these values and stream-load data, the denudational rate, *D,* expressed in inches per thousand years, can be determined by simple arithmetic according to the formula

$$D = .0052 \ L$$

where *L* is equal to the total load in tons per square mile per year.

Calculations of denudational rates are based on techniques of sampling suspended and dissolved loads used for many years by engineers, geologists, and hydrologists, but this does not mean that errors are wholly absent. Moreover, sampling programs must be extended over a sufficiently long time to assure maximum reliability of results. Because the suspended load increases dramatically with increasing discharge (Fig. 11.9), it is important to have data for many years so that high discharge values that occur only once in a period of many years will not be omitted.

Among the errors connected with determining denudational rates from river-load data are those related to man's activities. The clearing of vegetation from the land surface in the development of urban areas and the construction of highways adds considerable suspended load to streams draining those areas. The sediment production of agricultural land is ten times greater than forested land. Land under construction produces twenty to thirty times more sediment than land with a natural forest cover. Because many forested areas have been changed to agricultural production or have succumbed to urbanization, the suspended loads of rivers affected by these changes will yield denudational rates that are erroneous when extrapolated over a period of a thousand years.

Another error is introduced in the denudation rates by assuming that all of a river's dissolved load is derived from the solution of rock and regolith of the earth's surface. It is known, for example, from the chemical analyses of rainfall along the Atlantic coastal states as well as in Wyoming and New Mexico, that rainfall may be directly responsible for as much as 50 percent of a river's dissolved load. Rainfall flushes soluble air pollutants of industrial origin from the atmosphere, and these substances end up in rivers as part of the dissolved load, thus leading to a value of chemical denudation that is too high. One geologist estimates that the denudational rates of the Atlantic states based on

Table 11.b Estimated Rates of Denudation for the United States*

Drainage Region	Drainage Area Thousands of Square Miles	Load (Tons per Square Mile per Year)			Denudation Rate (Inches per 1000 Years)
		Dissolved	Solid	Total	
North Atlantic	148	163	198	361	1.9
South Atlantic and Eastern Gulf	284	175	139	314	1.6
Mississippi	1250	110	268	378	2.0
Western Gulf	320	118	288	406	2.1
Colorado	246	65	1190	1255	6.5
Columbia	262	163	125	288	1.5
Pacific Slopes, California	117	103	597	700	3.6
Total United States†	2627	121	340	461	2.4

* After Judson, S. and D. F. Ritter, 1964, Rates of Regional Denundation in the United States, *Jour. Geophysical Research*, v. 69, pp. 3395–3401.

† Exclusive of Great Basin, St. Lawrence, and Hudson Bay drainages.

stream-load data are too large by a factor of 2 because studies in New Hampshire and North Carolina indicate that much of the dissolved loads in these rivers represent material derived from the atmosphere and a direct contribution by the activities of man. Moreover, the suspended load of rivers draining the Atlantic states may be four to five times what they were before the European settlers began farming the land 200 years ago. Denudational rates based on such data will therefore be too high.

The portion of a river's dissolved load contributed by groundwater should be subtracted from denudational rate computations because the dissolved load of groundwater does not necessarily represent material that has been removed from the earth's *surface*. The separation of that part of the dissolved load contributed by overland flow and that part contributed by base flow is not possible because data are not uniformly available or readily obtainable. This represents a further source of error in determining denudational rates.

Having thus forewarned the reader of the pitfalls involved in calculating denudational rates, attention can be now directed to Table 11.b which gives the estimated denudation rates for some of the drainage areas identified in Figure 11.2. These values have not been corrected for the errors previously mentioned, and the load data for some of the areas are based on records from too short a period for reliable results. For this reason, the rates given in Table 11.b should be taken as estimates rather than as firm values.

River Deposits

Rivers carry their loads ultimately to the oceans, but during the transit time from land to sea, both bed load and suspended load may come to rest, temporarily, during times of low discharge. River deposits are made up of detrital grains that range from clay and silt, to sand, gravel, cobbles, and even boulders.

Alluvial Deposits. Rivers tend toward the establishment of meandering courses. Once meanders are established they shift their positions by eroding the channel banks on the outside of a river bend where the velocity

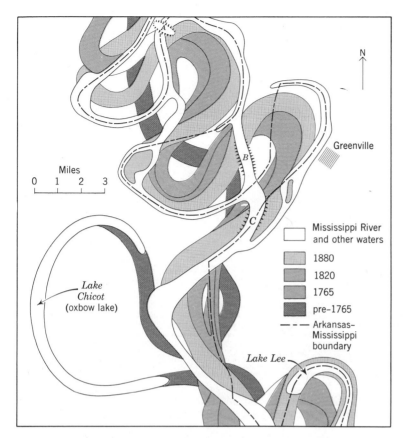

Figure 11.12 Map of a segment of the Mississippi River on the Arkansas-Mississippi State boundary showing previous courses of the river, an artificial fill (*A*), and cutoffs (*B* and *C*) made by the U. S. Corps of Engineers. The Arkansas-Mississippi boundary does not follow the modern course of the river, but is fixed, in part, by the natural channel that existed prior to the artificial cutoffs of 1935 (*B*) and 1933 (*C*). (Based on Fisk, H. N., 1944, *Geological investigations of the alluvial valley of the lower Mississippi River,* Mississippi River Commission, Vicksburg, 78 pp.)

is highest and by depositing sandbars on the inside of the bend where the velocity is slower.

The shifting of meanders eventually produces a broad valley floor called a *floodplain*. A floodplain is not produced by rivers in flood, but is so named because flood waters spread over it when a channel cannot accommodate the high discharge produced by the spring runoff from melting snow, or the heavy rains during other times of the year. A river floodplain is made up of *fluvial* (river) deposits collectively known as *alluvium*.

Some meanders become isolated or cut off from the main channel by a breakthrough during flooding of the stream across the narrow neck of land separating two adjoining meanders. The meander cutoff or *oxbow lake* fills in with silt, clay, and organic matter in the course of time. Oxbow lakes occur in great profusion along the lower Mississippi River (Fig. 11.12).

When a river spills over its channel banks

Figure 11.13 River terraces along the Clutha River, east of Arthur's Pass, South Island, New Zealand. (Photograph by Tad Nichols.)

during flood stage, part of the suspended load is deposited on the floodplain near the channel banks because the water velocity diminishes there. Gradually, low ridges of silty river sediment are formed parallel to the river channel. These *natural levees* confine the river during nonflood stages and tend to raise it above the general level of the floodplain.

River Terraces. Once established, a floodplain continues its function as a catchment for flood waters. If, however, a river begins to incise its channel because of base-level lowering, uplift of the land, or any other cause capable of upsetting the river's equilibrium, the floodplain is no longer within reach of flood waters. It becomes a *river terrace* (Fig. 11.13). River terraces are thus old floodplains and they are arranged in steplike fashion along the river valley. River terraces provide a basis for the reconstruction of a river's history in terms of the changes in the parameters that govern its capacity to erode or alluviate its channel. They include changes in base level, climatic changes on the watershed, and uplift of the land.

Deposits at River Mouths. A river entering a natural or man-made lake, or the ocean, deposits its detrital load as a *delta* (Fig. 11.14). As a delta expands seaward, the river extends its length and branches into several channels called *distributaries* that are flanked by natural levees. At any one time in the delta's history one or more distributaries carries more discharge than adjoining ones, and delta growth is shifted to other dis-

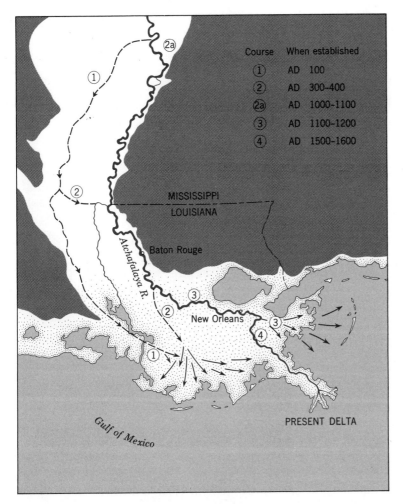

Figure 11.14 Map of the lower Mississippi River Valley showing different courses of the river during the past 2000 years, and the areas of delta building. The next natural change in course would have been through the Atchafalaya River, but in 1963, the U. S. Corps of Engineers finished a control dam at the threatened diversion site, thus ending the possibility of a new route to the sea for the Mississippi River. (After Fisk, H. N. *et al.,* 1952, *Geological investigations of the Atchafalaya Basin and the problem of Mississippi River diversion,* Mississippi River Commission, Vicksburg, 145 pp.)

tributaries so that the entire deltaic area grows seaward over a period of several hundred years.

Alluvial Fans. A river that debouches abruptly from a mountainous area onto a plain builds a body of alluvium that has a fan-shape (Fig. 11.15). These alluvial fans are very common in arid regions, but they are known also in humid regions where the topographic situation is conducive to their formation. Although running water is mostly responsible for fan deposits, it is not unusual for those in arid regions to carry mudflows under certain conditions. The drainage pattern on an alluvial fan is characterized by a distributary system in which the main channel divides into smaller channels in a downstream direction.

Figure 11.15 Oblique aerial photograph of alluvial fans in the Mojave Desert, California. (Photograph by J. R. Balsley, U. S. Geological Survey.)

The Cycle of Erosion

Denudation has two components, river channel erosion and hillslope downwasting by mass movement and overland flow. During the course of denudation from the time a landmass is uplifted by orogeny or epeirogeny until it is reduced to a near sea-level surface, there is reason to believe that the landscape passes through different evolutionary stages of development, each of which reflects a stage in the erosional history of the area.

The *cycle of erosion* is the sequence of events involved in the reduction of a land area to base level. An erosion cycle could begin with the uplift of a continental shelf, coastal plain, or a low-lying interior basin to an elevation high above sea level. Once set in motion by uplift, the cycle of erosion progresses through a series of identifiable stages until an end stage is reached.

The cycle of erosion concept was conceived in the late nineteenth century by the American geologist-geographer, William Morris Davis, and bears his name. According to the Davisian concept, each stage in the cycle of erosion (or geographical cycle, as Davis termed it) has certain diagnostic land forms that distinguish it from earlier or later stages. Theoretically, every segment of the earth's surface should fit into one of the main or intermediate Davisian stages, but because of the idealized form of the cycle, this is not the case. Nevertheless, even though the Davisian concept has been questioned on several grounds, it is a useful hypothesis that emphasizes change with time, which is a fundamental geologic principle.

Davis defines three main stages in an

"ideal" uninterrupted erosion cycle,[3] *youth, maturity,* and *old age.* These names have no meaning in terms of absolute years but, instead, reflect progressive conditions of a particular landscape during the time in which it is reduced from a position above sea level to base level. The main characteristics of each stage are given in the paragraphs that follow, but it should be borne in mind that the three main stages are only parts of a time continuum; transition stages between each are to be expected.

The Youthful Stage. A geologically new, more-or-less flat or nearly flat uplifted surface, is subjected to incision by streams vigorously eroding their channels. The profiles of these streams are steep, their competence is high, and they have no floodplains. Rapids and waterfalls are common and the valleys are V-shaped in cross section. Tributaries are extending themselves headward but do not reach back to the divides between adjoining drainage basins. Valley walls are steep and mass movement and erosion by overland flow are strongly operative.

Mature Stage. As the drainage pattern evolves, erosion by running water in the stream channels and by mass wasting and overland flow on the hillslopes narrows the flat areas between adjacent drainage basins until the divides are nearly as sharply defined as the valleys. A fully mature landscape contains no flat land except in the bottoms of the valleys where incipient floodplains begin to appear along the courses of the main streams. The drainage patterns are most fully developed in the mature stage and are well adjusted to the structure of the underlying bedrock.

Old Age Stage. The rivers at this stage of the cycle are characterized by extensive floodplains. The upland surfaces contain no features of sharp relief except a few isolated erosional remnants or *monadnocks,* rock hills that were unconsumed in the denuda-

---

[3] In modern scientific terminology Davis' ideal uninterrupted erosion cycle would be called a model.

tion process. Surfaces not occupied by valley alluvium are deeply weathered, and gentle slopes prevail everywhere. The ultimate base level surface that characterizes the end stage in an ideal erosion cycle was called a *peneplain* ("almost a plain") by Davis. All rock masses are worn down to a peneplain level so that no differential resistance to denudation is manifest in the old-age topography.

Using a denudation rate of 2.0 in. per 1000 years, a land surface standing at 2500 feet above sea level will be reduced to a peneplain in 15 million years. The denudation rate is considered as an average for the entire period, but in reality it would be higher than average during the early part of the cycle and lower than average toward the end of the cycle. Such an extrapolation also assumes that external conditions such as climate and vegetation remain constant and that no further uplift takes place during the time in which the cycle runs its course.

The geologic history of the earth reveals that climate cannot be considered to remain constant for even a period of one million years much less 15 million years. Furthermore, uplift is an ongoing process in mountainous regions, and sea level itself rises and falls a few hundreds of feet within a time span of 50 to 100 thousand years. Because the factors that control the cycle are always in a state of flux, one must conclude that the Davisian idealized cycle of erosion is an exception rather than the rule in the natural scheme of things, and that it is unlikely that it could continue from start to finish without interruption.

Interruptions of the Erosion Cycle. Davis recognized intuitively that the ideal cycle of erosion was a model rather than an expression of reality, mainly because he could not find an example of a modern peneplain anywhere on earth. No *bona fide* modern peneplain has yet been discovered, but remnants of dissected peneplains at high elevations have been described; not all geologists agree that these features are actually uplifted peneplains, however.

To account for interruptions of the ideal cycle of erosion, Davis described certain telltale criteria that would identify causes of *rejuvenation* in a landscape. For example, a meandering river in the Davisian scheme should be restricted to an alluvial valley during the late mature stage of the ideal cycle. Yet, certain rivers that flow in rockbound channels through deep canyons have meandering courses. Davis claimed that these *entrenched meanders* were inherited from a previous stage, but because of uplift, the river had regained its ability to deepen its channel through renewed downward erosion.

It can be seen as well that other processes such as volcanism and continental glacial deposition, can interrupt the erosional cycle by the addition of new material on the earth's surface.

Other Concepts. The Davisian philosophy of landscape development has been faulted by some geologists on the grounds that it is a model never approached in nature, that it really cannot explain the origin of certain landforms, or that it ignores conditions of dynamic equilibrium or steady-state relationships that some geologists believe are achieved in nature. These alternative views cannot be elucidated in detail here because they require more advanced concepts than can be introduced in an elementary treatment of this subject. However, it can be pointed out that no satisfactory alternative to the Davisian scheme of landscape evolution has been proposed that deals with the same span of geologic time envisioned by Davis. Alternative explanations are usually concerned with cause and effect relationships in landscape processes that are of much shorter duration than the millions of years required by the ideal Davisian model. Moreover, the Davisian terms of youth, maturity, and old age are useful terms that convey reasonably concise descriptions of fluvially formed terrains.

The difference between the Davisian view and the non-Davisian outlook is equivalent to two different views of the health of an individual. One view looks at the life history of the individual from birth to death. The other view deals only with a few years in the life of a healthy man. The first observes the gradual changes that accompany an individual as he gets progressively older and finally succumbs to the inexorable causes of death. The second sees a normal, healthy individual who eats enough to sustain his physical well-being without gaining or losing weight. A physician would proclaim such a person in good health and in equilibrium with his environment, but the same physician would also admit without hesitation that this healthy man is undergoing change that will lead finally to the end of his life cycle.

So it is with landscapes. On a short-term basis of the order of 1000 years or less a landscape may be in equilibrium with its environment, but over a period of geologic time of millions of years, however, the same landscape must undergo change. The logic of this statement is supported by the fact that the drainage system which produces a landscape is gradually losing energy as its surface is lowered closer to base level with each ton of suspended and dissolved load carried to the sea.

Rivers and Man

The economic well-being of mankind is intimately associated with the rivers of the world. The fertile delta of the Nile has been farmed for thousands of years, electricity is produced from turbines driven by falling water, cargoes are transported on barges that ply navigable rivers, and primary water supplies are drawn from one part of a stream, while municipal sewage and industrial waste are discharged at another.

These uses and many others are reminders that man has more than an academic interest in the rivers of the world. He needs to have a thorough understanding of them in order to achieve maximum benefits from them. This section will consider briefly some of the ways in which man and his activities relate to rivers in a highly civilized world. They will be discussed under the general headings of navigation, dams and reservoirs, floods, and water supply and pollution. Although an exhaustive treatment of them is beyond the scope of this book, a few remarks about each will serve to introduce some of the problems encountered when man and river meet.

Navigation. Rivers afford access by ocean-going vessels to ports inland from the seacoasts of the world, and river barges provide a cheap means of transportation for a variety of goods in many countries. The maintenance of a navigable channel involves stabilization of the river's course, dredging of channel bottoms, and regulation of the discharge.

In the United States, the Corps of Engineers of the United States Army is charged with the responsibility of maintaining navigable channels on certain rivers, among which is the Mississippi. Since the 1930's the Corps of Engineers has been engaged in shortening the course of the Mississippi River between Memphis, Tennessee and the mouth of the Red River through artificial cutoffs of meanders and the realignment of the channel (Fig. 11.12). Miles of concrete walls or revetments were placed along the outside bends of the river to reduce the tendency of the channel to migrate laterally. This program was only moderately successful because as soon as one segment of the channel was stabilized by this technique, other segments not protected by revetments would start shifting their positions. A laboratory study on a scale model of a river meandering in an alluvial channel convinced the engineers that a meandering course was a kind of equilibrium state needed by the Mississippi to remain in balance with its geologic and hydrologic environment. Hence the river engineers were able to devise a more rational plan for coping with the river's shifting channel.

Another problem confronting the engineers was the increasing discharge of the lower Mississippi through a parallel channel, the Atchafalaya River (Fig. 11.14). If this natural diversion had been allowed to increase, it eventually would have carried all the Mississippi's discharge to the Gulf of Mexico by 1975, and would have reduced the cities of Baton Rouge and New Orleans

to ghost ports. A geological study of the matter revealed that the Atchafalaya diversion was not a unique event because the lower Mississippi River had, in the last few thousand years, followed a number of different routes to the sea across its own delta (Fig. 11.14). Not wishing to block the flow in the function as an outlet for excessive flood waters, the engineers decided to maintain the status quo of the Atchafalaya channel by the construction of a regulatory barrier across its mouth, a task that was completed in 1963.

Dams and Reservoirs. Dams are structures built on rivers to cause upstream ponding of the water in a reservoir. Dams and their associated reservoirs are built for the purpose of water power, flood control, water supply (including irrigation), and aquatic recreation. Normally, a single dam cannot serve all of these interests efficiently. The reservoirs used for flood control must be kept partially *empty* during much of the year to provide room for flood waters, whereas reservoirs used for most other purposes need to be as *full* as possible all year long to maintain a steady supply of water. *Multipurpose* dams are those that are designed to store large quantities of water the year around and still have adequate storage for flood waters.

Dam sites are usually examined by competent geologists before the dam is designed to determine if the earth materials involved are adequate to support the dam, and to identify possible zones that might allow seepage of water from the reservoir when it is filled. Large seepage losses from reservoirs not only constitute a loss of water intended for surface storage but also are potential causes of dam failure if the seepage occurs beneath the base of the dam.

Rivers that supply water to reservoirs also bring in their sedimentary loads. Some reservoirs situated on turbid rivers have useful lives of 50 years or less, but Lake Mead, the reservoir behind Hoover Dam on the Colorado River, has such a large capacity that it will take more than four centuries to fill it with sediment. About 2000 million tons of sand, silt, and clay were deposited in Lake Mead in the first 14 years since Hoover Dam

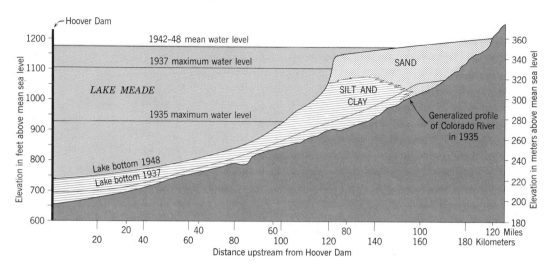

Figure 11.16 Cross section of Lake Mead behind Hoover Dam on the Colorado River showing sedimentation during the 14-year period 1935 to 1949. At the present rate of accumulation the reservoir will not be filled with sediment until the year 2350. (After Smith, W. D., C. P. Vetter, and G. B. Cummings, 1960, *Comprehensive survey of sedimentation in Lake Mead, 1948-49,* U. S. Geological Survey Professional Paper 295, Washington, D. C., 254 pp.)

was completed in 1935. This is within 2 percent of the predicted amount based on measurements of the suspended load carried by the Colorado River before the dam was built. Figure 11.16 is a longitudinal profile of Lake Mead showing the original profile of the Colorado River as it was in 1935, and the distribution of sediment in Lake Mead as of 1948. The sand fraction of the accumulated sediment lies partially above lake level because it is part of the delta formed by the Colorado River.

Floods. Rivers are able to accommodate the water delivered to them within the confines of their channels most of the time. At intervals that may be seasonal or random, however, rivers overflow their banks and inundate large areas of their floodplains. The disastrous consequences of such floods are well known (Fig. 11.17).

The magnitude of a flood can be subdued by the construction of dams to create reservoirs in which excessive flood waters can be stored and discharged downstream when the conditions that cause flooding have sub-

sided. Dikes and artificial levees are other effective means of combating flood waters, but when unusually large floods occur, these structures may be unable to contain them.

One solution to the flood problem is to avoid the installation of costly man-made structures on floodplains. It is unrealistic, however, to assume that this solution is always the most practical in a rapidly expanding municipality where all available land is sought for development. In planning the development of unused floodplains, for example, it may be possible to relegate them to a special use that would not be influenced adversely by a major flood every 5 or 10 years. A city park, for example, is a use of this type because it could be inundated frequently by flood waters and be restored to its former condition with minimum effort and expense. Even if buildings must be located on land subject to potential flooding, the designers of such structures can take into account the frequency of floods and the probable heights that the flood waters will reach. Such information is available for a number of

Figure 11.17 The Kentucky River in flood stage, March 1963. (Photograph by Billy Davis, *Courier-Journal and Louisville Times.*)

communities in the United States from the United States Geological Survey, a unit of the Department of the Interior that has the responsibility for collecting and disseminating this information. In newly developed areas where past records of river activity are not available, the delineation of the river's floodplain is the first step in intelligent planning of land use. This task can be performed by a competent geologist who can identify a river's floodplain by an on-site inspection.

Water Supply and Pollution. Rivers serve not only as a major source of water used in the world but also as the chief means whereby wastes of all kinds are disposed. About 95 percent of the major rivers in the United States are used mainly as a conveyor belt on which to send waste products to the sea.

The question in the industrialized and urbanized regions of the United States, Europe, Russia, and elsewhere, is not one of water *quantity* but one of water *quality*. The industrial revolution is now being followed by the population explosion, both of which aggravate the water quality of the rivers in highly populous areas.

In the early days of settlement, waste disposal in rivers did not materially affect the quality of river water because of the natural ability of flowing water to oxidize waste products. When the threshold of this ability is reached and all available dissolved oxygen is used by decaying organic wastes, pollution results. To partially meet this situation, towns and cities situated along rivers built sewer systems that treated waste products at a central point before being returned to the rivers. Many of these waste treatment plants, however, became too small as rapid urbanization outstripped their capacities. Also, many industrial plants did not bother with the additional expense of treating their waste products, and discharged enormous quantities of toxic substances into the rivers, thereby poisoning fish that were not already dead from the lack of oxygen. Waste waters from mines were also disgorged into the river channels and added further to the deteriorating open sewers that the rivers had now become. Additional sources of river pollution arose in the form of soil fertilizers, insecticides, pesticides, and hot water from nuclear power plants.

In many parts of the civilized world today—in England, Europe, the Soviet Union, and the United States—the foregoing description is one that describes the present, not the past. Although some steps are being taken to alleviate these conditions, a fantastic amount of cleanup remains to be done. Pollution abatement is a political-social-economic problem as much as it is an engineering or geologic problem. Techniques for treating most wastes are known, and their use awaits only the will of the people to pay the necessary costs for implementation.

Even though pollution abatement is not essentially a geological problem, the geologist can participate in a meaningful way in programs designed to restore rivers to a less-polluted state. Waters flowing in natural or artificial channels are part of the hydrologic cycle and thus lie within the broad framework of the earth sciences. Because geologists are familiar with the natural flow of river waters and their solid and dissolved contents, they can play a useful role in determining which components of a particular river are contributed by nature and which are introduced by man's activities.

References

* Bue, C. B., 1967, *Flood information for flood-plain planning*, U.S. Geological Survey Circular 539, Washington, D.C., 10 pp.

* Davis, W. M., 1909, *Geographical essays* (Chapters 13 and 14), Ginn and Co., Boston (reprinted by Dover Inc., New York, 1954, 777 pp).

Holeman, J. N., 1968, The sediment yield of major rivers of the world, *Water Resources Research*, V. 4, pp. 737-747.

Howard, A. D., 1967, Drainage analysis in geologic interpretation: a summation, *The American Association of Petroleum Geologists Bulletin*, V. 51, pp. 2246-2247.

Langbein, W. B., 1949, *Annual runoff in the United States*, U. S. Geological Survey, Circular 52, Washington, D. C., 14 pp.

LaMarche, V. C., 1968, Rates of slope degradation as determined from botanical evidence, White Mountains, California, *U.S. Geological Survey Professional Paper 352-I*, U.S. Government Printing Office, Washington, D. C., pp. 341-377.

Leopold, L. B., M. G. Wolman, and J. P. Miller, 1964, *Fluvial Processes in Geomorphology*, W. H. Freeman, San Francisco, 522 pp.

* Leopold, L. B., and W. B. Langbein, 1966, River meanders, *Scientific American*, V. 214, No. 6 (June), pp. 60-70. (Offprint 869, W. H. Freeman and Co., San Francisco.)

Livingstone, D. A., 1963, Data of geochemistry, 6th edition, Chapter G. Chemical composition of rivers and lakes, *U. S. Geological Survey Professional Paper 440-G*, U. S. Government Printing Office, Washington, D. C., 64 pp.

Meade, R. H., 1969, Errors in using modern stream-load data to estimate rates of denudation, *Geological Society of America Bulletin*, V. 80, pp. 1265-1274.

* Morisawa, M., 1968, *Streams, their dynamics and morphology*, McGraw-Hill Book Company, New York, 175 pp.

* Nace, R. L., 1964, Water of the world, *Natural History*, V. 73, No. 1, pp. 10-19.

Nordin, C. F., Jr. and J. P. Beverage, 1965, Sediment transport in the Rio Grande, New Mexico, *U. S. Geological Survey Professional Paper 462-F*, U. S. Government Printing Office, Washington, D. C., 35 pp.

Rankama, K., and Th. G. Sahama, 1950, *Geochemistry,* Chap. 6, Geochemistry of the hydrosphere, University of Chicago Press, Chicago, 912 pp.

Schumm, S. A., 1963, The disparity between present rates of denudation and orogeny, *U. S. Geological Survey Professional Paper 454-H,* U. S. Government Printing Office, Washington, D. C., 13 pp.

Smith, W. D., C. P. Vetter, G. B. Cummings, 1960, Comprehensive survey of sedimentation in Lake Mead, 1948-49, *U. S. Geological Survey Professional Paper 295,* U. S. Government Printing Office, Washington, D. C., 254 pp.

Wolman, M. G., 1955, The natural channel of Brandywine Creek, Pennsylvania, *U. S. Geological Survey Professional Paper 271,* U. S. Government Printing Office, Washington, D. C., 56 pp.

* Recommended for further reading.

12 Wind Action and Deserts

. . . , boundless and bare
The lone and level sands stretch far away.
Shelley

Wind is a geologic agent responsible for a number of features on the earth's surface. The geologic activity of wind, be it one of erosion, transportation, or deposition, is called an *eolian* process. Optimum conditions for eolian processes on terrestrial areas are found in the arid regions of the world where the lack of moisture results in a sparse plant cover that leaves rock, regolith, and soil unprotected from the force of the wind. Desert regions are not the exclusive domain of eolian processes, however; parts of the more humid regions are also susceptible to modification by blowing winds.

The first part of this chapter will cover some basic principles of wind action exclusive of wind-generated waves on oceans and lakes (Chapter 14). The second part will be devoted to desert regions per se and some of the special geologic features characteristic of them, including desert landforms by noneolian processes.

Desert sand near the Kharga Oasis, Western Desert, Egypt. (Photograph by Tad Nichols.)

Wind: Movement of the Atmosphere

The earth is surrounded by an envelope of gases called the atmosphere. The movement of the atmosphere parallel to the earth's surface is *wind,* and vertical movements are *air currents.* The science that deals with atmospheric circulation and patterns of wind movement is meteorology. Meteorologists are interested in the forces that produce winds and wind patterns, while geologists deal with the effects of the wind on the surface materials of the earth. Before proceeding with examples of wind-produced features, it will be necessary to review a few basic terms related to wind movement.

Wind Velocities. The speed of winds ranges from less than one mile per hour to well over 100 miles per hour. The ferocity of hurricanes and the havoc produced by them along the Gulf and eastern coasts of the United States and islands of the Caribbean have resulted in the widespread destruction of man-made structures and the loss of human lives. Much of the geologic work of the wind, however, is accomplished by winds of more moderate velocities. Winds that blow in the 20- to 25-mile-per-hour range are capable of transporting large quantities of sand and dust particles. Moreover,

winds in this range of intensity blow sand grains against rock outcrops, boulders, and cobbles that project above the general level of the ground surface, and produce modifications in the original form of these objects.

Wind Direction. The pattern of atmospheric circulation is fairly well established for most land areas of the earth. Even remote places such as the Antarctic are now under surveillance by manned weather stations or orbiting weather satellites. *Prevailing winds* for any given locality are those that come from one direction for the longest cumulative period during a single season or the entire year. Prevailing winds are not necessarily the strongest because local variations in atmospheric conditions may produce winds from another quarter of shorter duration but of higher velocity. Thus, geologic features related to wind action are produced by prevailing winds as well as by local storm winds.

Mechanics of Wind Action

Eolian activity can be divided into three categories: *erosion, transportation,* and *deposition.* All three are interrelated and depend on wind velocity, the nature of materials

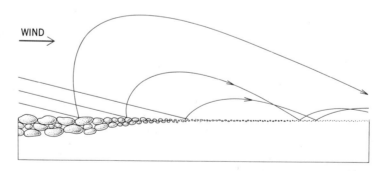

Figure 12.1 Paths of sand grains driven along the ground by wind. This type of movement is called saltation. Particles moved in this manner constitute the bed load of wind-borne sediments. (After Sharp, R. P., 1949, Pleistocene ventifacts east of the Big Horn Mountains, Wyoming, *Jour. Geol.,* v. 57, pp. 175-195; and R. A. Bagnold, 1941, *The physics of blown sand and desert dunes,* Methuen and Co., London, 265 pp.)

at the earth's surface, and vegetational cover.

The material moved by the wind falls into two categories, *bed load* and *suspended load*. The bed load consists of particles too large or too heavy to be lifted more than a few feet above the ground. The wind moves this part of its load either by rolling or by a process known as *saltation,* a kind of a jumping motion in which sand grains are dislodged from a position of rest and momentarily lifted above the ground. During the brief time aloft, the particles are bodily carried downwind before falling back to the ground (Fig. 12.1). When a particle strikes the ground, another particle is dislodged so that the process continues as long as the wind velocity is sufficiently high to keep the sand particles moving. Experimental work in wind tunnels indicates that a wind velocity of 11 miles per hour is necessary for fine dune sand, 0.25 millimeter in diameter, to be moved.

The suspended load consists of particles small enough that they can be kept constantly in suspension by the turbulence of the wind. The size of the fragments that comprise the suspended load ranges from 0.02 millimeter to extremely small dust particles. Finely disseminated dust originating from volcanic eruptions may remain in suspension for months thousands of feet above the ground; and top soil, unprotected by vegetation, is borne aloft for days and weeks before it eventually settles back to earth.

Wind Erosion

Two types of erosive action of the wind are *deflation* and *abrasion*.

Deflation. This process involves the removal of loose particles from an area, and leaves a denuded surface covered with coarse material too large for wind transport. Eventually, the residual cover of coarse gravel and stones becomes so concentrated that it prevents any further removal of fine-grained particles. Such a surface is covered with a *lag gravel,* and it may involve several hundred square miles or only a few acres.

Large tracts of lag gravels are characteristic of many true deserts.

The deflation process is impeded by wet sand and patches of vegetation, the roots of which tend to bind the soil particles together so they cannot be dislodged by the wind. Where the surface materials contain no coarse particles of gravel or stones, the deflation process may continue until a depression is formed. These features are called *blowouts,* and they occur where dry sand deposits are attacked by the wind. Blowouts are not usually much deeper than a few feet, but occasionally they attain depths of ten or more feet and are several acres in extent. Some blowouts are transformed to lakes when the water table rises as a result of an increase in rainfall brought about by a climatic change.

Deflation is the major process in denuding agricultural lands during extended periods of drought. The infamous "dust bowls" of Kansas, Oklahoma, and adjoining states during the 1930's and the 1950's resulted when crops failed to grow because of the lack of rainfall. The bare top soil, deprived of its vegetational cover, was thus easy prey, even for moderate winds.

Such disasters probably cannot be prevented entirely, but they can be averted, in part, by the use of strip farming. This method involves tilling parallel strips of land with intervening tracts of equal area planted to grass, thereby creating a surface only partially vulnerable to attack by the wind. Planting rows of trees along field boundaries also materially reduces wind velocity and protects loose top soil from deflation.

Abrasion. Bed-load particles are propelled by the wind against natural or artificial objects. Hard grains of quartz sand, driven by the wind, sandblast telephone poles, fence posts, stones, and rock outcrops, causing wearing or abrasive action on these objects. Because the bed load is concentrated within a few feet of the ground, wind abrasion is restricted to the parts of obstructions near ground level.

Stones and boulders subjected to wind abrasion are smoothed and polished if they

are composed of one of more minerals of uniform hardness, and are pitted or etched if they consist of grains or crystals of unequal hardness. Wind abraded stones are known as *ventifacts,* and they usually occur in association with lag gravels (Fig. 12.2).

A common misconception is that wind abrasion produces "wind caves," indentations along bedrock cliffs in arid regions. Although these features do occur in regions subject to strong winds, their origin is more likely the result of concentrated physical or chemical weathering due to the local concentration of moisture. At most, abrasion by wind only exerts a modifying influence on such features. Some "wind caves" may develop through deflation of a vertical cliff of sandstone whose grains have been loosened by weathering or were not strongly cemented during lithification.

Some of the most intensive wind abrasion occurs in the ice-free areas of the Antarctic. The total lack of vegetation makes a bouldery regolith very susceptible to wind attack, and the extreme aridity assures an ample supply of dry sand as the abrasive material. Large boulders and rock outcrops bear witness to the intensity of wind abrasion. Pitted and polished surfaces are commonplace, and where a bouldery regolith has been under wind attack for many thousands of years, all rock projections above the general ground level have been worn down by a combination of physical weathering and wind abrasion.

Deposition by Wind

Bed load and suspended load materials eventually settle out of the atmosphere. Where bed-load particles accumulate, *sand dunes* are formed. Suspended load materials are distributed over large areas to

Figure 12.2 Wind-faceted pebbles (ventifacts) from Rock Springs, Wyoming. About one-half normal size. (Photograph by M. R. Campbell, U. S. Geological Survey.)

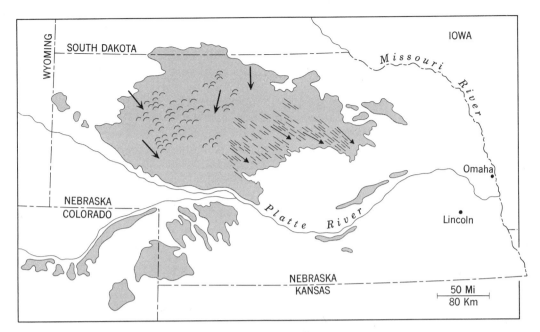

Figure 12.3 Map of the Sand Hills of Nebraska showing stabilized barchans and longitudinal dunes. The heavy arrows show the prevailing
wind direction during the time of formation of the barchans, and the
light arrows indicate the direction of the winds that produced the
longitudinal dunes. (Based on the *Map of Pleistocene eolian deposits*
of the United States, Alaska, and parts of Canada, published by the
Geological Society of America, 1952; and Smith, H. T. U., 1965,
Dune morphology and chronology in central and western Nebraska,
Jour. Geol., v. 73, pp. 557-578.)

form blanket deposits that may cover hundreds or thousands of square miles.

Sand Dunes. Sand dunes fall into two
categories, *active* dunes and inactive or
stabilized dunes. Active dunes are dunes
that are actively in the process of formation
and occur in true desert regions and along
coastal areas. These dunes change their
shapes and positions on the ground in
response to prevailing winds. Stabilized
dunes are those whose shape and position
have become fixed because of a vegetational cover. Stabilized dunes reflect a change
from an arid climate to one in which grass
or even trees become established on the
dune surface because of increased rainfall.

Shapes of Dunes. Sand dunes occur in a
variety of shapes and sizes. Dune shape is
a function of wind direction and degree of
stabilization by vegetation. The shapes of

active dunes are generally different from the
forms of the stabilized group. Contours of
stabilized dunes are somewhat softer than
those of active dunes, especially along
the ridge crest which separates the windward and lee slopes. The orientation and
surface configuration of active dunes reflects
existing wind directions, and the form and
orientation of stabilized dunes can be used
to reconstruct wind patterns of a past geologic period. For example, the stabilized
dunes in the Sand Hills of Nebraska reflect,
at least, two different wind directions during
two episodes of dune formation during the
last several thousand years (Fig. 12.3).

Active dunes include (1) *barchans,* cres-
cent-shaped dunes with the convex side
facing toward the wind; (2) *transverse dunes,*
dune ridges extending at right angles to the
wind directions; (3) *longitudinal* dunes,

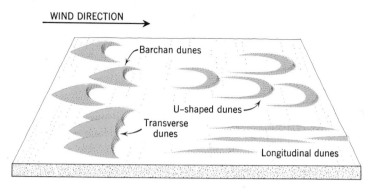

Barchan dunes

U-shaped dunes

Transverse dunes

Longitudinal dunes

Figure 12.4 Block diagram showing the various types of sand dunes.

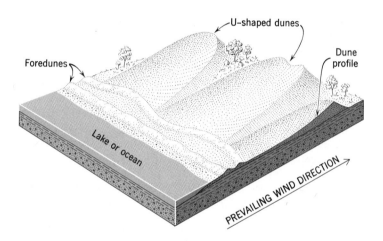

U-shaped dunes

Dune profile

Foredunes

Lake or ocean

PREVAILING WIND DIRECTION

Figure 12.5 U-shaped dunes are common to coastal areas of large lakes and oceans. The sand that nourishes these partially stabilized dunes is derived from the beach area. The dunes nearest the beach are transverse dunes (foredunes) that evolve into U-shaped dunes.

sand ridges elongated parallel to the wind direction; and (4) *complex* dunes of irregular shapes (Fig. 12.4). The barchans and transverse dunes are both strongly asymmetrical in cross section with the steep side (lee side) facing away from the wind direction and inclined at an angle of about 30 to 33 degrees with the horizontal.

A common form of stabilized or partially stabilized dune is the U-shaped dune, often called a *parabolic* dune. Unlike the barchans, the U-shaped dunes have the open end facing toward the wind, and the steep lee slope of the convex side facing in a downwind direction.

Some coastal dunes are U-shaped as a result of the formation of blowouts on the windward side of preexisting dune forms (Fig. 12.5). Vegetation holds the rest of the dune in place while the central blowout zone is further eroded by wind. A study of dunes along the southeastern shore of Lake Michigan shows that a regular evolutionary sequence of dune shapes develops, begin-

Figure 12.6 These partially stabilized dunes along the eastern shore of Lake Michigan are being modified by wind action that produces a blowout in the central part of each. View is toward the east, prevailing winds are from the west. (Courtesy of Hann Photo Service, Hartford, Michigan.)

ning with the original transverse dunes (foredunes), passing through a parabolic stage (Fig. 12.6), and finally ending with longitudinal dunes.

Other coastal dunes occur in Washington, Oregon, and California where ocean beaches provide the source of sand which is blown into a variety of dune shapes and patterns.

Most dune sand consists of quartz grains that have lost their shiny appearance because of constant collision with each other during saltation. When viewed under a magnifying glass, grains of dune sand look like rounded grains of frosted glass, a diagnostic characteristic that distinguishes eolian sands from fluvial or beach sand.

Figure 12.7 Dust arising from the floodplain of the Delta River in central Alaska is deposited as loess on the surrounding uplands. The loess deposits of the central United States may have originated in a similar manner from dust storms on the floodplains of the Mississippi and Missouri rivers during the Pleistocene (Ice Age). (Courtesy of Troy Péwé, Arizona State University.)

An unusual occurrence of nonquartz sands are those which form the "white sands" near Alamogordo in New Mexico. These active dunes consist entirely of white gypsum grains derived from local gypsum beds.

Loess

The suspended load wafted by winds consists mainly of silt and dust particles. As the suspended load settles out of the air suspension, it builds a blanket deposit of silt known as *loess* (pronounced as lōs, lō-ess, lŭs, lĕrs, or lĕs). These deposits are generally calcareous, unstratified, and consist mainly of silt grains, 0.002 to 0.05 millimeter in diameter. The narrow range of particle size gives all loess deposits a remarkable homogeneity, and reflects the efficacy of sorting by wind action.

Loess deposits are derived from unconsolidated surficial materials that are only sparsely vegetated or completely barren of plants. Some deserts provide a source of silt particles that are transported to peripheral semiarid regions and deposited as loess. For example, the thick loess of western China was derived from the deserts of central Asia.

In humid regions, broad floodplains are replenished seasonally with new layers of fine sand and river silt that are excellent sources of silt for suspended-load wind transport. Figure 12.7 shows a dust storm rising from the river deposits of the Delta River in central Alaska. Silt of similar origin

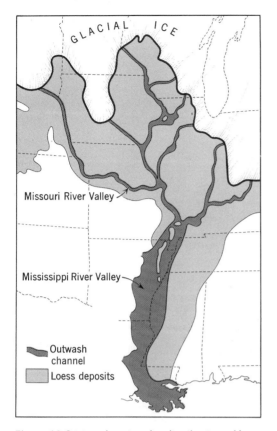

Figure 12.8 Map showing the distribution of loess deposits in the central United States, and their relationship to rivers that carried the meltwaters of the retreating continental glaciers of the Pleistocene (Ice Age). (Based on a diagram by Leighton, M. M. and H. B. Willman, 1950, Loess formation of the Mississippi Valley, *Jour. Geol.,* v. 58, p. 604.)

geologists concur that this loess was derived from floodplain silts that were deposited by rivers fed by silt-laden glacier meltwater (Fig. 12.8).

Loess and Agriculture. The loessial soils of the central United States, especially Iowa, Illinois, and Indiana, are among the most fertile in the world. The rich soils of southern Russia and parts of Europe are also developed on a loessial parent material. Their silty texture makes them easy to plow and endows them with a high capacity for holding moisture. Moreover, a great variety of minerals in the loess parent material provides an adequate supply of elements required for many kinds of field crops. Not the least among these is calcite ($CaCO_3$) which, because of its solubility in water, is readily available for plant use.

Loess and Engineering. Loess deposits that occur in cold climates require special treatment when highways or airfields are built on them. Because of the high susceptibility of silty sediments to frost heaving, it is necessary to remove the loess to a depth below the level of frost penetration and to replace it with a coarser material, thereby preventing the accumulation of ice lenses which develop during freezing. Loess that occurs in regions underlain by permafrost (Chapter 8) presents special problems in construction because of the presence of ice wedges and other irregular subsurface clear ice masses.

occurs around Fairbanks, Alaska as a blanket deposit ranging in thickness from 10 to 100 feet on the hilltops, and more than 300 feet in the valleys. The Delta River and others like it in central Alaska are fed by melting glaciers that supply the load of sand and silt to the rivers. This load is deposited on the floodplains during high stages and is later exposed to the wind when the rivers return to their low stages.

This direct observation of river-deposited silt being lifted by the wind corroborates the widely held theory that the loess deposits of the central United States and parts of Europe and Russia are of eolian origin. Most

DESERTS

Traditionally, the word desert connotes hot sun, drifting sand, and endless miles of barren wasteland inhabited by scattered shrubs or spine-covered cactus struggling for survival in soils of low fertility. Although these descriptions fit many desert regions of the earth, they are not characteristic of all. The world desert is derived from the Latin, *desertis,* which means barren or deserted. Desert is a type of climate that is characterized by aridity, the lack of moisture. In this sense deserts can be hot or cold and are distributed geographically from the equa-

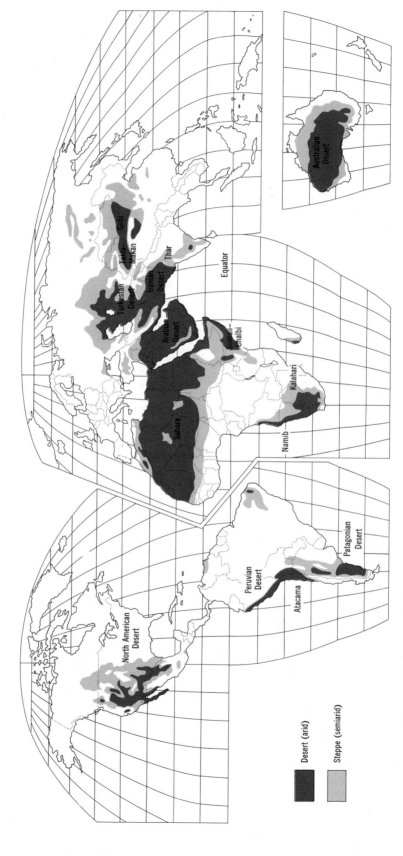

Figure 12.9 Map of the world showing the distribution of arid lands. [After McGinnies, W. G., B. J. Goldman, and P. Paylore (eds.), 1968, *Deserts of the world*, University of Arizona Press, Tucson, 788 pp.]

torial regions to the poles. Some writers refer to the Antarctic ice cap as a desert because of its very low precipitation and extremely low relative humidity. Not all geographers, however, are willing to include the dry polar areas in a classification of deserts because of the presence of ice caps (Chapter 13). Nevertheless, the ice-free areas of Antarctica exhibit almost all features normally associated with warm deserts. These include soils containing evaporite accumulations and barren surface areas with lag gravels and ventifacts in a variety of shapes and sizes. The major differences between the ice-free Antarctic deserts and those found in other parts of the world are the intensive frost action in the former and the occurrence of rain in the latter. Rainfall, even though it occurs infrequently in hot deserts, is absent in the Antarctic deserts. For this reason, the ice-free areas of the Antarctic continent do not contain the fluvially derived landforms that are common to the deserts of lower latitudes. Hence, polar deserts will be excluded in the discussion that follows.

Origin and Distribution of Deserts

The worldwide patterns of atmospheric circulation produce high pressure zones in the general region of the tropics. The high altitude air in these belts is dry and cold, but as it descends to lower altitudes it warms adiabatically,[1] so that by the time it reaches the surface of the earth it is hot and dry. An air mass without moisture is incapable of producing precipitation. Indeed, it is characteristic of desert air masses to produce high rates of evaporation, so much so that one way of defining a desert climate is to say that the evaporation exceeds the precipitation. The greater the excess of evaporation over precipitation, the higher the aridity. To differentiate areas of different degrees of aridity, arid lands are classified as extremely arid,

[1] The change in temperature of a gas when it expands (causing adiabatic cooling) or compresses (causing adiabatic heating) without any subtraction or addition of heat.

arid, and semiarid (Fig. 12.9). The latter climatic term is synonymous with *steppe* (see Table 8.a).

Most of the world's deserts are concentrated in two world-encircling belts. The *tropical deserts* occur between the latitudes of 15° and 35° in the Northern and Southern Hemispheres, and the *mid-latitude deserts* lie between 30° and 50° north and south of the equator in the continental interiors. Approximately 20 percent of the earth's land area is classified as desert.

Temperature and Precipitation

The classification of climates given in Chapter 8 (Table 8.a) provides a brief description of rainfall and temperature conditions of the various kinds of deserts around the world. The thermohyet for In Salah, Algeria in the Sahara Desert (Fig. 8.3) is typical of an extremely arid climate in a tropical desert, and the one for Las Vegas, Nevada (Fig. 8.4) is typical for a mid-latitude desert. Mid-latitude deserts have much wider annual temperature ranges than tropical deserts. The Mongolian Desert, for example, a mid-latitude type, has subfreezing temperatures in winter and summer temperatures that reach 100° F. Maximum summer temperatures in tropical deserts commonly exceed 110° F. At Yuma, Arizona, for example, the highest summer temperature on record is 123° F.

Desert rainfall is very meager and sporadic. In the extremely arid deserts no rain may fall in a full 12-month period, while in the semiarid deserts, annual precipitation is about 10 inches (250 mm). However, when rainfall does occur in deserts, the runoff is often torrential because there is little vegetational cover to retard overland flow, and soils are cemented with caliche which renders them almost impervious to infiltration by surface water. During infrequent storms that produce high intensity rainfalls, dry washes are transformed into raging torrents in a matter of minutes, and mudflows laden with boulders and other debris ooze down steep slopes causing considerable destruction of man-made structures that happen to lie in their paths. Fluvial features in deserts are more commonplace than is generally realized because most casual observers of desert landscapes see them only when the surface water courses are dry.

The Desert Landscape

The same geologic processes that operate in humid regions also function in deserts, but these same processes give rise to different terrain features because of the factor of aridity.

In humid regions both physical and chemical weathering are active but in arid regions, chemical weathering is less intense. The desert soils, aridisols (Table 8.d), have poorly developed horizons and are lacking in organic material. Secondary deposits of "lime" ($CaCO_3$) and other salts that cement the mineral grains together occur in both soil and regolith. These cemented soils or *caliches* impede the infiltration of rainwater into the groundwater reservoir and materially increase runoff during high intensity rainfalls.

Desert stream channels are ephemeral as well as influent (Fig. 11.1). The first impression of the desert traveler is that running water is not an important geologic agent in the shaping of the desert landscape but, in fact, running water is the most effective agent in desert denudation.

Desert Erosion. The desert landscape is a product of the interaction between bedrock

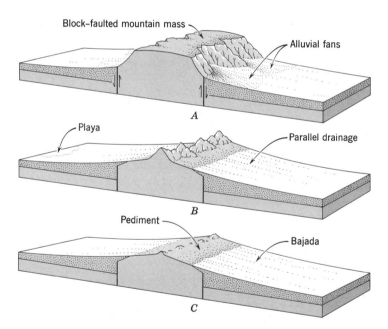

Figure 12.10 Sequential diagrams showing the erosional evolution of a pediment. *A:* a mountain range produced by block faulting is dissected by stream erosion. Alluvial fans are formed along the flanks. *B:* continued erosion causes further dissection of the mountain range and the development of an inclined erosional surface that merges imperceptibly with the adjacent depositional surface. Playas occur where closed bains occur between adjoining block mountains. Parallel drainage channels attest to the presence of surface runoff, which is the chief agent active in the process of pedimentation. *C:* in the very late stages of pedimentation, the pediment surface is covered by a thin veneer of debris so that the boundary between the edge of the pediment and the adjoining alluvial fill is unidentifiable at the surface. The sloping pediment and the alluvial surface form a bajada.

lithology and structure, and the forces of weathering and erosion. The lack of moisture in soils reduces the tendency for weathered products to creep in a downslope direction as is the case in more humid climates. For this reason the desert topography is more angular than a humid topography. Whereas the contours of a humid terrain are softened because of the presence of a relatively thick regolith that becomes distributed downslope by creep, weathered products in the desert are usually thin and do not accumulate on hillslopes because they are washed into the valleys and low areas by runoff. Changes in slope from hillside to valley floor are gradual in humid regions but tend to be abrupt in the desert.

Desert landforms produced by erosion are as varied in type as any found in humid regions. One of the most prominent and best-developed types is the *pediment*. A pediment is an inclined bedrock erosion surface that fringes a mountain range in the desert (Fig. 12.10). It separates the rugged bedrock relief of the mountains from the flat alluvial floor of the valley. Pediment surfaces are so uniform in shape that their surface geometry can be very closely described by a mathematical formula.

The exact manner in which a pediment is formed is not clearly understood, but the fact that it has a concave-upward profile suggests that running water is a primary agent. Pediments are found on different kinds of bedrock but they seem to reach their best expression on coarse-grained granitic rock

Figure 12.11 Badlands topography in the Death Valley National Monument, California-Nevada. The dissection of this landscape is caused by ephemeral streams that are dry washes most of the time. (Photograph by George Grant, National Park Service.)

types. The granular debris resulting from the physical weathering of coarse granite is moved downslope by sheet wash during periods of high intensity rainfall. The pediment surface grows in areal extent as the mountain front retreats under the combined assault of physical weathering and slope wash until it is literally buried in its own debris (Fig. 12.10C).

Sedimentary rocks with a horizontal attitude give rise to certain landforms that are exceptionally well developed in arid and semiarid regions. If the rocks are fine-grained clastics such as fine sandstone, siltstone, or shale, *badlands* are produced (Fig. 12.11). If the sedimentary rocks contain alternating layers of hard and soft rocks, *mesas* and *buttes* are formed (Fig. 12.12). The rock layer forming the top of one of these erosional forms is more resistant to erosion than the rocks underneath. In desert regions these *cap rocks* may consist of limestone, strongly cemented sandstone, a resistant lava flow, or an exhumed sill.

Eolian Deposits. Only about one-fourth of the extremely arid and arid lands are covered by active sand dunes, but where these

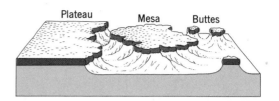

Figure 12.12 Block diagram of a mesa and butte. The flat tops of these desert landforms are capped by horizontal strata that are resistant to weathering and erosion by running water. See also Figure 19.8.

occur they are impressive elements of the landscape. Photographs taken from high-flying aircraft and from manned spacecraft reveal much about the distribution of large tracts of desert sand (Fig. 12.13). Some of these massive sand areas contain dunes that clearly reflect the direction of prevailing winds, but others consist of highly complex dunes that are difficult to interpret in terms of prevailing winds.

The largest expanse of dune sand in the world is found in the Sahara Desert of northern Africa. Approximately 200,000

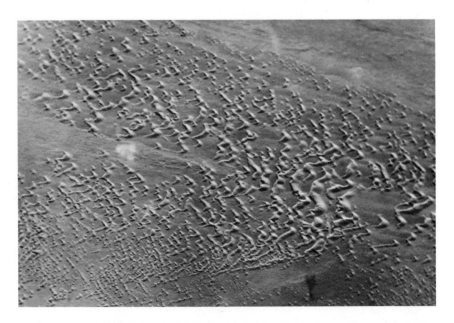

Figure 12.13 Aerial photograph of a field of barchan dunes in the Spanish Sahara, Africa. Wind direction is from right to left. (Courtesy of H. T. U. Smith, University of Massachusetts.)

square miles of shifting sand blankets the Libyan sector of the Sahara. Many of the dune areas cover irregular shaped tracts 2 to 3 square miles in area and are separated by intervening flat ground of about equal size. The dune sand is piled into sharp crested ridges, some of which radiate from a central peak that rises several hundred feet above the adjacent flat areas.

The sand dune areas of the Arabian Desert spread from the Persian Gulf across the Rub' al Khali toward the Red Sea. Some of the dunes are so large that they are referred to as "sand mountains." The different kinds of dunes and their orientation provides a basis for establishing the general direction of wind transport for much of the Arabian Peninsula (Fig. 12.14).

Loessial deposits are not widespread in the extremely arid deserts because the winds are too strong to allow the settling of silt and dust. Furthermore, these regions do not have a cover of vegetation that can trap airborne silt particles once they have reached the ground. Some desert areas act as a source area of silt which is transported downwind

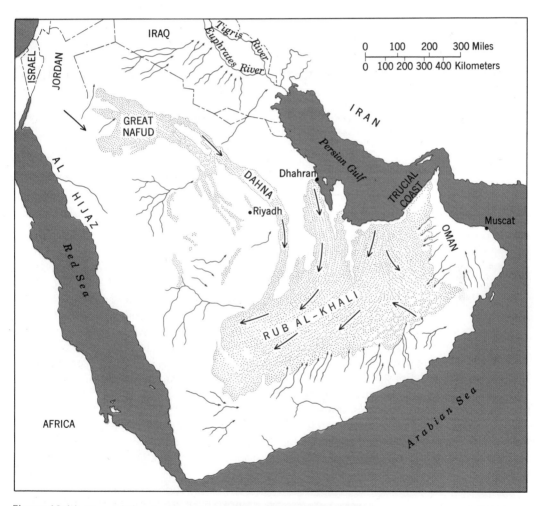

Figure 12.14 Map of the Arabian Peninsula showing major desert areas (stippled). Notice that most of the surface drainage lines flow toward the desert region where they terminate. The heavy arrows show wind directions as inferred by the orientation of sand dunes. (After Holm, D. A., 1960, Desert geomorphology in the Arabian Peninsula, *Science*, v. 143, pp. 1369-1379.)

Figure 12.15 A playa in Smith Creek Valley, Nevada. This flat surface is formed by the accumulation of sediments in the playa lake which occupies the area from time to time. These features are the basins of the Basin and Range Province of the arid southwestern United States. Other playas are visible in the background. (Photograph by James T. Neal, United States Air Force Academy.)

and deposited in marginal semiarid regions or steppes where sufficient vegetation, usually grass and shrubs, effectively aids in the accumulation of a loess blanket.

Noneolian Depositional Features. In some desert regions such as those of the western United States the desert landscape consists of three major topographic elements: mountain ranges, flanking pediments, and alluvial basins. The interrelationship of these three components is relatively simple. The debris eroded from the mountains is transported across the pediments and is deposited in the basins. The alluvial fill in some intermontane basins is a thousand or more feet thick and consists of alternating layers of sediment ranging from coarse gravel to fine silt and clay.

Some of the basins in which alluvial fill accumulates over long periods of geologic time are closed; that is, there are no surface drainage channels leading away from them. Consequently they trap great quantities of detrital sediment, and from time to time are occupied by very shallow lakes called *playa lakes*. These water bodies are ephemeral and are obliterated by evaporation soon after they are formed. The dry lake floor or *playa* is exposed to wind deflation and other subaerial processes of weathering and erosion. Material dissolved in the waters of a playa lake remain as salty residues on the playa surface after evaporation consumes the lake waters.

Part of the floor of Death Valley, California is a classic example of a playa that has undergone successive periods of flooding from runoff originating on the flanks of the surrounding mountains during periods of high precipitation. Similar playas are commonplace all through the Basin and Range Province (Fig. 12.15) of the western United States.

Alluvial fans are fluvial deposits previously described in Chapter 11. They are a widely occurring landform of the desert landscape.

References

Bagnold, R. A., 1941, *The physics of blown sand and desert dunes,* Methuen and Co., London, 265 pp.

Cooper, W. S., 1967, Coastal dunes of California, *Geological Society of America Memoir 72,* 169 pp.

* Hadley, R. F., 1967, Pediments and pediment-forming processes, *Jour. Geol. Education,* V. 15, pp. 83–89.

Holm, D. A., 1960, Desert geomorphology in the Arabian peninsula, *Science,* V. 132, pp. 1369–1379.

Howe, G. M. and others, 1968, *Classification of world desert areas,* Technical Report 69-38-ES, U. S. Army Natick Laboratories, Natick, Mass., 104 pp.

* Logan, R. F., 1968, Causes, climates, and distribution of deserts, in *Desert Biology,* V. 1, pp. 21–50, Academic Press, Inc., New York.

* McGinnies, W. G., B. Goldman, and P. Paylore (eds.), 1968, *Deserts of the World,* University of Arizona Press, Tucson, Ariz., 788 pp.

Péwé, T., 1955, Origin of upland silt near Fairbanks, Alaska, *Geological Society of America Bulletin,* V. 66, pp. 699–724.

Sharp, R. P., 1949, Pleistocene ventifacts east of the Big Horn Mountains, Wyoming, *Jour. Geology,* V. 57, pp. 175–195.

* Smith, H. T. U., 1968, Geologic and geomorphic aspects of deserts, *in Desert Biology,* V. 1, pp. 51–100, Academic Press, New York.

* Recommended for further reading

13 Glaciers and Glaciation

*Some vast store of ice beyond seemed
to take advantage of the break in the
mountain chain, and to pour down in one
great river of ice to the sea.*
Robert F. Scott

A glacier is a flowing mass of land ice derived from snowfall. Glacier ice contains more than 2 percent of all the water on the globe (Table 10.a) and covers about 10 percent of the earth's land area. At various times during the last two or three million years, however, glaciers covered nearly one-third of the land area and contained an estimated 8 percent of the earth's water volume.

In one sense glaciers are capacious storehouses of solid moisture removed temporarily from the hydrologic cycle. Glaciers originate when annual snowfall in a given region exceeds the amount of snow melted each year. A succession of such years causes glaciers to expand in size with a concomitant lowering of sea level, the source from which the snow is derived. Conversely, a general shrinking of the world's glaciers produces a corresponding rise in sea level.

In another sense, glaciers are geologic agents capable of modifying the land surface

Large outlet glacier flowing from the continental ice sheet of Antarctica to the Ross Ice shelf. (U.S. Navy Photograph.)

that they occupy. Glaciated landscapes are among the most spectacular in the world, as those who have seen the Alps, Himalayas, Greenland, Alaska, or Antarctica will testify. This chapter is concerned not so much with the inspiring grandeur and natural beauty of glaciers but, instead, with the dynamics of existing glaciers and the geologic effects of modern and ancient ones.

Glacier Dynamics

Glaciers occur in the high latitudes of both hemispheres and at high elevations at any latitude where climatic conditions for glacial nourishment are favorable. Glaciers exist even in the equatorial regions of Africa, South America, and New Guinea.

The chief requirement for the maintenance of a glacier is a climate conducive to the *accumulation* of snow, the source of nourishment for all glaciers. Glaciers can originate only where more snow accumulates annually than wastes away by melting or, as in the case of Antarctica, by direct discharge to the ocean. Each year a layer of residual snow is added to the accumulation of past years until a sizable thickness is attained. As older snow layers are buried beneath subsequent layers, the deeper snow is transformed into a granular material called *firn*. Firn has a density greater than newly fallen snow but less than pure ice. Ultimately, firn is metamorphosed by further compaction and *recrystallization* into glacier ice.

Glacier Mass Balance. A glacier normally can be divided into a *zone of net accumulation* and a *zone of net wastage* (Fig. 13.1). Each winter a layer of fresh snow covers the entire glacier surface. During the summer months some of the snow in the accumulation area is lost by melting, and in the wastage zone all of the previous winter's snowfall plus some glacier ice formed by older snows is lost. The boundary between the zones of net accumulation and net wastage is called the *firn limit* or *annual snow line*.

If the mass of the annual snow surplus is

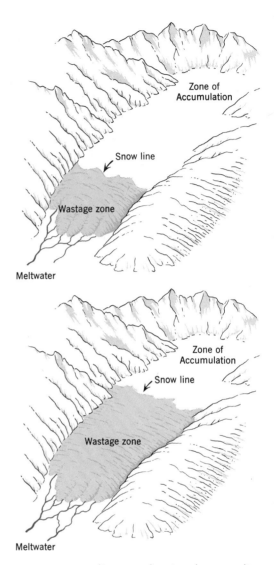

Figure 13.1 A: diagrams showing the snow line and the zones of accumulation and ablation on a valley glacier at the end of the melt season for two different years. In the upper diagram, accumulation exceeds wastage so the glacier is said to have a positive mass balance for that year. In the lower diagram, wastage exceeds accumulation, a condition that results in a negative mass balance for the year. Notice that the elevation of the snow line increases as the area of accumulation diminishes. B: a longitudinal profile through a valley glacier at the end of the melt season. Colored arrows show direction of ice flowage along flow lines. (From Sharp, R. P., 1960, *Glaciers,* Condon Lectures, Oregon State Systems of Higher Education, Oregon, 78 pp.)

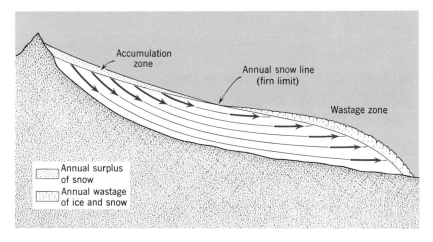

Figure 13.1 *B*

equal to the mass of annual net wastage, the glacier is in equilibrium; it has neither gained nor lost any of its mass during the year. Flowage of ice from the accumulation zone to the wastage zone continues during both summer and winter, so that the position of the glacier's *snout* or *toe* remains more or less the same, if the glacier is in equilibrium.

The relationship between net wastage and net accumulation can be expressed in quantitative terms because techniques have been devised to measure both of these quantities. Accumulation and wastage values are both stated in volumes of their water equivalents.[1] If A_c = net annual accumulation, A_w = net annual wastage, and A = the mass balance of the glacier, then

$$A = A_c - A_w$$

When $A > 0$, the glacier is said to have a *positive mass balance,* and when $A < 0$, the glacier has a *negative mass balance.* If $A = 0$, the glacier has a *zero mass balance.*

The twelve-month period during which accumulation and wastage measurements are made on the glacier is called the *budget*

year. The relationship of accumulation and wastage to a budget year of a glacier in the Northern Hemisphere is shown in Figure 13.2. Observe that the budget year does not coincide with the calendar year.

The determination of a glacier's mass balance, for a number of successive years, provides an interesting analysis of the glacier's "state of health." For example, a record of glacier mass balance has been kept for the Hintereisferner of the Tyrolean Alps in Austria. Between the budget years of 1952–1953 and 1967–1968, this glacier experienced a positive mass balance for five of the 16 years of record (Fig. 13.3). In other years the glacier lost more of its mass by wastage than it gained by accumulation. Notice also in Figure 13.3 the general relationship of the elevation of the annual snow line and the area of the glacier surface to the mass balance for each budget year.

Glacier Movement. Flowage is a phenomenon ordinarily associated with liquids, but under certain conditions solids will flow also. Some kinds of earth materials (Chapter 9) exhibit characteristics of flowage, and flowage in foliated metamorphic rocks is manifested in the orientation of platy or elongate mineral grains.

The proof that glaciers flow in response to the pull of gravity was provided in the

[1] The water equivalent is the volume of water contained in a given volume of snow, firn, or glacier ice.

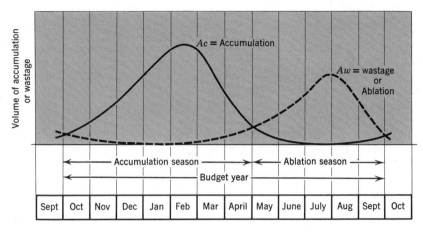

Figure 13.2 A graph showing the relationship between accumulation and ablation on a glacier during the budget year. In this example the total accumulation is greater than the total ablation during the budget year. The glacier represented by this diagram has a positive mass balance. [After Meier, M., 1964, *Glaciers,* Section 16, p. 18 in Chow, T. Ven (ed.), *Handbook of hydrology,* McGraw-Hill Book Co., New York.]

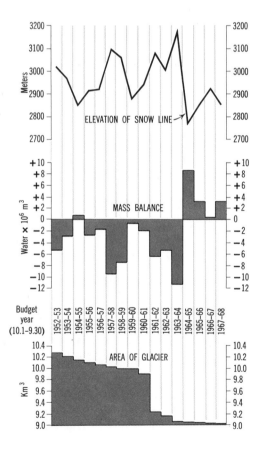

eighteenth century when a Swiss scientist set a straight line of stakes across the axis of a glacier and observed their change of position over a number of years. All the stakes moved in a down valley direction, but those near the center of the glacier moved farther than those near the edge. Since that first experiment, the surface movement of many glaciers all over the world has been measured. Surface veolocites are generally on the order of less than one foot per day but some glaciers are known to flow at rates in excess of 50 feet per day.

Stake measurements reveal only the surface velocity of a glacier. The determination of ice velocity between the glacier surface and the rock floor over which it flows is a

Figure 13.3 This diagram shows the mass balance of the Hintereisferner in the Tyrolean Alps (Austria) for the budget years of 1952-1953 to 1967-1968. In years when this glacier had a negative mass balance the snow line was higher than years in which there was a positive mass balance. (Courtesy of H. Hoinkes, University of Innsbruck, Austria.)

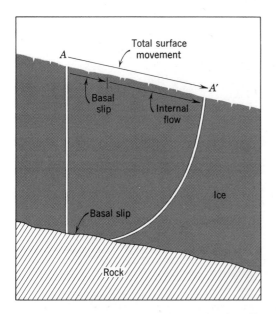

Figure 13.4 Cross section of a longitudinal profile of a valley glacier. A borehole drilled vertically at "A" deforms into a curved line at A' after a few years. The total surface movement, AA', is made up of two flow components, basal slip and internal flow. (After Sharp, R. P., 1960, *Glaciers*, Condon Lectures, Oregon System of Higher Education, Eugene, Oregon, 78 pp.)

more difficult task that requires the drilling of a vertical hole through the glacier. By measuring the change in the inclination of a borehole with time, the *vertical velocity profile* of a glacier can be determined. Figure 13.4 is an example of the results of a borehole experiment. Note that part of the ice movement is by slippage of the glacier over the rock floor, but that most of the movement is by internal flow.

The mechanics of internal flow (that is, deformation) of a crystalline solid are not clearly understood. Inasmuch as ice is a mineral, the water molecules of which it is composed occur in an orderly geometric arrangement. An ice cube dropped on a hard surface will shatter, yet ice deep within a glacier will flow without disrupting. Glaciologists have learned from laboratory experiments that single ice crystals can be deformed under certain conditions of stress without losing their solidity or coherence. These studies show that ice crystals deform by internal gliding along definite crystallographic planes. Ice crystals are made up of layers of the H_2O molecule, and deformation occurs when these layers slide past one another along the planes that separate them.

Crevasses. The ice near the glacier surface is brittle and does not deform in the same manner as ice deeper in the glacier. The upper part of a glacier is thus carried along by the deeper ice which acts as a sort of conveyor belt. The less plastic surficial ice layer on a glacier is generally 100 to 150 feet thick and contains cracks and fissures known as *crevasses* (Fig. 13.5). Crevasses are abundant in areas where the slope of the glacier bed steepens markedly, thereby causing the velocity of the basal ice to increase. Under these conditions the surface of the glacier becomes an *icefall* that is characterized by a chaotic system of crevasses and jagged masses of glacier ice. An icefall presents an imposing barrier to mountaineers who use glaciers as a means of gaining access to high peaks in glaciated mountains.

Crevasses can occur in either the zone of accumulation or the wastage zone. In the latter, they are visible during the ablation season and can be avoided, but in the former, they are usually covered by a snow layer called a *snow bridge*. Some snow bridges are strong enough to hold the weight of a man but many are not. For this reason cautious mountaineers do not attempt the crossing of snow bridges alone unless they are on skis or are roped together in parties of at least three people.

Surging Glaciers. A glacier with a positive mass balance is ordinarily expected to advance. That is to say, the increasing mass of the glacier will be reflected in an increase in length of the glacier in a down-valley direction. Some glaciers, however, advance very rapidly without any apparent increase in their mass. These glaciers are called *surging glaciers* (Fig. 13.6). Very little is known

Figure 13.5 Crevasses on the Tasman Glacier, South Island, New Zealand. The glacier is moving from upper right toward lower left, more or less at right angles to the orientation of the crevasses. (Photograph by Tad Nichols.)

about the causes of glacier surges, but it appears that some mechanism triggers the glacier into a phase of more rapid flowage, thereby causing a spectacular advance down the valley. Considerable research remains to be done before the peculiar behavior of surging glaciers can be explained.

Classification of Glaciers

Glaciers can be separated into three general groups, (1) *valley glaciers,* (2) *piedmont glaciers,* and (3) *ice sheets.* Valley glaciers are literally rivers of ice because they flow between valley walls in mountainous regions. Piedmont glaciers develop wherever one or several valley glaciers spread out at the foot of a mountain range. Ice caps or ice sheets cover broad areas of continental proportions and flow outward in all directions.

Valley Glaciers. In mountainous regions where conditions are favorable for snow accumulation, a network of valley glaciers may occupy a conspicuous part of the landscape. These ice streams may radiate from a central volcanic peak such as Mt. Rainier in Washington, or they may form a tributary system feeding into a main valley glacier such as the Yentna Glacier of Alaska (Fig. 13.7).

The glacier surface usually contains rock debris worn from the valley walls. The debris forms dark bands parallel to the flow

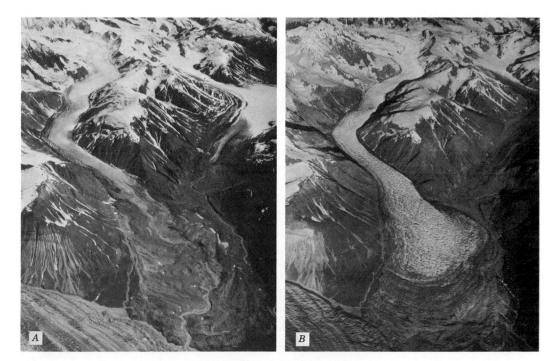

Figure 13.6 These two photographs of the Variegated Glacier in Alaska (center) were taken one year apart. *A:* on October 29, 1964 the snout of the glacier appeared to be either in equilibrium or retreating. *B:* a year later on October 22, 1965 the glacier snout had a dramatically different appearance, having advanced down valley a considerable distance. This kind of rapid movement is known as a glacier surge and cannot be explained by an increase in mass balance during a one-year period. The glacier immediately to the right of the Variegated Glacier presumably was subjected to the same climatic factors as the Variegated Glacier, and it did not surge. Glacial surges are caused by factors that are not fully understood by glaciologists. (Photographs by Austin Post, U. S. Geological Survey.)

direction of the glacier and are known as *moraines.* When viewed from the air or when seen on aerial photographs, the moraines are plainly visible and their relationship to the glacier system is readily apparent. Three kinds of moraines can be identified. Those that occur along the sides of the glacier are *lateral moraines.* At the confluence of two valley glaciers, the lateral moraines of each join to form a *medial moraine as* seen in Figure 13.7. At the glacier snout or terminus the debris contained in the lateral and medial moraines, as well as stones and other detritals carried in the glacier, are released from the melting ice and heaped into an *end moraine.*

Piedmont Glaciers. If the snout of a valley glacier spreads out in a bulbous form, or if the snouts of several valley glaciers coalesce into a single broad mass, the result is a piedmont glacier. These types are relatively rare. One of the best-known and largest piedmont glaciers in North America is the Malaspina Glacier in Alaska. A piedmont glacier is shown in Figure 13.8.

Ice Sheets

Only two major sheets exist today, Greenland and Antarctica, but many smaller ones occur in Iceland, Norway, Baffin Island, and

Figure 13.7 Yentna Glacier, Alaska. The dark linear features are moraines. (Photograph by Austin Post, U. S. Geological Survey.)

the Canadian Archipelago. Together, Greenland and Antarctica contain about 99 percent of all the glacier ice in the world.

The Greenland Ice Sheet. Greenland is a large elongate landmass between North America and Europe which contains an ice sheet of 666,400 square miles which represents about 83 percent of the area of Greenland itself (Fig. 13.9). The mean elevation of the surface of the ice sheet is 7000 feet above sea level, and much of the base of the ice sheet lies below sea level (Fig. 4.1).

The ice flows from the two dome-shaped high points toward the margin where it spills through valleys in the coastal mountains as *outlet glaciers* (Fig. 13.10). In other places it debouches directly to the sea where massive portions break off and become *icebergs*.

The mean thickness of the ice sheet is 4900 feet, but the maximum thickness is nearly 10,000 feet. The total volume of glacier ice on Greenland is 620,000 cubic miles which, if melted and distributed over the world's oceans, would raise sea level about 20 feet (6 m).

Figure 13.8 The Bering Piedmont Glacier, Alaska. (Photograph by Austin Post, U. S. Geological Survey, 1966.)

Measurements of accumulation and wastage indicate that Greenland is gaining in mass, and the ice sheet is expanding. Wastage takes place by direct melting around the margins and the breaking off of large masses of glacier ice directly into the ocean, a process called *calving*. The total annual loss by calving is estimated to be about 220 cubic kilometers of water equivalent. Losses by direct melting are estimated to be about 250 cubic kilometers annually, whereas net accumulation is considered to be about 550 cubic kilometers per year. Thus, if $A_c = 550$ km³, and $A_w = (220 + 250)$ km³, then $A = A_c - A_w = +80$ km³ annual surplus.

Antarctic Ice Sheet. The surface area of the Antarctic ice sheet is 5.2 million square miles (Fig. 13.11). Ice thickness measurements by seismic methods[2] reveal an average of about 7500 feet. The greatest ice thickness reported anywhere on earth occurs about 300 miles inland from the Russian base of Mirny. At that point the ice surface is 9840 feet above sea level, and the ice is 13,000 feet thick. A large part of the bottom

[2] A small charge of dynamite is exploded at the surface of the ice. The time required for the sound waves thus generated to travel to the base of the ice and back is measured. Half the travel time measured at a point on the glacier surface multiplied by the speed of sound through glacier ice equals the ice thickness at that point.

Figure 13.9 Map of the Greenland ice sheet and smaller ice caps on Iceland and the Canadian Arctic. [After Bader, H., 1961, *The Greenland ice sheet,* in Cold regions science and engineering (F. J. Sanger, ed.), U. S. Army Cold Regions Research and Engineering Laboratory Monographs. Part I, Section B2, Hanover, New Hampshire.]

Figure 13.10 These outlet glaciers descending to Søndre Strømfjord on the west coast of Greenland are forms of valley glaciers. (Photograph by United States Air Force.)

of the Antarctic ice sheet lies below sea level because the continental crust of the earth is depressed about 1000 feet for each 3000 feet of ice piled on the land.[3]

The total volume of glacier ice in the Antarctic ice sheet is probably in excess of 6.3 million cubic miles, which amounts to enough water to raise the level of the oceans about 195 feet (59 m).

There is insufficient evidence to permit an unequivocal statement about the relationship between wastage and accumulation for the Antarctic ice sheet. However, available measurements of accumulation and reasonable estimates of wastage suggest that the Antarctic ice sheet is increasing in

mass. More precise measurements of both wastage and accumulation are needed before the magnitude of this positive mass balance can be made.

Fringing the Antarctic continent are floating masses of glacier ice called *ice shelves* (Fig. 13.11). They are attached to the main ice sheet and represent seaward extensions of continental ice sheets. The two largest ice shelves in the world are the Ross Ice Shelf (200,000 sq. mi.) and the Filchner Ice Shelf (160,000 sq. mi.). The Ross Ice Shelf is about the size of Spain and ranges in thickness from less than 300 feet near the seaward margin to 2300 feet in the interior. It is floating on seawater 1000 feet deep beneath its base. Icebergs that calve from ice shelves are flat-topped, and some of them are tens of miles long and hundreds of feet thick. Measurements near the edge of the Ross Ice Shelf show that much of it moves seaward at the rate of about 0.5 miles per year.

[3] This ratio of 1 to 3 derives from the ratio of the density of ice, 0.9, to the density of the earth's crust, 2.7. The ratio of 1 to 3 is further based on the assumption that perfect isostatic balance exists.

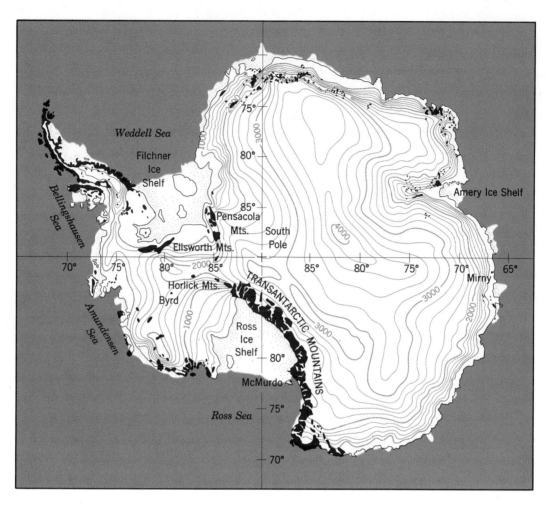

Figure 13.11 Map of Antarctica. This continent is 5.2 million square miles in area and is almost entirely covered with an ice sheet with an average thickness of about 7500 feet. The greatest thickness of ice anywhere on earth occurs about 300 miles inland from the Russian Base of Mirny. At that point the glacier surface is 9840 feet above sea level, and the ice sheet is 13,000 feet thick. (Based on Zumberge, J. H., and C. W. Swithinbank, 1962, *The dynamics of ice shelves, Antarctic Research, Geophysical Monograph No. 7,* American Geophysical Union, p. 198, and Bentley, C., 1964, *Antarctic Map Folio Series,* Folio 2, The Ice Sheet, American Geographical Society, New York.)

Geological Work of Glaciers

Glacial Erosion. Glaciers carry a load of detrital material in their basal ice layers. As a glacier moves over a bedrock surface, this basal load abrades the rock surface and forms *glacial polish,* scratches or *striations* (Fig. 13.12), and even large grooves. The direction of movement of former glaciers is as-

sumed to have been parallel to the scratches and grooves, and when these directions are plotted on a map, it is possible to reconstruct the overall pattern of flow of former glaciers.

Glacial abrasion is far less effective in eroding bedrock than glacial *quarrying* or *plucking.* This process involves the lifting and removal of blocks of bedrock by the moving

Figure 13.12 This jointed rock surface in Devil's Postpile National Monument in California has been smoothed by a former glacier moving over it. The small grooves and striations were also formed by glacial abrasion, and they are oriented parallel to the direction of ice movement.

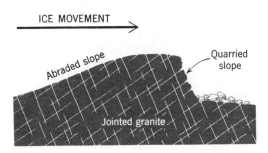

Figure 13.13 Cross section of a bedrock outcrop shaped by glacier erosion. Both an abraded slope and a quarried slope are shown.

ice. Glacial quarrying operates most effectively in rocks with closely spaced joints that define the incipient blocks to be removed by the plucking action.

A combination of quarrying and abrasion of a projecting rock surface produces an asymmetrical rock hill with the steep side facing in the direction of ice movement (Fig. 13.13). The terrain produced by ice-cap erosion consists of rounded and smoothed bedrock surfaces with less relief than the preglacial topography. Good examples of bedrock that are essentially the product of ice-cap erosion can be seen in the lake country of eastern Canada and in many parts of Scandinavia and Finland.

Glacially Eroded Valleys. A valley glacier is a more effective agent of erosion than the stream that originally formed the valley. Instead of the V-shaped cross section produced by stream erosion, a glaciated valley is a deep U-shaped trough (Fig. 13.14). Tributary valleys occupied by smaller glaciers are less deeply eroded than the main valley and, hence, are left "hanging" above the trunk valley. With the disappearance of

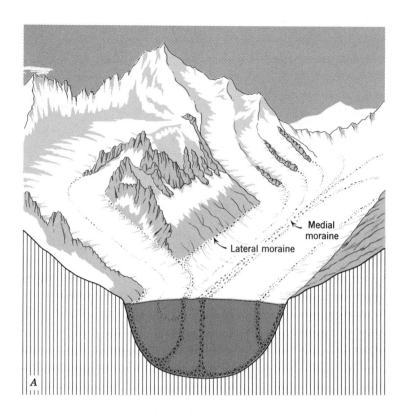

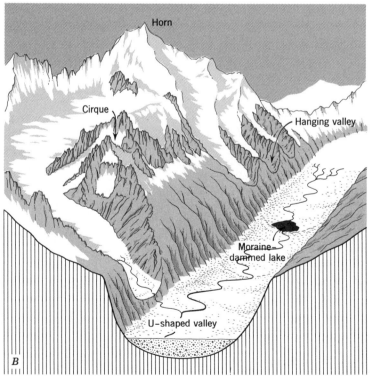

THE DYNAMIC EARTH

Figure 13.15 Yosemite Valley in Yosemite National Park, California is a classic example of a glaciated valley. Its steep walls and tributary hanging valleys attest to the presence of a valley glacier in the recent geologic past. Bridalveil Fall plunges over the lip of a hanging valley. (Photograph by F. E. Matthes, U. S. Geological Survey.)

the glaciers, the *hanging valleys* are reoccupied by streams that discharge into the main valley via waterfalls over precipitous cliffs, such as the Bridalveil Fall in Yosemite National Park (Fig. 13.15). In the headward reaches of a glaciated valley, bedrock basins, called *cirques,* formed by glacial erosion contain lakes, for example, Iceberg Lake in Glacier National Park. Other glacially formed lakes are the result of end moraines that act as dams in glaciated valleys.

In general, the terrain sculptured by valley glaciers undergoes a pronounced increase in ruggedness of relief. Direct action by valley glaciers plus the effects of intense frost action on the adjacent unglaciated peaks produces sharp divides, steep slopes, and pyramidal peaks called *horns* (Fig. 13.14).

Figure 13.14 Block diagrams showing changes in a landscape brought about by mountain glaciation. *A:* a system of valley or alpine glaciers occupies the former stream valleys and causes deep erosion. Frost action causes weathering of the unglaciated slopes and provides a source of debris for lateral moraines. *B:* climatic change causes cessation of glaciation and the uncovering of a telltale topography that includes hanging valleys, U-shaped valleys, lakes, end moraines, horns, and cirques.

Figure 13.16 Till exposed in a road cut in southern Michigan. The heterogeneous sizes of particles and the lack of stratification are characteristic of tills. Hammer handle near the center of the photograph is 12 inches long.

Glacial Deposits

The landscape uncovered by glacier retreat contains various deposits formed beneath the ice and in front of the ice margin. These deposits are visible long after the ice has disappeared. The general term *drift,* used to designate all deposits associated with glaciers, is a holdover from the days when the loose, unconsolidated deposits found in Europe and North America were attributed to the results of the Noachian deluge described in the Scriptures. The huge boulders and heterogeneous assortment of stones were thought to have originated from floating icebergs that "drifted" through the Noachian seas and dropped their load of debris when melting destroyed them. This notion is no longer tenable as a general explanation for the origin and distribution of glacial drift, because it is now clear that drift was transported and deposited by the action of former ice caps that covered one-third of the land surface during the Pleistocene (The Ice Age). In the oceans around Greenland and Antarctica, however, sediments accumulating on the sea floor do contain fragments released from melting icebergs that "drifted" into the polar seas after being calved from glaciers.

There are two general types of glacial drift, *nonstratified* and *stratified.* Nonstratified drift is distinguished chiefly by its unsorted texture and lack of bedding. It represents an accumulation of material deposited directly by the ice without the intervening action of water. Stratified drift, on the other hand, is deposited by melt water from the glacier and accumulates in streams issuing from the ice, in temporary lakes along the ice border, or in the sea marginal to the glaciated regions.

Nonstratified Deposits. All nonstratified glacial deposits are generally referred to as *till.* Textually, till ranges from a hard dense clay with intermixed sand, stones, and boulders, to a collection of boulders with very little intermixed fines (Fig. 13.16). Till occurs in two topographic forms, *ground moraine* and *end moraine.*

Ground moraine is characteristically a

Figure 13.17 Aerial photograph showing drumlins in Charlevoix County, Michigan. Ice movement was toward the south, parallel to the drumlin axes. (U. S. Department of Agriculture photograph.)

gently rolling to nearly flat terrain. Stream-lined ridges called *drumlins* characterize many ground moraine surfaces (Fig. 13.17). They are ice-molded hills developed beneath the glacier as the till is plastered onto the ground. The long axis of a drumlin is parallel to the direction of ice movement, and swarms of them provide excellent evidence of ice movement.

End moraines are belts of hills that mark the former position of the ice front. End moraines deposited by the continental glaciers during the Pleistocene Epoch are traceable for many miles in the Great Lakes region and provide the basis for reconstructing the position of the terminus of the ice sheet at various times during its retreatal history (Fig. 13.18).

Stratified Deposits. Sand and gravel de-

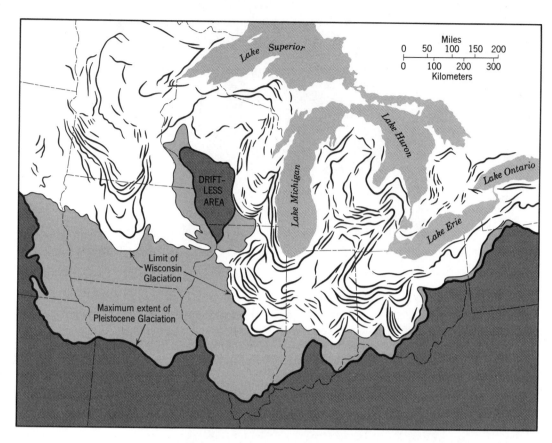

Figure 13.18 Map of the Great Lakes Region showing the general distribution of end moraines formed during the Wisconsin or last stage of the Pleistocene. The end moraines marking the successive retreatal positions of the ice front are shown in heavy black lines. The area between the Wisconsin drift boundary and the line marking the maximum extent of Pleistocene glaciation is covered by glacial deposits of pre-Wisconsin age. Many of them are mantled with loess or modified by stream erosion so that their original topographic forms are obscured. (Based on Flint, R. F., 1959, *The glacial map of the United States east of the Rocky Mountains,* The Geological Society of America, Boulder, Colorado.)

posited by glacial meltwater streams is commonly referred to as *outwash.* Many topographic features are made of outwash. Streams emerging from the ice front carry a great deal of bed load and suspended load. The bed load of several coalescing streams forms a broad *outwash plain,* and outwash deposited down valley from the snout of a valley glacier is a *valley train.* Residual blocks of stagnant glacier ice buried in outwash become ice-block pits when they melt. Some ice-block pits contain lakes (Fig. 13.19).

Much sand and gravel outwash collects close to the ice front and adjacent blocks of dead ice, which is ice thinned by melting so that movement ceases. Such outwash features are known as *ice-contact* deposits. Although stratification is visible where roads have been cut through those features or where gravel pits have been opened in them, the bedding shows signs of slumping and

THE DYNAMIC EARTH

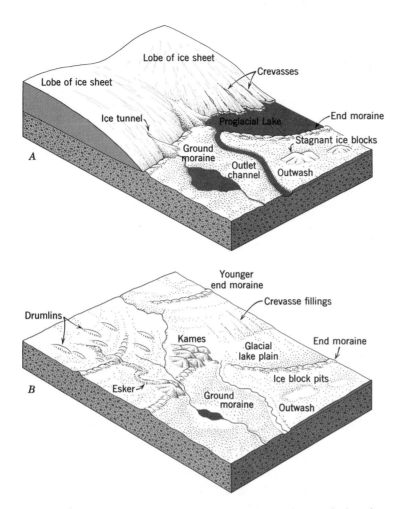

Figure 13.19 Block diagrams showing the origin of several glacial landforms produced by continental glaciation. *A:* a proglacial lake is formed between the ice front and an end moraine. Stagnant ice blocks are partially buried in outwash, and a large lake occupies an irregular depression in ground moraine. Kames are formed where sediment-laden meltwater emerges from an ice tunnel or flows off the glacier surface. *B:* the same area shown in *A* after the ice has retreated. An esker marks the position of the ice tunnel, and drumlins reflect the flow direction of the lobe that produced them. Kames are slumped hills of sand and gravel and a younger end moraine marks a pause in the retreat of the ice. The stagnant ice blocks have melted to leave ice block pits, and the proglacial lake has drained causing exposure of its former bottom which is now a glacial lake plain.

distortion produced when the supporting walls of ice melted away. *Kames* are irregularly shaped knolls or hummocks of outwash commonly associated with end moraines. *Kame terraces* are flat, linear, and pitted outwash patches that form between dead ice in a valley and the adjoining valley walls. *Eskers* are sinuous gravelly ridges deposited by subglacial streams in ice tunnels. *Crevasse fillings* are similar to eskers but form in an ice valley rather than in an ice tunnel.

Figure 13.20 A commercial gravel pit in glacial outwash near Ann Arbor, Michigan. (Photograph by William Kneller.)

Most of the glacial outwash deposits are suitable sources of commercial sand and gravel (Fig. 13.20). If large quantities of shale, soft sandstone pebbles, or even small amounts of chert particles are present, however, the outwash makes a poor concrete aggregate. Shale and soft sandstone pebbles are not durable, and the chert reacts chemically with the cement, resulting in a poor quality concrete. Some large gravel operators find it profitable to remove the deleterious stones by various methods, thereby producing a high quality aggregate that can be marketed at a premium.

Lake Deposits. Glacial meltwater impounded around the retreating ice front provides collecting basins for sediment. Lakes fringing the ice border are called *proglacial* lakes, and their deposits are widespread around the Great Lakes in Canada and the United States. Predecessors of the modern Great Lakes covered much of the land fringing the present water bodies, as is indicated by the flat topography of the old lake bottoms in the Great Lakes states.

Lake sediments are known as *lacustrine* deposits, and they consist of deep water clays and silts, and shallow water sands and beach gravels. Much of the deep water sediments consist of *rock flour,* a finely ground material produced by the abrasion of debris-laden ice moving over bedrock surfaces. Glacial meltwater streams carry this sediment in suspension to the proglacial lakes where it eventually settles to the bottom in the deeper, quiet waters. Some of these deposits are finely laminated or banded, indicating a cyclic type of deposition. Each cycle is represented by a relatively coarse layer of very fine sand a fraction of an inch thick, on top of which is a thinner layer made up of fine silt and clay. A pair of those bands makes a *varve,* which represents the sediments of one year. The coarse layer was deposited during the spring and summer months at the height of the melting season, and the fine layer was laid down during the fall and winter months after the lake had frozen. Varve series have been used by glacial geologists to construct chronologies of glacier recession during the

Pleistocene in Sweden, Canada, and the United States.

A Brief History of Pleistocene Glaciation in North America

The great ice sheets or continental glaciers that covered much of North America, northern Europe, and parts of Russia during the Pleistocene Epoch reached their maximum extent sometime within the last 2 or 3 million years of earth history. The former presence of these immense ice sheets is well documented by the deposits that they laid down and the characteristic landscapes that they produced.

The North American ice sheet reached a maximum southern limit roughly equivalent to the Missouri and Ohio rivers (Fig. 13.18). One main center of original snow accumulation was in the highland regions of Labrador and eastern Quebec. Other local centers may have developed later on, as a result of the shifting of storm tracks and other meteorological factors. A second major center of glaciation persisted in the Canadian Rockies.

The ice advanced in lobate fashion, each lobe being guided by major topographic lowland areas, especially large river valleys. Eventually, however, the entire land area north of the Ohio and Missouri rivers was engulfed in glacier ice. A tract in southwestern Wisconsin known as the "Driftless Area," may have escaped glaciation altogether, although some geologists believe that it was invaded by early ice advances of the Pleistocene but escaped glaciation during the last or Wisconsin stage of the Pleistocene.

The classic area of Pleistocene glacial deposits of North America is the Upper Mississippi Valley. This area has received the attention of glacial geologists for more than 75 years, and is the type area for most of the Pleistocene drift sheets.

A major stride toward a better understanding of the Pleistocene was the discovery of multiple glaciation. Until the late nineteenth century, geologists assumed that one major advance of the ice was followed by one major retreat. But recognition of two drift sheets separated by *interglacial deposits* led to the concept that more than one advance of the ice had taken place, and that each major advance or *glacial age* was separated in time by an *interglacial age*. In fact, fossils from one of the interglacial deposits near Toronto indicate that the climate there was warmer than today before the last major advance of the ice in that region.

The study of drift deposits in the Great Lakes region has led to the identification of at least four major glacial and three interglacial ages during the 2 to 3 million years of Pleistocene time. It is by no means certain, however, that only four major glacial advances occurred during the Pleistocene, but as yet, the evidence from the glacial deposits in North America supports only four major glacials and three interglacials. The names assigned to each are given in Table 13.a.

Table 13.a Names of Glacial and Interglacial Ages of the Pleistocene Epoch in North America

Glacial Ages	Interglacial Ages
Wisconsin or Wisconsinin	
	Sangamon
Illinoian	
	Yarmouth
Kansan	
	Aftonian
Nebraskan	

The last glacial age, the Wisconsin, began about 70,000 years ago and ended about 10,000 to 11,000 years ago. The basis for these figures comes from radiocarbon dating (see page 314) of organic remains associated with relatively minor advances and retreats of the Wisconsin age glaciers. Peat deposits and logs from trees that grew along the border of Wisconsin ice sheet were buried beneath till laid down by the advancing glacier. Radiocarbon dates on such deposits as these provide the means for assigning absolute dates to the Wisconsin glacial age.

The area covered by the Wisconsin ice sheet is distinguishable from pre-Wisconsin landscapes by the uneroded character of the glacially constructed landforms such as end moraines, kames, eskers, and the like. Moreover, the landscape produced by the Wisconsin ice sheet does not have an integrated drainage system because there has been insufficient time for one to develop, and the soils developed on the Wisconsin drift are generally younger, in terms of development of the solum, than soils on pre-Wisconsin drift material. The thousands of lakes, countless swamps, and inefficient surface drainageways that abound in Minnesota, Wisconsin, and Michigan, are all characteristic of a very youthful glacial topography.

In contrast, the older glacial drift surfaces do not possess the youthful topographic features of the Wisconsin drifts. Drainage patterns are well developed, many of the end moraines have been eroded beyond recognition, and the upper parts of the pre-Wisconsin drifts have been deeply weathered. Where buried beneath younger glacial deposits, weathered zones are a valuable tool for the reconstructing of Pleistocene history. One of the most persistent and readily identifiable buried weathered zones is the one that occurs at the upper boundary of the Illinoian drift. This weathered zone was produced during the Sangamon interglacial age, and is generally referred to as the *Sangamon soil*. The buried Sangamon soil and associated nonglacial deposits have a more strongly developed solum than the soils that occur on Wisconsin age drifts of the modern land surface. From this comparison of Sangamon and post-Wisconsin soil profile development it can be inferred that the Sangamon interglacial age lasted much longer than the 11,000 years that have lapsed since the disappearance of the last Wisconsin ice sheet.

The Origin and History of the Great Lakes

The five Great Lakes of Superior, Michigan, Huron, Erie, and Ontario and their con-

necting channels constitute the largest inland waterway system in the world. With an area of 95,000 square miles and a shoreline of 11,000 miles, these lakes are important in the economic well-being of the states and provinces that border them. The people inhabiting the Great Lakes region look to them not only as a primary source of water for cities and industries but also as disposal areas for municipal and industrial wastes, transportation routes for commercial shipping, a source of energy for the generation of hydroelectric power, an arena for commercial and sport fishing, and a retreat for the small-boat sailor and the summer shore-dweller.

Because millions of people have a direct interest in the lakes, or are directly or indirectly affected by the chemical and biological changes that the lakes have experienced since the industrial revolution, it is appropriate to consider the origin and history of these lakes in some detail at this point.

The Ancestral Great Lakes. The presence of vast areas of old lake bottoms and ancient shorelines in the vicinity of the present Great Lakes is the chief evidence for the former existence of larger and deeper ancestral water bodies in those regions (Fig. 13.21). These ancestral Great Lakes had their inception during the retreat of the lobate front of the Wisconsin ice sheet (Fig. 13.22A). Ice lobes protruding from the main glacier mass filled the lowlands now occupied by the Great Lakes. These lowlands, except possibly for the Superior lowland, were large river valleys formed by a pre-Pleistocene drainage system. The exact configuration of these preglacial rivers is only vaguely known, but there is little doubt that they had eroded valleys sufficiently large to influence the general pattern of advance and retreat of the Wisconsin ice.

At the beginning of the retreat of the Wisconsin glacier about 15,000 years ago, the meltwater issuing from the ice front flowed away from the glacier margin, but as the lobes shrank, lower land was uncovered, and the meltwaters became impounded in front of the ice margin. These water bodies grew in extent as the various lobes diminished in size. To differentiate these early *proglacial lakes*

LAKE
HURON

0 ½
Mile

Figure 13.21 Aerial photograph of beach ridges in Huron County, Michigan. These ridges were produced by waves of older and higher lake stages in the Lake Huron basin during the late Pleistocene. (U. S. Department of Agriculture photograph.)

from the modern water bodies, geologists have assigned names to them. Glacial Lakes Chicago and Maumee were associated with the early retreat of the ice, and came into existence about 14,000 years ago (Fig. 13.22A).

Each ancestral lake possessed an outlet channel, just as many modern lakes do, but different outlets were used at different times because, as deglaciation progressed, new and lower outlets came into existence. Lower outlets caused a drop in lake level which, in some extreme cases, amounted to several

hundred feet but was usually less than 25 feet. An outlet of sufficient duration caused the lake level to persist at a single elevation long enough to build a well-defined shoreline or beach ridge. Closely spaced beach ridges like those visible in Figure 13.21 suggest a steplike lowering of the outlet.

In the early stages of deglaciation the outlets of the proglacial lakes converged toward the Mississippi River which flowed into the Gulf of Mexico. As deglaciation continued, however, the drainage shifted to an eastern course, first through the Mohawk-Hudson

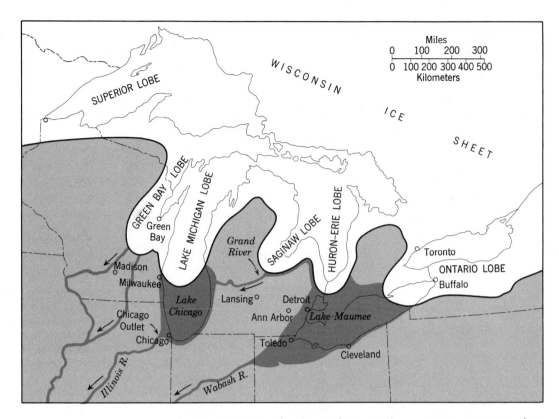

Figure 13.22A. The ancestral Great Lakes came into existence when
the lobate ice margin began its final retreat in late Wisconsin time.
Lake Chicago and Lake Maumee were the first of the pro-glacial
lakes to form. They drained to the Gulf of Mexico via outlets shown
(about 14,000 years ago). B. Further retreat of the ice front opened
an eastern outlet to the Atlantic Ocean via the Hudson River. Lake
Chicago and the newly formed Lake Duluth still drained to the Gulf
of Mexico (about 10,500 years ago). C. Niagara Falls came into existence when the Ontario basin became ice free about 8,000 years ago.
At the same time, Lake Algonquin existed in the Huron-Michigan
basins and for a time had two functional outlets. One was the Chicago
outlet and the other was the St. Claire-Detroit River channel near
Detroit. (After Jack L. Hough, 1963, *The prehistoric Great Lakes
of North America,* American Scientist, 51, 84-109.)

valleys (Fig. 13.22B) and later through the
St. Lawrence lowland via Niagara Falls (Fig.
13.22C).

The Niagara River flows from Lake Erie
into Lake Ontario over the edge of a resistant dolomite layer, forming the Niagaran
escarpment (Fig. 17.7). This drainage route
came into existence when the shrinking Wisconsin ice uncovered the escarpment and
the Ontario basin less than 10,000 years ago.
The waterfall thus created has since mi-

grated upstream creating a gorge of scenic
beauty. The average rate of retreat has been
about 4 feet per year, and geologists once
thought that they could calculate "post-
glacial" time simply by dividing the total
gorge length by the retreatal rate. Evidence
brought to light later, however, indicated
that not only was the gorge history more complex than formerly realized but also that the
discharge of the Niagara River had varied
greatly since its inception. In view of these

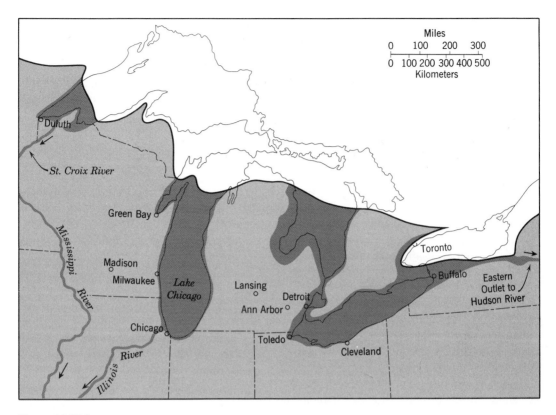

Figure 13.22*B*

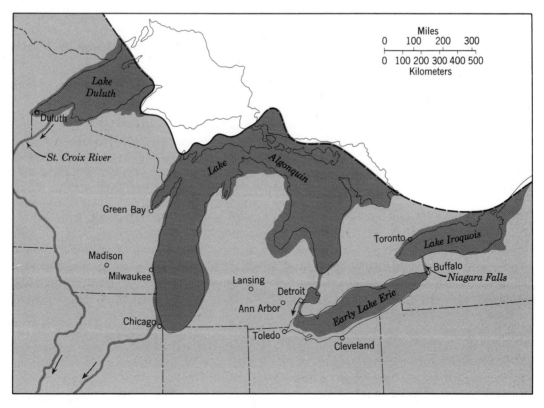

Figure 13.22*C*

complicating factors, attempts to use the retreat of Niagara Falls as a geologic clock have been abandoned.

The Modern Great Lakes. The modern Great Lakes reached their present levels about 2000 years ago. Since that time their outlets have remained unchanged and their water levels have fluctuated only slightly compared to the changes in level that occurred during the development of their ancestral stages (Fig. 14.12).

The water quality in all the Great Lakes except Lake Superior has not remained constant, however. The use of the lakes for municipal and industrial waste disposal, plus the runoff from agricultural lands subject to intense chemical fertilizations has produced substantial increases in phosphates, nitrates, sulfates, and other chemicals in the lake waters. Phosphates and nitrates provide nourishment for intense algal growth such as occurs in Lake Erie during the summer months. When these overabundant algal masses die, the process by which they are decomposed depletes the dissolved oxygen in the lake waters to a level that is lethal for many species of fish that were previously very abundant.

Other sources of pollution in the Great Lakes are hot waters discharged by thermonuclear electrical generating plants that use lake water for cooling purposes. Solid particles from a number of sources are also carried into the lakes where they increase the turbidity of the water.

The quality of the waters of the Great Lakes will continue to decline until much of their value is depreciated, unless firm measures are taken to reduce all sources of pollutants.

The Cause of "Ice Ages"

An "Ice Age" is a period of geologic time during which existing glaciers were greatly expanded over what they are today, and ice sheets of continental proportions came into being where they do not exist today or where they did not exist previously. The underlying cause of glacier growth to ice-age dimensions, such as those which pre-

vailed during the Pleistocene, are generally attributed to climatic cooling on a scale of worldwide proportions. But to say that glaciation is caused by climatic deterioration is to beg the question, because what we really want to know is what causes climatic changes that can initiate an ice age?

Before considering some of the speculative answers to this question, let us examine the phenomenon of a glacial age a little more closely so that we can appreciate more fully exactly what requirements must be satisfied in any answer that might be proposed.

1. Ice Ages occurred at least three times in the geologic record, late Precambrian, Permo-Carboniferous, and Pleistocene.
2. Pleistocene continental glaciation was a worldwide event marked by at least four glacial episodes separated by interglacial conditions when the climate may have been warmer than it is today. The difference in mean annual air temperature between glacial and interglacial ages was on the order of 10° F. in the tropics and 15° F. in higher latitudes.
3. The waxing and waning of Pleistocene ice sheets was probably synchronous in the Northern and Southern Hemispheres.
4. The distribution of land and sea was not markedly different in the Pleistocene than today.
5. The pole positions during the Pleistocene were nearly the same as today.

All hypotheses proposed to explain climatic changes that will account for these facts fall into three categories: terrestrial, extraterrestrial, or a combination of terrestrial and extraterrestrial causes.

Among the hypotheses that invoke terrestrial causes are changes in the elevation of continental areas, changes in the relative positions of continents with respect to the ocean basins, and changes in atmospheric composition. The increased elevation of a continental mass could cause a nonglaciated area to become glaciated because increased elevation would produce more snowfall as a result of colder temperatures at higher elevations, but whether such a mechanism can account for worldwide glacial growth is doubtful. Moreover, this hypothesis by itself can hardly explain the interglacials that occurred between glacial times. Changes in the relative positions of continents could bring about ice age conditions in pre-Pleistocene times. The Permo-Carboniferous glaciation of the Southern Hemisphere (Fig. 7.8), is well accounted for by this means, but continental drift is not a valid hypothesis to explain the advance and retreat of Pleistocene glaciers. By Pleistocene time the continents were in their present positions. Atmospheric changes that conceivably could cause worldwide changes in climate involve changes in the amount of carbon dioxide and volcanic dust. Carbon dioxide produces a "greenhouse" effect on the earth; the more there is the warmer the temperature because less heat escapes from the earth. Although this premise is not questioned, no one has been able to demonstrate that the magnitude of the change in carbon dioxide content of the atmosphere would be sufficient to produce the climatic changes required to bring about ice-age conditions. The volcanic dust content of the atmosphere has also been ascribed by some theorists to explain glaciation on a worldwide scale. An increase in dust would reduce the amount of solar energy reaching the earth and thereby would result in a general cooling. Even though it is known that volcanic dust can be widely disseminated in the atmospheric envelope, it is difficult to envision volcanism of the order of magnitude necessary to cause the mean annual air temperature to drop by 10° F. or more, to say nothing of explaining alternating glacial and interglacial conditions.

Among the extraterrestrial hypotheses that have been proposed are those that take account of the earth's cyclic change with respect to the sun, and the possibility that the radiant energy of the sun itself changes with geologic time. The earth does, in fact, periodically change in position with respect to the sun, and these changes could cause variations in the amount of energy that reaches the earth. The major defect of this mechanism to explain glacials and interglacials, however, is that glacial ages would not have been synchronous in the Northern and Southern Hemispheres, as they apparently were.

The outpout of solar radiation varies by a few percent from time to time, but to what extent these variations could account for major glaciations at various times in the history of the earth is uncertain. The same is true for the hypothesis that cooling of the earth in past geologic time was caused by passage through cosmic dust clouds.

No hypothesis by itself provides a convincing explanation for the repeated birth, growth, and decline of the huge glaciers that covered nearly a third of the earth's land surface during the Pleistocene. Most likely, more than one cause is involved, and it is conceivable that a combination of both terrestrial and extraterrestrial events is needed to explain the phenomena that characterizes an ice age, not only of the Pleistocene but also of the more ancient ones that occurred in late Precambrian and late Paleozoic times.

References

Bentley, C. R., and others, 1964, Physical characteristics of the Antarctic ice sheet, *Antarctic Map Folio Series, Folio 2*, American Geographical Society, New York, 10 pp., 10 plates.

Crowell, John C., and L. A. Frakes, 1970, Phanerozoic glaciation and the causes of ice ages, *Amer. Jour. Science*, V. 268, pp. 193–224.

* Dyson, J. L., 1962, *The world of ice*, Alfred A. Knopf, New York, 292 pp.

* Embleton, C., and C. A. M. King, 1968, *Glacial and periglacial geomorphology*, St. Martins Press, New York, 608 pp.

Flint, R. F., 1971, *Glacial and Quaternary geology*, John Wiley and Sons, New York, 892 pp.

* Fristrup, Borge, 1966, *The Greenland ice cap*, University of Washington Press, Seattle, 312 pp.

Gow, A. J., 1965, The ice sheet, *in Antarc-*

tica (T. Hatherton, ed.) Chapter 9, Methuen and Co. Ltd., London, 502 pp.

Kelley, R. W., and W. R. Farrand, 1967, *The glacial lakes around Michigan,* Michigan Geological Survey, Bulletin 4, Lansing, Michigan, 23 pp.

Meier, M. F., and Austin Post, 1969, What are glacier surges? *Canadian Journal of Earth Sciences,* V. 6, pp. 807–817.

Meier, M. F., 1964, Ice and Glaciers *in Handbook of Applied Hydrology* (Ven Te Chow, ed.) Sec. 16, McGraw-Hill Book Co., New York.

Mellor, Malcolm, 1964, *Snow and ice on the earth's surface,* Cold Regions Science and Engineering (II-C1), U. S. Army, Cold Regions Research and Engineering Laboratory, Hanover, N.H., 163 pp.

* Robin, G. de Q., 1962, The ice of the Antarctic, *Scientific American,* V. 207 No. 3 (September) pp. 132–146. (Offprint 861, W. H. Freeman and Co., San Francisco.)

* Sharp, R. P., 1960, *Glaciers,* Condon Lectures, Oregon State System of Higher Education, Eugene, Oregon, 78 pp.

Wright, H. E., and D. G. Frey, 1965, *The Quaternary of the United States,* Princeton University Press, Princeton, N. J., 922 pp.

*Recommended for further reading.

14 Oceans and Shorelines

And I have loved thee, Ocean! and my joy
Of youthful sports was on thy breast to be
Borne, like thy bubbles, onward; . . .
Byron

Continents and ocean basins constitute two major physiographic elements of the planet Earth. The subject matter of the preceding five chapters has dealt with geologic processes operating on land areas; this chapter will deal with large lakes and the ocean basins.

The oceans occupy nearly 72 percent of the earth's surface and contain over 97 percent of all of the water in the hydrologic cycle (Table 10.a). The science of oceanography has become a professional field of great importance to the future of mankind. Some oceanographers study the biological world of the ocean, and others concentrate their efforts on the physical aspects of seawater. Still others are interested in the ocean floors and the sediments that accumulate there, and a fourth group deals with the chemistry of seawater.

This chapter is divided into two sections, the ocean shorelines and the ocean basins. The section on shorelines deals with the

Hawaiian surfers. (Photograph by Emil Kalil.)

interface between land and sea where waves not only cause shore erosion but also are responsible for the accumulation of sand in the form of beaches, off-shore bars, and other depositional features. In the section on ocean basins the reader is introduced to the major topographic elements of the sea floor and the nature of the sediments that accumulate there.

OCEAN SHORELINES

Where land and sea meet along the thousands of miles of coastline, the attack of wind-driven waves on the shore provides a variety of geologic phenomena within the reach of man's direct observation. Wave action can produce geologic changes in a relatively short time so that the cause and effect relationships are apparent after a period of a few tens of years. Some changes are so rapid that they are almost catastrophic, but others are gradual and develop over much longer periods of time.

This section deals in particular with the way in which shorelines along the oceans and large lakes are modified by the action of water waves. Most water waves are wind generated, but some of the most disastrous are *tsunamis* that result from a sudden displacement of the sea floor. Both types are considered in the paragraphs that follow.

Mechanics of Wave Action

Origin and Description of Waves. Waves result from the friction of wind passing over a water surface. Wave size depends on (1) wind velocity, (2) the length of time the wind continues to blow from a single direction, and (3) the length of the open water over which the wind blows. The highest point on the wave is its *crest* and the adjacent low point is the *trough*. The distance from crest to crest or trough to trough is the wavelength *L,* and the difference in elevation between the crest and the trough is the wave height *H* (Fig. 14.1). A standard method of describing a single wave is by means of the ratio of length to height, *L/H*. The heights of storm waves in the open sea are often exaggerated in reports from observers at sea, but the average height is about 30 to 40 feet, and individual waves of 75 feet have been recorded.

Wave Motion. The motion of waves in the open sea is somewhat analogous to the surface of a field of grain being swept by the wind. Only the wave *shape* is transmitted during the process of wave motion. The paths taken by various surface water particles are shown in Figure 14.2. The time necessary for a wave crest to travel one wavelength is the wave period, *T,* and the velocity, *V,* of the wave is equal to the wavelength divided by the wave period, as expressed in a simple equation,

$$V = \frac{L}{T}$$

Wave motion dies out with depth because the orbital paths of the water particles become smaller and smaller. The depth of water below which no wave motion occurs is equal to one-half the wave length, ½*L,* and is known as the *wave base* (Fig. 14.2); below this level winds can cause no water motion. A submarine can easily ride out a

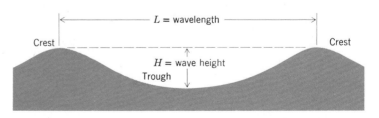

Figure 14.1 Cross section of a deep-water wave.

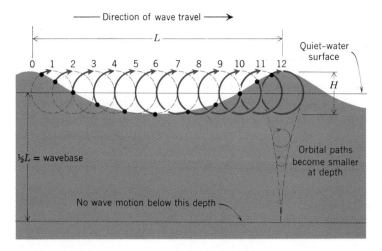

Figure 14.2 Cross-sectional diagram of a water wave showing motion of water particles (black dots) at 12 different points on the wave surface. Solid arrows show the direction and distance traveled by each point on the water surface as the wave crest reaches that particular position. When the water particle at position 12 has made one complete circuit in its orbital path, the wave crest will have traveled one wavelength. The orbital paths decrease in diameter down to a water depth equal to ½ L. (After Kuenen, P. H., 1950, *Marine Geology*, John Wiley and Sons, New York, 568 pp.)

storm when submerged, while a surface vessel of the same size is wildly tossed about.

Waves produced by high winds in the open ocean may travel hundreds or even thousands of miles away from the storm center. During this period of travel a single wave decays, that is, its height is reduced and its length is increased. For example, by the time a 15-foot wave with a length of 300 feet travels 2000 miles across the open sea, its height decreases to 2½ feet and its length increases to 1300 feet. It will be noted that the *L/H* of the original wave is 20 whereas the *L/H* of the same wave after 2000 miles is 520. Waves with *L/H* ratios between 10 and 35 are storm waves, those with an *L/H* ratio between 35 and 70 are intermediate waves, and those whose *L/H* ratio is greater than 70 are the well-developed *swells* that reach seacoasts many miles from the place where they were generated.

Waves in Shallow Water. In the near-shore areas where the water is shallow the incoming waves undergo a marked change in shape because of the effect of the sea bottom. The wave height increases and the length becomes shorter, but the period remains more or less constant. Eventually the wave becomes so high that the crest topples over and the wave is said to *break* (Figs. 14.3 and 14.4). The energy of waves breaking on shore produces geologic changes on almost all shorelines.

Distribution of Wave Energy. A wave approaching an irregular shore does not reach wave base everywhere along its crest at the same time. The part of the wave that "feels bottom" first is slowed down, thereby causing a flexure in the crest line (Fig. 14.5). If a single deep-water wave is divided into equal units along the crest line, each unit contains the same amount of energy. As the wave is bent or *refracted* when wave base is reached along part of its crest, the lines of equal

OCEANS AND SHORELINES   271

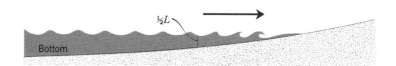

Figure 14.3 Cross section of waves breaking on shore. Wave crest becomes distorted on reaching a water depth of ½ L. Wave height increases and wavelength decreases. The period remains about the same.

Figure 14.4 Breaking waves on the coast of Maine. (Photograph by M. R. LaMotte, Maine Department of Economic Development.)

energy are concentrated on the headlands, whereas the waves reaching the shore of the bay are less potent and act more as a transporting agent than a force of erosion.

The material eroded from the headlands is carried toward the bay by a process of *beach drifting*. As the waves strike the shore obliquely, sand and larger grains are thrown diagonally up the beach, but when the water flows back down the beach slope, it carries some sand and pebbles with it. The wash and backwash action of successive waves produces a net movement of beach materials parallel to the shore through a series of curved paths, as shown in Figure 14.6.

Geologic Work of Waves

Wave Erosion. Coastal promontories and headlands are particularly susceptible to wave attack. Waves striking a shore are capable of doing great damage to man-made structures, and are able to produce permanent changes in the configuration of the shore itself.

One of the chief results of wave erosion on a headland is the landward retreat of the shore. The wave attack is concentrated in the zone between low tide and the highest point reached by storm waves. Material is removed by abrasion, solution, or simply

272   THE DYNAMIC EARTH

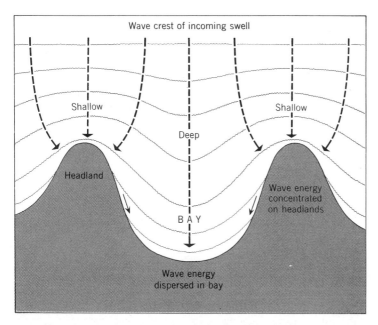

Figure 14.5 Diagrammatic map view of waves striking an irregular shore. Wave crests are bent (refracted) when one part of the wave strikes shallow water before the rest of the wave. Dashed arrows are lines of equal energy. They are equally spaced along the crest of the incoming swell in deep water but converge toward the headlands as the water shoals off the headlands. Wave energy is thus concentrated on the headlands and dispersed in the bays. Short solid arrows show direction of beach drifting. See Figure 14.6.

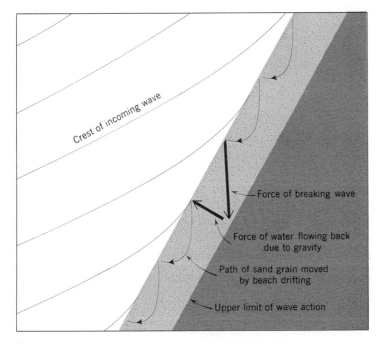

Figure 14.6 Map showing the curved paths followed by sand particles moved by beach drifting.

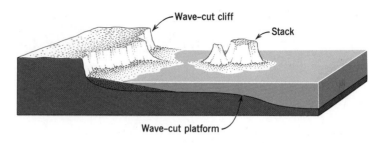

Figure 14.7 Block diagram of a wave-cut cliff and stack. The base of the cliff is eroded by storm waves but is not reached by smaller waves in calm weather.

Figure 14.8 *A:* wave-cut cliff near St. Joseph on the eastern shore of Lake Michigan, November 1952. Landsliding was initiated by waves attacking the base of the cliff during high lake levels of the early 1950's (Fig. 14.12). Unconsolidated glacial deposits are exposed in the cliff face. *B:* same area as in *A,* October 1954. Continued landsliding caused retreat of the cliff and endangered houses, many of which were evacuated and wrecked or moved. (Hann Photo Service, Hartford, Michigan.)

by the hydraulic action of the waves as they smash into poorly cemented sediments or loosely jointed rock. A land surface sloping gently seaward is undermined by wave action near its base and is eventually transformed into a *wave-cut cliff* (Fig. 14.7). The rock is planed off just below water level and becomes a *wave-cut* platform. Residual rock columns, isolated by wave erosion from the retreating sea cliff, are called *stacks* because of their vague resemblance to smoke-stacks. *Sea caves* are undermined notches at the base of a sea cliff.

Secondary results of wave erosion are landslides. A landslide occurs when the base of a wave-cut cliff is removed by wave

Figure 14.9 Sea walls near South Haven on the eastern shore of Lake Michigan, December 1952. Note the ineffectiveness of the offshore "fence" as a protective measure against wave erosion. (Hann Photo Service, Hartford, Michigan.)

attack so that the cliff becomes unstable and slides into the water. Buildings and other installations perched near the edge of a wave-cut cliff face possible destruction from landslides (Fig. 14.8).

Protection against extensive shore erosion is expensive. *Sea walls* built parallel to the shore are especially difficult to construct properly (Fig. 14.9), but less expensive and equally effective are *groins,* impervious "fences" of heavy timber or concrete installed at right angles to the shore. The groins trap the sediment transported by beach drifting and thereby establish a protective beach at the base of the cliff (Fig. 14.10). Groins are also an effective measure against extensive loss of beach sand where no wave-cut cliffs exist.

Wave Deposition. The transportation of sand along the shore by beach drifting accounts for many of the beaches, bars, and spits common to coastal areas. Although a complex terminology for these features has been devised by coastal engineers and geologists, a simplified system of nomenclature is used in this book. A *bar* is a submerged ridge of sand or gravel lying offshore more-or-less parallel to the mainland. A *beach* is a zone of unconsolidated shore material lying between the waterline and the upper limit of normal storm waves. *Beach ridges* above the upper limit of fair-weather swells are products of storm waves. Several parallel storm beaches are produced by waves of different storms varying in intensity. A *spit* is a low, sandy ridge project-

Figure 14.1 Groins constructed to protect the highway along the top of the wave-cut cliff near St. Joseph, Michigan, December 1954. (Hann Photo Service, Hartford, Michigan.)

ing into a body of water from the shore, and it may have a curved tip that is shaped by currents (Fig. 14.11).

All variations of these three basic features exist, as for example, the *offshore* or *barrier bars* that occur extensively along the Atlantic coast of the United States from Atlantic City to Miami. The crests of these ridges are *above* water, a condition not easily accounted for by normal wave action. Possible explanations are that they originated by the extension of spits through beach drifting, or that they are beach ridges isolated from the mainland by a rising sea level.

Shoreline Evolution. Shorelines are dynamic features that are undergoing profound changes of both the short- and long-term varieties. Shorelines, therefore, must

be evaluated in terms of their development with the passage of time. This point needs repeated emphasis because it is the very heart of geologic reasoning and distinguishes the science of physical *geology* from physical *geography*. The latter is more often concerned with the description of earth features as they now appear, whereas the geologist places the emphasis on the *origin* of landforms, not merely their size, shape, and distribution. The origin of any landscape cannot be accomplished without consideration of the factor of time.

Applying this fundamental concept to the geologic interpretation of a particular shoreline, be it on the seacoast or the Great Lakes, it seems self-evident that a shoreline evolves to its present condition through a series of

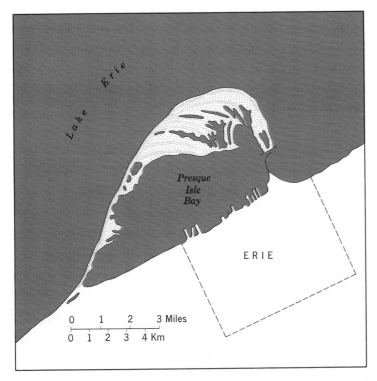

Figure 14.11 A compound spit is responsible for Presque Isle Bay in Lake Erie near Erie, Pennsylvania. It was formed by currents from the southwest. Stippling shows progressive stages in the growth of the spit. (From U. S. Geological Survey, Erie Quadrangle, edition of 1900.)

gradual changes. What we observe today may be only a stage in its developmental history, not necessarily the end result.

One of the main factors that influences shoreline evolution is the relative changes in sea level or lake level with time. On the Great Lakes the changes in elevation of the water surfaces amount to several feet during a decade or so, but significant changes in the level of the ocean require centuries or even thousands of years.

Lake Michigan, for example, had a water surface elevation of nearly 581 feet above sea level in 1952, 575.4 feet in 1964, and 579.8 feet in 1969 (Fig. 14.12). These changes are simply due to greater or lesser amounts of precipitation on the drainage basin. During the high water stages of 1952 and 1969, the waves attacked the shore and caused severe damage, loss of property, and extensive destruction of sand beaches. During the low water stages (1964, for example) the beaches were generally restored by wave action.

Ocean shorelines generally change more slowly than lake shores because of the longer time required for changes in sea level. The *relative* position of sea level with respect to the land surface may be caused by a decrease or increase in volume of the water in the oceans, by waxing and waning of ice sheets, for example, or by uplift or subsidence of the land surface. Sea level "rise" or "fall" is a relative rather than absolute movement of either the ocean surface or landmass.

Coastal areas that are experiencing a relative rise of sea level are said to be *submer-*

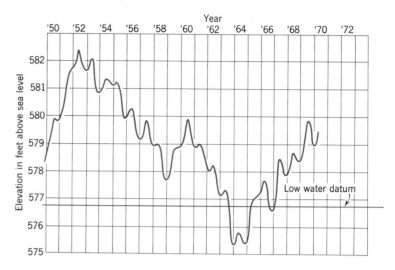

Figure 14.12 Graph of water levels of Lake Michigan plotted against time from 1950 to 1970. Water levels are higher during years of heavier precipitation. (Based on reports by the Lake Survey, U. S. Corps of Engineers, Detroit, Michigan.)

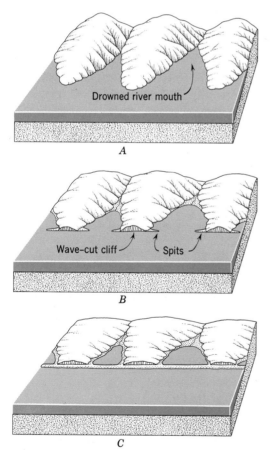

gent, and those where sea level has fallen relative to the coastal area are *emergent.* Where a coastline is clearly emergent or submergent, certain morphological features of the shoreline are produced. Submerged shorelines, for example, are characterized by drowned river mouths and extensive bays and inlets produced by the encroachment of the sea on low-lying coastal areas. A recently submerged shoreline is therefore highly irregular in map view during the early stages of submergence (Fig. 14.13*A*). As time passes, however, the initial irregularities are modified by wave erosion on the headlands and the building of spits across the bays, a combination of processes that eventually produces a straighter shoreline (Fig. 14.13*B* and C).

Emergent shorelines, in contrast, are generally straightened by the relative withdrawal of the sea because the previously

Figure 14.13 Block diagram showing the evolution of a shoreline of submergence. The shoreline is irregular to begin with but becomes straighter as wave action cuts back the headlands and builds spits across the mouths of drowned rivers.

submerged sea floor is, under ideal conditions, a gently sloping surface with only minor surface irregularities. Evidence of emergence is also found in old beaches and strand lines now lying beyond the reach of waves on lake shores or seacoasts (Fig. 13.21).

A classification of shorelines based solely on recent submergence or emergence is far too simple, however. Some shores bear evidence of both emergence and submergence and others exhibit characteristics related to some other geologic agent such as glaciation, volcanism, or coral reef growth. It is thus best to evaluate the origin of a particular shoreline on the basis of all factors involved. These factors include not only the evidence of submergence or emergence and the stage of development (that is, the passage of time), but also the influence of geological materials and processes that have influenced the origin and history of the shoreline.

Tsunamis

The most destructive of all water waves are not produced by wind but by large submarine landslides, subocean earthquakes, or volcanic eruptions beneath the sea. Any of these events results in a rapid displacement or dislocation of the ocean bottom which, in turn, generates a series of water waves, or *wave trains,* capable of traveling for thousands of miles across the open ocean until they strike distant shorelines with tremendous destructive force.

In popular descriptions these waves are called tidal waves, but they have nothing to do with the tide. The technical term is *tsunami* (tsoo-nah-mee), a Japanese word now employed more-or-less universally for these large types of waves. A more precise and better descriptive term is *seismic sea wave.* Tsunamis are one of the great hazards of man's physical environment; hence, it is appropriate to dwell briefly on some historical examples of this frightening phenomenon.

In 1883 the eruption of Krakatoa, a volcano in the East Indies, produced a train of about a dozen seismic sea waves that traveled at a velocity of three to four hundred miles per hour. They reached such distant points as South Africa (4690 miles away), Cape Horn on the southern tip of South America (7280 miles distant), and Panama (11,470 miles).

The Alaskan earthquake of March 27, 1964 produced a tsunami that smashed into several ports on the Alaskan coast, causing the loss of many lives and tens of millions of dollars in damage, and that raced onward to wreck other towns and villages along the west coast of Canada and the United States. This tsunami was recorded as far away as the Argentine Islands in the Antarctic, Chile, New Zealand, Australia, Japan, and the Hawaiian Islands.

The Hawaiian Islands are especially susceptible to tsunamis generated around the margin of the Pacific Ocean basin. Many tsunamis have struck the Hawaiian Islands with severe violence. One of the most disastrous events was in April 1946 when a submarine landslide of the Aleutian submarine trench produced a tsunami that killed several hundred people on Hawaii. Again, on May 22, 1960, a Chilean earthquake sent a wave chain speeding across the Pacific toward the Hawaiian Islands. At Hilo on the Island of Hawaii one of the larger waves rose to a height of 35 feet above sea level and smashed into the city where it demolished part of the business district, killed 61 people, and injured 282 others.

The scientific studies of the Hilo disaster in May 1960 are available in documented reports which clearly show that the residents of all the Hawaiian Islands were warned of the impending tsunami about 12 hours before it struck. The warning was made possible by the fact that seismograph stations recorded the earthquake waves produced by the Chilean quake, thereby fixing its geographic coordinates and exact time of occurrence. Based on this information and the knowledge of the distance and depth of wa-

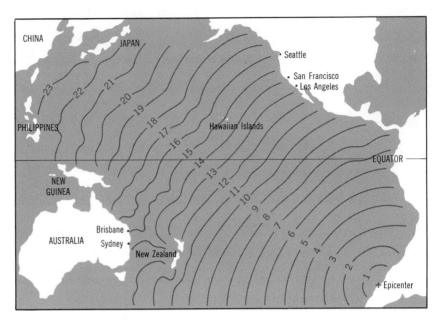

Figure 14.14 The numbered lines show the position of the seismic sea wave (tsunami) for each hour after it was generated by the Chilean earthquake of Mary 22, 1960. (After Wadati, K., T. Hirono, and S. Hisamoto, 1963, On the tsunami warning service in Japan, *Proceedings of the meetings associated with The Tenth Pacific Science Congress,* Monograph 24, Int'l. Union of Geodesy and Geophysics, Paris, pp. 138-146.)

ter between Chile and Hawaii, a prediction of the time of arrival of the tsunami was calculated. Actually, the main wave traveled the 6600 miles from its source along the Chilean coast to Hilo in 14 hours and 56 minutes at an average speed of 442 miles per hour (Fig. 14.14).

With such information at hand, the alert sounded by the United States Coast and Geodetic Survey Observatory in Honolulu was based on more than a mere hunch. Yet, it appears that nearly two-thirds of the residents of the stricken area failed to take it seriously enough to move to higher ground, an act that they had ample time to accomplish before disaster struck. Clearly, better public education in the urgency of these warnings is needed if further loss of life from tsunamis is to be avoided, not only in the Hawaiian Islands but elsewhere in the coastal regions of the world.

THE OCEAN BASINS

Ever since man learned to build ships that could carry him to new lands beyond the sea's horizon, he has viewed the ocean as a mysterious place whose fury in storm was to be avoided but whose serenity in calm had no equal. The attention of early explorers and seafarers of all lands was focused mainly on the ocean surface. Only in the shallow waters of harbors and coastal inlets was any concern given to the sea floor because of the need to navigate through treacherous reefs or rock shoals. Once at sea, however, a ship's captain could concentrate on his navigational problems and cease to worry about the water depth. Thus it remained for centuries that the floor of the ocean was virtually unknown because there was no compelling reason to seek further knowledge about it.

In the nineteenth century a few deep soundings were made in the open sea. Between 1872 and 1876 the British ship, *Challenger*, made a worldwide scientific cruise during which time a number of depth soundings were made by the laborious process of lowering a weight to the sea floor. Although many soundings were made they were so far apart that much of the topographic detail of the ocean bottoms was missed. For half a century the ocean floors were believed to be "flat, featureless plains."

The technique of measuring water depths by echo sounding devices is now commonplace. A ship equipped with an echo sounder is able to record, continuously, the depth of water along the course it is sailing. It was not until the middle of the twentieth century, however, that detailed bathymetric charts could be made with great accuracy. This was made possible by the systematic cruises of research vessels from a number of oceanographic institutions in the United States and other countires. It is now possible, therefore, to differentiate certain major features of the ocean basins even though the topography of the ocean floors is still imperfectly known. The three major divisions of the ocean floor are (1) the continental margin, (2) the ocean-basin floor, and (3) the mid-ocean ridge. (In the next several sections in which these features are discussed, frequent reference should be made to Plates II and III.)

The Continental Margin

All continents are fringed by a relatively shallow submerged zone that terminates at a point where there is a marked increase in the slope of the sea floor. This fringing zone is the *continental shelf* and the steeper sloping surface forming its seaward margin is the *continental slope*. These two elements form the continental margin.

Continental Shelves. The continental shelves of the world show a great variation in width and considerable diversity in water depth at their outer edges. Some continents have long stretches of coastal area where the sea floor plunges to great depths immediately offshore. Other seacoasts are bordered by shelves several hundred miles in width.

The outer edges of continental shelves occur in water depths as shallow as 65 feet or as deep as 1800 feet but the average depth is about 430 feet. Nominally, the edges of continental shelves are considered to be about 100 fathoms (600 feet) deep, but this generalization reflects the information on navigational charts that were based on such widely spaced soundings that the precise definition of the break in slope between continental shelf and continental slope could not be made.

Continental shelves are not easily accounted for by any simple combination of geologic processes. Exploration of a number of them around the world attests to their complexity of origin and diversity of form. The majority of the world's continental shelves are built of sediment derived from the adjacent continents. Various kinds of geological obstructions paralleling the coastlines keep the land-derived sediments from being washed farther seaward. Long, narrow fault blocks, deep marginal trenches, and extensive coral reefs are examples of the "geological dams" that are responsible for the occurrence of thick layers of sediment on many continental shelves.

During the Pleistocene build-up of glacier ice, sea level was considerably lower than today. A lower stand of the sea would expose many of the continental shelves to a subaerial environment. For example, the last major lowering of sea level occurred about 15,000 years ago (Fig. 14.15). Evidence for a lowering of about 425 feet (130 meters) below the present level has been discovered in the form of land fossil plants and animals on the present-day continental shelves off the Atlantic and Gulf coasts of North America as well as the shelves bordering Japan and Europe. Accurate age determinations by carbon-14 dating techniques of freshwater peats that now occur below sea level pro-

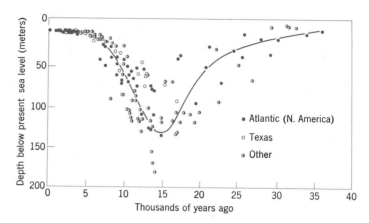

Figure 14.15 Graph showing worldwide sea-level changes during the last 40,000 years. (After Emory, K. O., 1969, The continental shelves, *Scientific American*, v. 221, no. 3, September, pp. 106-122.)

vide the basis for the sea-level changes of the last 40,000 years of earth history. The evidence for a lower stand of the sea around 15,000 years ago is corroborated by independent evidence of the advance of continental glaciers on land at the same time.

An interesting corollary to the relationship between sea-level lowering and the advance of continental glaciers is the probable location of the shore of the Atlantic coast of North America if all of the ice on Greenland and the Antarctic should melt (Fig. 14.16). Estimates of the magnitude of this rise in sea level are based on the knowledge of the volume of water represented by glacier ice in Greenland and the Antarctic. If this volume of ice should melt and be returned to the ocean basins, sea level would rise about 215 feet[1] and would more than double the area of the continental shelf from Florida to Cape Cod. All major seaports of the world would also be inundated. A rise in sea level due to the melting of ice on Greenland and Antarctica is not predictable, but if it should happen, it would take place on the same time scale as the sea level rise and fall of the last

[1] If this figure is corrected for isostatic adjustment, the net sea level rise would be only about 140 feet.

40,000 years and, hence, would not be catastrophic.

Resources of Continental Shelves. Economic exploration of the world's oceans has been concentrated on the continental shelves for a number of reasons. Relatively shallow waters permit the use of oceanographic techniques that are less expensive than the ones needed for the deep ocean, and nearness to coastal ports provides for greater economy in the extraction of resources on the continental shelves as compared to the open ocean. Moreover, the discovery of recoverable bodies of petroleum and natural gas in offshore areas of the Gulf of Mexico, the Pacific Coast, the North Sea, and Alaska has given impetus to further exploration of other continental shelves around the world. Other mineral deposits such as sand and gravel, phosphates, diamonds, tin, and gold are known to occur in quantities along various stretches of continental shelves. Some favorable areas for the exploration of mineral resources on the Atlantic continental shelf off the coast of the United States are shown in Figure 14.17.

One of the most valuable resources of the continental shelves is the commercial fisheries industry. About 90 percent of the world's marine food resources are extracted from these waters. Some species of fish taken in commercial quantities from continental-

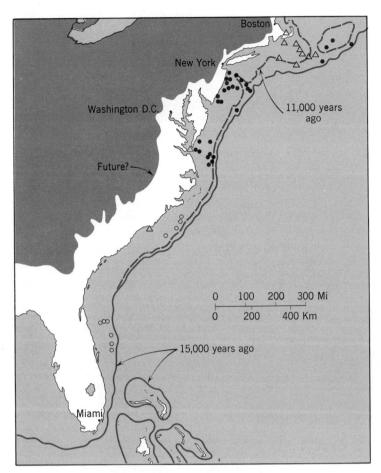

Figure 14.16 This map of the Atlantic coastal area shows the shoreline 15,000 and 11,000 years ago. Areas of the continental shelf that are now submerged were dry land at those times as confirmed by the discoveries of elephant teeth (dots), freshwater peat (triangles), and shallow-water deposits called oolites (circles). The white area landward from the present coastline shows the terrain that would be inundated if the present ice sheets of Greenland and Antarctica were to melt. (After Emery, K. O., 1969, The continental shelves, *Scientific American*, v. 221, no. 3, September, pp. 106-122.)

shelf fisheries are already on the decline, but in other areas, especially in South American waters, the fisheries can be expanded many fold before they reach maximum development.

Continental Slopes and Submarine Canyons. The outer edge of the continental shelf is marked by a break in the slope of the sea floor. The sloping sea floor from the edge of the continental shelf to the deep ocean basin is called the continental slope. The inclination of the continental slopes around the world shows considerable variation. Some, like those off the Atlantic coast of North America, are inclined only 4 to 5 degrees, while others, such as the slope off the coast

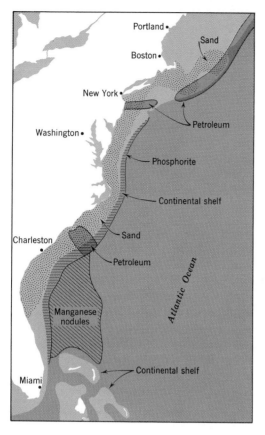

Figure 14.17 Potential mineral resources from the continental shelf off the east coast of the United States. (After Emery, K. O., 1965, *Some potential mineral resources of the Atlantic continental margin,* United States Geological Survey Professional Paper 525-C, Fig. 1.)

The canyons are V-shaped in cross section, have winding courses, and terminate at the base of the continental slopes thousands of feet below sea level. Their geometry bears a strong resemblance to deep valleys cut by rivers on land. The similarity between river valleys and submarine canyons led to a hypothesis that explained submarine canyons as submerged river valleys that were formed when sea level was much lower than it now is. Although one or two of the existing submarine canyons, notably the one off the mouth of the Congo, appear to be seaward extensions of modern rivers, most of the canyons bear no apparent relationship to modern river mouths. A greater objection to this hypothesis, however, is the fact that there is no evidence to support a sea-level lowering of the magnitude required to expose all the continental slopes to subaerial conditions, a condition implicit in the hypothesis that submarine canyons are caused by river erosion.

The most likely explanation of submarine canyons is that they are produced by *turbidity currents* and other submarine processes such as landsliding, creep, and tidal currents that have been operative for very long periods of time, perhaps thousands or even millions of years. Turbidity currents are bottom currents containing mud, silt, and sand that flow independently of the surrounding water mass. They have been produced in laboratory experiments and are known to exist on continental slopes.

It is almost certain that the last word on the origin of submarine canyons is yet to be spoken, but with the availability of deep-sea diving vessels from which oceanographers can observe directly the details of canyon morphology and the forces acting in these deeply submerged "valleys," additional evidence to account for their origin is to be expected.

The Ocean Basin Floor

The ocean floors beyond the continental margins are more variable in relief than the

of West Florida, are more like 25 degrees. (Care must be used in studying the continental slopes on Plates II and III because the highly exaggerated vertical scale greatly distorts the slope angle.)

Possibly the most intriguing aspect of the continental slopes is the presence of deep gashes that slice across them. These *submarine canyons,* as they are known, are characteristic features of continental slopes but they are by no means commonplace. Several occur off the Pacific coast of North America, some off the Atlantic coast, and others are known to exist off the mouth of the Congo River, along the coast of Japan, and along the French Riviera.

surface of the continents. Ocean floor topographic features include clusters of volcanic peaks, long straight fracture zones, deep arcuate trenches, and abyssal hills and plains. The geologic significance of some of these features in terms of global tectonics has already been discussed in Chapter 7. There, it was pointed out, for example, that the deep trenches are regarded by many geologists to be linear zones where crustal material is forced downward when two crustal plates are thrust together.

Seamounts, Guyots, and Atolls. Whereas the ridges, trenches, and fracture zones are produced by essentially horizontal movements, other features of the ocean floors provide proof of vertical movement. Many of the submerged volcanic peaks or seamounts that rise from the floor of the oceans have never been above sea level and have retained their characteristic conical shape. Other seamounts, however, are flat-topped, in which case they are called *guyots*. The

submerged flat surface of a guyot is ascribed to erosion by waves that planed off the peak of a volcanic island when it was above sea level. The presence of a wave-planed surface submerged as much as 3000 to 4000 feet (900 to 1200 m) below sea level is more likely due to subsidence of the crust rather than a rise in sea level.

Further evidence of crustal sinking beneath the oceans comes from the *coral atolls* of the equatorial Pacific Ocean. An atoll consists of a chain of low, elongated islands that lie on the fringes of a shallow lagoonal platform. The atolls are made of live coral colonies below sea level along with the debris from dead corals which occur both above and below sea level. Corals can live in water that is not much colder than 65° F. and not much deeper than 150 feet, a fact that prompted Charles Darwin to suggest that coral atolls rested on a submerged volcanic platform. Darwin's theory called for a slowly sinking volcanic peak

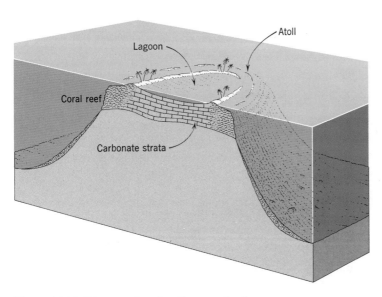

Figure 14.18 Diagram showing the growth of an atoll on the wave-planed top of a subsiding volcanic foundation. Drilling on Eniwetok Atoll in the Pacific has confirmed this hypothesis which was originally proposed by Charles Darwin. (From a diagram by Dietz, R. S., and J. C. Holden, 1966, Miogeosynclines in Space and Time, *Jour. Geol.,* v. 74, p. 570.)

that originally projected above sea level sometime in the geological past. Coral reef growing on the flanks of the volcano just below sea level continued to expand upward as the volcano itself subsided. Eventually the volcano sank to a great depth, but the rate of sinking was slow, and the coral reefs maintained themselves in a shallow water environment by growing upward and outward until an atoll was formed.

Darwin was challenged by the American geologist, Daly, and the Austrian geologist, Penck, both of whom argued that the atolls formed by a rising sea level attending the melting of the ice sheets at the end of the Pleistocene. Darwin's hypothesis required a considerable span of geologic time for atoll development, while the Penck-Daly scheme called only for the length of time since the retreat of the last Pleistocene ice caps.

After World War II scientists of the United States drilled into the atolls of Bikini and Eniwetok in the Pacific Ocean. The drill on Bikini penetrated 2500 feet of coral without reaching the volcanic basement, but on Eniwetok, holes on both sides of the atoll encountered lava rock at somewhat over 4000 feet. The fossil corals at this depth were on the order of 60 m.y. old and had been growing in place when they died. This evidence favors a subsidence hypothesis similar to Darwin's and does not support the glacial-control hypothesis of Penck and Daly. Darwin was wrong only in the idea that the coral growth started as a fringing reef. The evidence suggests that the corals originally attached themselves to a subsiding wave-planed volcanic platform (Fig. 14.18). If the volcanic platform was surrounded by warm seas and if its subsidence rate was slow, atolls resulted. If the platform was either surrounded by cold water or was sinking rapidly, guyots resulted instead. This explanation nicely accounts for the distribution of atolls in the warm equatorial waters of the Pacific and the swarm of guyots in the colder waters off the coast of Alaska as well as the many guyots in the Emperor Seamount Chain extending south from the western tip of the Aleutians.

The Mid-Ocean Ridge

The Mid-Ocean Ridge system is a world-encircling submarine mountain range some 40,000 miles in length that can be traced through the Atlantic, Indian, and Pacific Oceans (see Chapter 7). One of the best-known segments of this undersea mountain system is the Mid-Atlantic Ridge, a 1000-mile belt of peaks and ridges rising some 2 miles above the surrounding ocean floor. A few of its highest peaks, such as the Azores and Ascension Island, rise above sea level to form islands, and Iceland is a volcanic landmass lying right on the Mid-Atlantic Ridge. The entire Mid-Ocean Ridge system is characterized by a rift or valley that occupies the central portion of the range, and more-or-less defines the axis or backbone of the system. This median rift is narrow—only 20 to 80 miles wide—compared with the width of the Mid-Ocean Ridge, and ranges in depth from ½ to 1½ miles. In some places the rift extends into the continents, as for example, the Gulf of California, the Red Sea, and the East African rift valleys.

The Mid-Ocean Ridge is offset in a number of places because of movement along fracture zones which show up as straight linear troughs bearing at an angle to the ridge axis. Earthquake foci along these fracture zones indicate that movement is still occurring. This is in accordance with the general concept of sea-floor spreading and plate tectonics as already discussed in Chapter 7.

Ocean Sediments

The sea floor is the "final" resting place of a great variety of sediments that are delivered to it from sources outside the oceans or produced by processes in the oceans themselves. Ocean sediments originating outside the oceans are *terrestrial* and *extraterrestrial,* and sediments produced in the oceans are *biogenous* and *authigenic.*

Terrestrial sources include the detrital and chemical loads of rivers, volcanic ash, dust derived from deserts by deflation, and detrital particles carried by glaciers that enter the sea directly. Extraterrestrial sediments are those that originate outside the planet Earth and are generally categorized as *meteoritic dust.*

Biogenous sediments are produced in the ocean by marine plants and animals that secrete calcareous or siliceous substances to form a part of the organism, either as skeletal material or as an external shell. Many of these organisms are microscopic while others, such as corals, are larger. Authigenic minerals are those that form in place by chemical precipitation from seawater, as for example the manganese nodules that occur over large areas of the sea floor.

It is rare for any sediment derived from a single source to occur as a "pure" deposit on the floor of the open ocean far from land. Sediments from many sources are usually mixed together in the open sea because of the ability of ocean currents to disperse fine particles over a large area, and the tendency for wind-borne dust to be widely disseminated over the entire earth's surface. Only when rapid deposition takes place very close to a source area does a marine sedimentary deposit occur in relatively pure form. For example, nearly pure volcanic ash beds may form in the sea around a volcanic island, and in the lagoonal areas associated with coral reefs nearly pure coral sand may accumulate. By and large, however, deep-sea sediments are mixtures of substances that have been derived from more than one source, a fact that complicates the classification of deep-sea deposits.

Classification of Deep-Sea Sediments

Two broad categories of deep-sea sediments provide the framework for the classification of material deposited in the ocean depths (that is, outside the continental shelves). They are *pelagic* and *terrigenous.*[2]

[2] This classification scheme is based on the one proposed by F. P. Shepard in *Submarine Geology,* 2nd ed., Harper and Row, New York, 1963, p. 402.

Pelagic sediments, according to Shepard, are ". . . those sediments of the deep part of the open ocean that settled out of the overlying water at a considerable distance from land and in the absence of appreciable currents so that the particles are either predominantly clays or their alteration products or consist of some type of skeletal material from plants or animals. The clays may be derived from the land either by water or by wind or may come from volcanic dust or meteorites."

Terrigenous sediments are those derived directly from a land source. Table 14.a gives a brief classification of deep-sea sediments based on the two major categories, and the paragraphs that follow provide additional comments on the origins and characteristics of the major types. Figure 14.19 shows the distribution of deep-sea sediments.

Brown Clay. This pelagic sediment is a soft very fine-grained, low carbonate sediment that comes from several sources. These sediments were formerly called red clays, but subsequent studies have shown that most of the deposits in this category are chocolate brown in color rather than red. Studies on countless bottom samples taken from many cruises by oceanographic ships reveal something of the composition of brown clays. Although there is still much to be learned about the origin of these sediments it appears that they consist mainly of clay derived from atmospheric dust and fine-grained terrigenous sediments distributed over the oceans by currents. Some of the airborne dust is undoubtedly of meteoritic and volcanic origin. In the South Pacific Ocean the major mineral constituent of the brown clays is *phillipsite,* a mineral believed to be formed by the chemical reaction of seawater with volcanic ash.

Authigenic Deposits. One of the most curious of all deep-sea deposits are the manganese "nodules." Actually, manganese nodules are only one form of manganese-bearing minerals on the sea floor. Manganese oxide (MnO_2), along with other elements such as iron, occurs not only in nodular form but also as coatings or encrustations on hard objects lying on the bottom of the

Table 14.a Classification of Deep-Sea Sediments*

I. Pelagic Sediments

    A. Brown clay (chiefly composed of inorganic particles of very small size. Some biogenous material may be present).

    B. Authigenic deposits (minerals formed by crystallization in seawater).

    C. Biogenous sediments (chief constituent is material derived from marine organisms, either calcareous or siliceous in composition. Nonbiogenous material may also be present).

        1. Pelagic oozes (chiefly microscopic organisms).

            a. Calcareous ooze (contains abundant remains of calcareous skeletons of one-celled animals, called Foraminifera).

            b. Siliceous ooze (contains abundant remains of the siliceous single-celled marine plants called Diatoms, or one-celled animals called Radiolaria, or a mixture of both).

        2. Coral reefs and associated calcareous sands and muds.

II. Terrigenous Sediments

    A. Terrigenous muds (contain abundant silt and clay derived chiefly from land areas).

    B. Turbidites (produced by turbidity currents).

    C. Glacial marine sediments (contain clastics of various size ranges derived from glaciers or icebergs).

* After Shepard, F. P., 1963, *Submarine Geology,* Harper and Row, New York, p. 402.

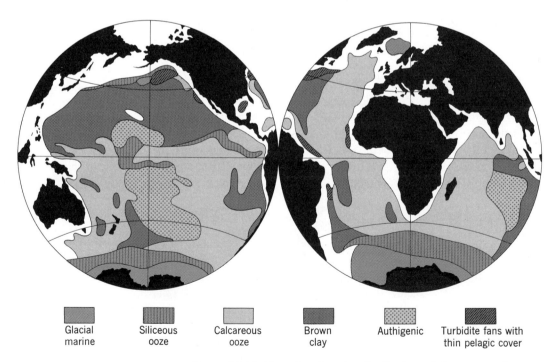

Glacial marine   Siliceous ooze   Calcareous ooze   Brown clay   Authigenic   Turbidite fans with thin pelagic cover

Figure 14.19 Map showing the distribution of deep-sea sediments. [After Shepard, F. P., 1967, *The earth beneath the sea* (revised), The Johns Hopkins Press, Baltimore, p. 166.]

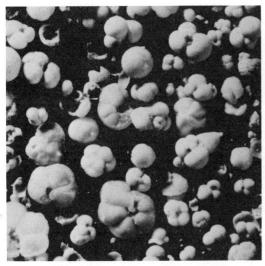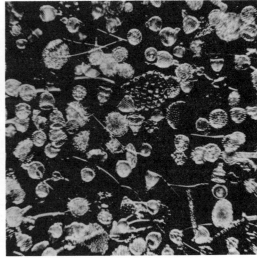

Figure 14.20 Photograph of Radiolarian (right) and Globigerina oozes (left). (From Dunbar, C. O., and K. M. Waage, 1969, *Historical geology,* John Wiley and Sons, Inc., New York, p. 523.)

ocean. In all cases, it is certain that the manganese encrustations form in place on the ocean floor and are, thus, authigenic in origin. Some occurrences are close to land while others are far away, so it is possible that the manganese is derived both from land areas and submarine volcanic eruptions. Large portions of the sea floor are literally covered with iron-manganese nodules or paved with manganese encrustations, and it is conceivable that some day they will be mined as a source of manganese and other metallic elements such as nickel and cobalt.

Pelagic Oozes. These deposits of the deep sea are derived from planktonic organisms that live in the upper regions of the open ocean. The skeletal remains of these unicellular plants and animals consist of silica or calcium carbonate, and when these organisms die, their remains sink slowly to the sea floor and accumulate as sediment. If the resulting deposit is made up primarily of calcium carbonate, it is called a *calcareous ooze,* and if the chief constituent is silica, the term *siliceous ooze* is applied.

One of the most common of the calcareous oozes contains the small calcareous shells from the one-celled animal known as *Globigerina* (Fig. 14.20). Siliceous oozes commonly form from the skeletons of one-celled animals called *Radiolaria* (Fig. 14.20) or the remains of small, free-floating plants called *Diatoms.* Most pelagic oozes also contain varying amounts of brown clay.

Globogerina oozes are especially useful in reconstructing past conditions in the ocean because they are temperature sensitive. Certain forms are found in cold waters while others are restricted to warm waters, and when alternating layers of warm and cold water forms are found in successive layers of sediment on the sea floor, it is assumed that the upper levels of the ocean waters were alternately warm and cold in response to the climatic changes that were responsible for the glacial and interglacial periods of the Pleistocene.

Terrigenous muds. Land-derived silts and clays accumulate as muds in the deep ocean along the continental margins. These sediments differ from the brown clays in that they contain silt particles, a component that is absent in the brown clays. The color of terrigenous muds varies according to conditions of deposition and the nature of the source of the sediment. Red colors reflect an oxidizing environment and darker colors of green and black are the result of either the lack of oxygen or the abundance of organic compounds.

Turbidities. These sediments are composed of silt, sands, and coarser grains that are delivered to the deep sea floor, especially the abyssal plains, by turbidity currents. In some places the turbidites are overlain by a thin veneer of terriginous mud or pelagic ooze. The important thing about turbidites is that they have characteristics of a shallow-water origin but are actually found in a deep-water environment. This knowledge has led many geologists who are concerned with the study of ancient rock layers to reexamine previous interpretations of their origins. Before turbidites were known and their origin understood, similar sedimentary rocks of ancient geologic vintage were thought to have been deposited in shallow seas on the order of a few hundred feet deep. The occurrence of modern turbidites on *abyssal plains* in water depths measured in thousands of feet now provides an alternative explanation for the origin of those rock layers.

Glacial Marine Deposits. In polar areas where glaciers discharge directly into the oceans or where icebergs are calved from their snouts, coarse sedimentary particles, released from the melting ice, accumulate on the ocean floor. Glacial marine sediments were more extensive during the glacial periods of the Pleistocene than today. Extensive glacial-marine sediments are accumulating today off the coasts of Antarctica and Greenland. The study of the physical characteristics and organic content of these sediments is important in the recognition of similar deposits found in older geological strata.

Significance of Deep Sea Sediments. Before the era of modern oceanographic studies, the interpretation of old sedimentary strata exposed on the continents was based on inadequate knowledge of modern sedimentary processes in the ocean. If the doc-trine of uniformitarianism is a realistic basis for the interpretation of ancient geologic events, then the knowledge of modern sedimentary processes in the oceans of the world ought to provide the kind of data needed to unravel the geologic record as it is revealed in marine strata deposited in the seas during past geologic ages.

On this note, we now proceed to the second part of this book which deals with the history of the earth as recorded in the rock strata formed during the various geological ages of the past.

References

* Bascom, Willard, 1959, Ocean waves, *Scientific American,* V. 201 No. 2 (August) pp. 75–84. (Offprint 828, W. H. Freeman and Co., San Francisco.)

Cox, D. C. (ed.), 1963, *Proceedings of the tsunami meetings associated with the tenth Pacific Science Congress, Aug.-Sept., 1961.* Monograph No. 24, International Union of Geodesy and Geophysics, Univ. of Hawaii, Honolulu, 265 pp.

Dietz, Robert S., 1964, Origin of continental slopes, *American Scientist,* V. 52, pp. 50–69.

Keen, M. J., 1968, *An introduction to marine geology,* Pergamon Press, Ltd., Oxford, 218 pp.

Marine Science Affairs, 1969, *Third report of the president to the Congress on marine resources and engineering development,* U. S. Government Printing Office, Washington, D. C., 251 pp.

* *Scientific American,* 1969, V. 223, No. 3. This issue devoted entirely to the oceans.

Shepard, Francis P., 1963, *Submarine geology,* (2nd ed.), Harper and Row, Inc., New York, 557 pp.

* Shepard, Francis P., 1967, *The earth beneath the sea* (rev. ed.), The Johns Hopkins Press, Baltimore, 242 pp.

* Turekian, Karl K., 1968, *Oceans* (Founda-

tion of Earth Sciences Series) Prentice-Hall, Inc., Englewood Cliffs, N. J., 120 pp.

Weyl, Peter K., 1970, *Oceanography: An introduction to the marine environment,* John Wiley and Sons, Inc., New York, 535 pp.

Wiegel, Robert L., 1956, *Waves, tides, currents, and beaches, glossary of terms and lists of standard symbols.* The Engineering Foundation, Council on Wave Research, 113 pp.

Wilson, Basil W., and A. Tørum, 1968, *The tsunami of the Alaskan earthquake, 1964: Engineering Evaluation,* U.S. Army Corps of Engineers, Coastal Engineering Research Center, Tech. Memorandum No. 25, 401 pp.

* Recommended for further reading.

Part Two The Geological Story

SANDY

COASTAL PLAIN OF

PENEPLANE ON CRYSTALLINE PRECAMBRIAN RO

UPPER CAMBRIAN

LIMY COASTAL PLAIN OF LOWER ORDOVICIAN

15 The Key to the Past

There rolls the deep where grew the tree.
Oh Earth, what changes hast thou seen!
Tennyson

In the preceding chapters attention has been directed largely to the dynamic aspects of contemporary geologic processes. It has been shown that the present face of the earth is undergoing constant change because of the many geologic forces that are continually in operation, both from within and without. Now the perspective will be shifted from the present to the past in a survey of the geologic history of the earth. One of the principal goals of geologic investigation is the deciphering of earth history. The chief objective of the remaining chapters of this book will be to reconstruct the past events of earth history into a coherent and logical chronology deduced from observations made over many years by geologists.

It is from the rocks, their internal characters and relationships to one another, that the facts about our earth have come to light during the course of man's never ending search for truth. During the last hundred years' study of history, errors have been made, erroneous ideas have been propa-

The distribution of land and sea in North America during the Ordovician period as reconstructed from the distribution of various kinds of sedimentary rock layers. (From Kay, M., 1951, North American Geosynclines, *Geological Society of America, Memoir 48, Boulder, Colorado, 143 pp., Plate I.)*

gated, and faulty interpretations of field and laboratory observations have produced unsound hypothesis. Nevertheless, progress has been made toward unlocking the secrets of the earth's diary, and we now have a fascinating account of the major changes the earth has experienced during its life span. Some pages of this autobiography still remain undeciphered, because they are written in a language still poorly understood or, in some aspects, a language that is yet unknown; and other pages, even whole chapters, have been lost forever, destroyed by one or more geologic processes.

The record is then incomplete, but the search goes on. Geologists are probing the far reaches of the earth with a vigor equaling that of the earliest investigators in quest of additional data concerning the earth's past history. Perhaps through the reading of the chapters to follow, the student can participate vicariously in this intriguing study, and will enjoy the thrill of discovering something new about his planetary home.

Uniformitarianism

During the nineteenth century Baron Georges Cuvier (1769–1832), a prominent French scientist and statesman, promoted the idea that the history of the earth was punctuated by sudden and violent catastrophic events. Followers of Cuvier formed what has been called the "catastrophic school" of geologists. In simple terms, this belief held that most geologic features were to be explained as a result of violent upheavals of very short duration. Opposed to this view was Sir Charles Lyell (1797–1875), a contemporary of Cuvier. Lyell, expanding on the views developed earlier by James Hutton (1726–1797), held that earth changes were gradual, taking place at the same uniform slowness that they are today. Hutton and Lyell are thus credited with the propagation of the premise that has guided geologic thought ever since, namely that the present is the key to the past. In essence, the Hutton-Lyell doctrine of uniformitarianism holds that past geologic processes operated in the same manner and at the same rate that they do today.

This basic principle is the undergirding structure of historical geology, even though many geologic events constitute catastrophes in the strictest meaning of the word. Volcanic eruptions, earthquakes, landslides, tsunamis, and floods are catastrophic events from man's point of view, but within the framework of earth history they are but normal happenings of no greater or lesser magnitude today than they were a million years ago.

From a scientific point of view, it is unwise to accept uniformitarianism as unalterable dogma. As discussed in Chapter 1, man's experience with geologic processes is restricted to only a minute fraction of the total span of earth history. He should not close his mind to the possibility that conditions in past geologic time were different in some details than today, and that the doctrine of uniformitarianism may not apply in every case.

Fundamentals of Historical Geology

Relative Age of Sedimentary Rocks. Armed with the concept of uniformitarianism, we can now proceed to certain fundamental relationships between rock units on which geologic chronology is based. The first fundamental relationship is embodied in the *law of superposition*. This law states that, in any succession of sedimentary rock layers lying in their original horizontal position, the rocks at the bottom of the sequence are older than those lying above. Furthermore, in a single rock layer, such as a sandstone, the minerals at the bottom of the layer were deposited before those near the top. Hence, if the top or bottom of a single layer can be identified, the original succession can be unraveled even though the strata may be highly contorted or even overturned.

In instances, and they are very common, where sedimentary rocks are not in juxtaposition, their relative age may be deter-

NORTH

Figure 15.1 Aerial photograph of Duncan Lake and vicinity, Northwest Territories, Canada, showing rocks of three different ages. Black areas are lakes. The lightest grey areas are outcrops of a granite mass intruded into older metamorphic rocks. The linear belts (emphasized by narrow bays of the lakes) are outcrops of dikes intrusive into both the granite and metamorphic rocks. (Photograph by the Royal Canadian Air Force.)

mined by the application of simple logic. If one of the rocks in question contains fragments of the other, the former is younger than the latter.

Relative Age of Igneous Rocks. A second basic relationship between rock masses involves intrusive igneous rocks. *Igneous rocks are younger than the rocks that they intrude.* Figure 15.1 shows rocks of three different relative ages. The light gray tone represents a granite intruded into metamorphic rocks shown in darker gray. Both the metamorphic rocks and the granite are intruded by dikes. Simple logic requires that the dikes are the youngest rock, the granite is next oldest, and the metamorphics are oldest.

Age of Faults. Where rock units are displaced by a fault, the fault is younger than the youngest rock cut by the fault. And ap-

plying the principle of superposition it is clear that a fault is older than nonfaulted sedimentary rocks overlying it. One type of fault is responsible for a condition in which older rocks lie on top of younger rocks as shown in Figure 15.2. In fact, these circumstances have led some to question the validity of superposition and the whole edifice of historical geology. If what geologists call older rocks do lie above younger in "natural" order, the edifice crumbles. All such "exceptions" to superposition are found in mountainous areas of the earth where the forces indicated in Figure 15.2 are commonplace. Mountain-building forces have thrust the rock layers on one side of the gently inclined fault plane to a position on top of beds on the other side of the fault. This is an event subsequent to the period

THE KEY TO THE PAST 297

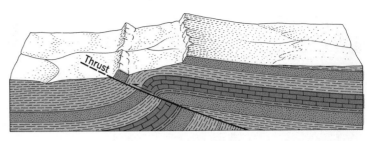

Figure 15.2 Block diagram showing how older beds may be thrust over younger beds along a thrust fault. Thrust faulting is associated with orogenic movements.

of sedimentation, so the relationship shown in Figure 15.2 is not an exception to the law of superposition.

In the examples cited, there is no evidence that would permit the establishment of the absolute ages of the rocks in question. The described relationships reveal only the *relative* ages. To determine the age in number of years requires other information and methods that are described later in this chapter.

Geologic Maps. The mutual physical relationships of rock units in a certain region may not be revealed unambiguously in a single exposure or outcrop. In such places as the Grand Canyon (Fig. 19.6), rocks are displayed on a grand scale and their relationships are generally clear. Such exposures are the exception, however. Therefore, the geologist may be obliged to observe the outcrops in many isolated places within the area he is studying, and the relative ages of the various rock units may not be obvious until lengthy field observations have been completed.

To assist him in drawing general conclusions about the rocks with which he is concerned, the geologist constructs a geologic map (Fig. 15.3). The geologic map is the principal tool of the geologist; the making of a geologic map is the first step in unraveling the geologic history of any area and is an art unique to the field geologist. Basically, a geologic map shows the areal distribution and geometric form of rock units, and the bound-

aries of those units drawn on a map of suitable scale. Most geologic maps portray the occurrence of geologic *formations* and their *contacts* with one another. (Geological maps are discussed on page 00.)

A formation is a rock layer or unit of sufficient size and with sufficiently distinct boundaries that a geologist can plot its distribution on a map. A formation, therefore, is a *mappable rock unit.* It could be a sandstone layer 50 or 300 feet thick, a 100-foot-thick conglomerate, a succession of alternating layers of shale and sandstone, or a massive plutonic intrusive rock such as granite. Formations are named for geographical localities where they were first described or are well displayed. A geologic contact may represent the normal sedimentary boundary between two rock layers, the juncture of a sedimentary layer and an intrusive or extrusive igneous rock, or a fault. A complete geologic map will contain a description of the formations shown on it and an indication of their relative ages.

Once the geologic map of an area has been completed, the job of interpretation can begin. Geologic cross sections that portray the geologist's interpretation of the underlying structure can be constructed, and the geologic history of the map area can be deduced. The reconstruction of geologic history from a geologic map is generally improved by laboratory examinations of the rocks of the area in order to provide more precise information on the origins of the rock masses. Both field

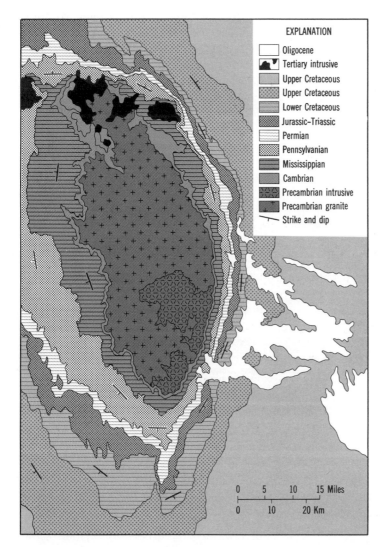

EXPLANATION

☐ Oligocene
⬛ Tertiary intrusive
▨ Upper Cretaceous
▨ Upper Cretaceous
▨ Lower Cretaceous
▨ Jurassic–Triassic
▤ Permian
▧ Pennsylvanian
▬ Mississippian
▨ Cambrian
▨ Precambrian intrusive
⬛ Precambrian granite
⌐ Strike and dip

Figure 15.3 Geologic map of the Black Hills of South Dakota. This well-known area is a structural dome as indicated by the pattern of strike and dip symbols and the arrangement of successively younger strata outward from the central Precambrian core. (Based on the Geologic Map of the United States, U. S. Geological Survey.)

The Meaning of Sedimentary Rocks

and laboratory studies are vital to a sound reconstruction of the geological history of an area.

The field geologist encounters many kinds of rocks during the course of his geologic mapping. The variety of rock types and number of formations he finds depend on the geologic complexity of the area in question. But whatever rock types are involved, the geologist ultimately will be required to gain from them an understanding of their origin and subsequent geologic history. His task is more difficult and his conclusions are less certain if he is dealing with a complex of highly metamorphosed rocks rather than a series of relatively undeformed sedimentary rocks.

It is, therefore, from the sedimentary rocks that the geologist receives the most accurate information, because the processes that produce sedimentary rocks are not beyond the reach of direct observation. Indeed, sandstones of future geologic ages are now being deposited in shallow waters of the coastal areas, and the muds collecting in the Gulf of Mexico will someday become hardened shales. The branch of geology called *stratigraphy* deals with the meaning that can be read into the many different kinds of sedimentary rocks in terms of their origins.

Factors that Control Sedimentary Rock Properties

If sedimentary rocks are to be used as a means of deciphering earth history, it is important that the factors which control sedimentary rock properties be understood. These factors are:

1. Kind of rock in the source area (provenance).
2. Environment of the source area.
3. Earth movements in the source area and in the depositional area (tectonism).
4. Environment of the depositional area.
5. Postdepositional changes of the sediment (lithification).

Kind of Rock in the Source Area (Provenance). Most of the sediments that come to rest in an area of deposition were derived from some other kind of rock on the earth's surface. For example, the sediments now accumulating in the Gulf of Mexico were derived by the weathering and erosion of rocks in the drainage basin of the Mississippi River and other streams. Thus, the mineralogical composition of the sediments must bear some relationship to the mineralogical composition of the rocks in the source region. It might be possible, therefore, to determine in a general way, by the study of certain mineral assemblages in the resulting sediment, the type of rock or rocks from

and temperatures, all of which, in turn, determine the kind and rate of weathering which they were derived. In a large drainage basin like the Mississippi River system, it is unlikely that an analysis of the Gulf of Mexico sediments would yield much diagnostic information about the rocks exposed in the source area because the source area encompasses rock types of great diversity. But where the source area is more restricted in its geographic extent, the rocks formed in adjoining depositional environments are likely to provide some information about the nature of the parent rock. This is especially true of sandstone which contains *accessory minerals* of higher specific gravity than the more abundant quartz and feldspar grains. These *heavy minerals* can be studied under the microscope in the laboratory after they have been separated from the lighter minerals. The mineralogical composition of the sediments thus yields information on the source area or *provenance.*

Certain heavy mineral suites are characteristic of definite source rock types, and it is possible for a geologist to specify the provenance from which certain sediments were derived. Usually he can state the provenance only in a general way, for example, acid igneous rocks, basic igneous rocks, metamorphic rocks, or reworked sediments. Heavy mineral analysis of a sediment does not always yield a detailed picture of the provenance, but it can be a valuable tool for the stratigrapher.

Environment of the Source Area. Although the term environment in the usual sense refers to the sum total of external conditions affecting the existence of some form of life, it can be used in a more general sense in reference to all conditions that prevail in an area, including the plants and animals themselves. Moreover, environment as used in this sense refers to the interplay of all external forces brought to bear on an area. Thus defined, it would include climate, topography, vegetational cover, animal population, and all the geologic processes at work during a specific span of time.

Climate involves rainfall, evaporation,

that will prevail. This is important because the way in which rocks weather influences the kind and amount of sediment that will be available for transportation by geologic agents.

Topography has a bearing on the competence and capacity of streams to carry the sediment to the depositional area. Vegetation influences the rate of runoff and the amount of organic matter added to the streams. The role of animal populations is extremely important in the formation of sedimentary rock. Some invertebrate animals such as reef-building corals concentrate vast quantities of calcium carbonate in the form of the shell material they secrete, and other microscopic animals with hard shells are added to marine sediments as the animals die and their shells fall to the sea bottom. Whole layers of organic limestone were produced in this fashion.

Foremost among environmental factors that prevail in the source area are the geologic processes themselves. The first part of this book is devoted to an understanding of these processes so there is little need for further elaboration here, except to emphasize the fact that the geologic processes active in the source area have a profound influence, not only on the kind of sediment produced but also on the manner in which the sediment is carried to the place of deposition, and on its distribution and special characteristics.

Earth Movements in the Source Area and in the Depositional Area (Tectonism). This term refers to the movement of any segment of the earth's crust during or between the periods of sedimentation. Tectonism includes not only the rise of a continental block constituting a source area from which sediments are derived but also the subsidence of the sedimentary environment in which the sediments accumulate.

During the middle of the last century, American geologists recognized the fact that great thicknesses of sediment—50,000 feet in some cases—appeared to have been deposited in relatively shallow seawater, probably on the order of 200 to 500 feet deep. This led to the conclusion that the basin or region in which the sediments were accumulating must have been constantly sinking during long periods of geologic time. Although no firm agreement exists among geologists as to the cause of subsidence (see Chapter 7), there is agreement that it has occurred in different places at different times during the history of the earth.

First let us consider tectonism of the depositional (negative) area, and second, the tectonism of the source (positive) area.

After geologists had agreed that most of the sedimentary rocks now lying on the continents were produced in relatively shallow seawater, it became necessary to classify sedimentary areas according to the degree of tectonic movement during major cycles of sedimentation. This led to the recognition of several different kinds of tectonic environments, three of which are of prime importance even to the casual student of geologic history. The first of these is the *geosyncline,* a linear belt of subsidence on a continent in which great thicknesses of sediments have accumulated (Fig. 7.2). The geosyncline may sink rapidly and continuously, or sporadically and at variable rates. The result, however, is a prodigious accumulation of sediments during long spans of geologic time.

The second negative tectonic element is the *shelf* (Fig. 7.2), a relatively stable, or perhaps mildly oscillating (but gradually subsiding) submerged continental platform receiving sediments. Such shelves are not unlike the present continental shelf bordering eastern North America (Plate II). In past geologic time, shelf areas were more widespread than at present, because greater areas of the continent were flooded by shallow inland seas.

The third major negative tectonic element is the *intrashelf basin,* a more or less isolated feature, moderately subsiding and surrounded by a more stable shelf area.

Turning now to positive tectonism, or the uplift that characterizes the source area, two kinds of movement are recognized. Vertical uplift or warping of the land is an *epeirogenic* movement, whereas *orogenic* movements are more intense and produce considerable uplift of the land as well as deformation of the rocks in the source area. Orogenic movements are now attributed to the interaction of crustal plates when they "collide" as explained by the hypothesis of plate tectonics (Chapter 7).

Environment of the Depositional Area. Tectonism provides the major framework of the depositional pattern, but on this larger picture are superimposed the various types of *sedimentary environments.* The classification of these environments is based on several factors, chief among which is the site of deposition with respect to sea level. The three major sedimentary environments are (1) *continental,* (2) *marine,* and (3) *transitional.*

The continental environments include the terrestrial deserts and glaciated regions as well as the aqueous environments of rivers (alluvial), lakes (lacustrine), and swamps. The marine environment is subdivided on the basis of depth of water and includes three zones: the *neritic* zone, extending from low tide to the edge of the continental shelf; the *bathyal* zone, from the edge of the shelf to 6000 feet in depth; and the *abyssal* zone which includes all oceanic depths greater than 6000 feet.

Transitional environments are mixtures of continental and marine zones and are classed as deltaic, lagoonal, or littoral (the zone between high and low tide).

Generally speaking, each of the sedimentary environments imparts certain characteristics to the sediments that accumulate in it, and depending on how carefully sedimentary rocks are examined, some understanding of the environment of deposition usually can be gained.

Postdepositional Change of the Sediment (Lithification). As sediment accumulates on the sea floor, it begins to undergo transformation of varying degrees. One significant change is *compaction,* owing to the weight of new sediment that is continuously being added. Compaction increases the density of the sediment by driving out water and reducing the pore space between particles. This is a far more important process in fine detrital materials like mud and silt than it is in sand. For instance, a mud at the bottom of the sea may contain more than 50 percent of voids or pore space. By the time

it is buried beneath 500 feet of additional mud the porosity will be reduced to 40 percent, and when that same mud layer is covered with 2000 feet of younger sediment only about 20 percent of its total volume will be pore space. Rocks composed of sand grains are less affected by the weight of overlying sediments, although a slight increase in density does take place during compaction.

Another change imposed on clastic sediments is *cementation,* the process whereby minerals form in the spaces between individual grains. This involves the precipitation of chemicals such as silica, calcium carbonate, or iron oxide, which bind the grains together like the cement in concrete and further reduces their porosity. Sands thus become sandstone and gravels become conglomerates.

Under certain conditions, newly deposited sediment will undergo some change in crystalline texture, a process called *recrystallization.* This is especially common in the chemical sediments such as limestone and dolomite. During recrystallization, some grains are dissolved while others grow larger because of redeposition of the dissolved material. Recrystallization usually increases the density of the sediment and when complete, tends to obscure the original sedimentary properties.

Many other changes take place in a sediment between the time it is first deposited on the sea floor and the time it is exposed as rock somewhere on the earth's surface. These changes may obscure or mask entirely the properties of the original sediments, thereby making the task of reconstruction of past environments more difficult, but not impossible. Table 15.a summarizes possible inferences that can be made from some sedimentary rocks.

Paleogeography

Paleogeography is a study of ancient land and sea relationships, the geographic distribution of areas of strong or mild tectonism, areas of sediment accumulation, the source

Table 15.a Possible Inferences Made from Various Sedimentary Rocks

Rock Type	Source Area			Depositional Area		
	Climate	Topography	Tectonism	Climate	Tectonism	Environmental Zone
Quartz ss.	Temperate	Moderately rugged	Stable	Variable	Mildly sinking	Shallow water
Eolian ss.	Temperate	Moderately rugged	Stable	Arid or coastal	Stable	Shore area or continental
Graywacke	Temperate	Rugged	Rapid uplift	Variable	Geosynclinal (rapid sinking)	Shallow to deep water
Chert	Temperate	Rugged	Moderate uplift	Variable	Geosynclinal	Moderately deep water
Turbidite	Temperate	Rugged	Moderate uplift	Variable	Geosynclinal or stable	Quiet water, shallow to deep
Evaporites (salt, gypsum)	Humid temperate	Moderate relief	Stable	Arid	Moderately sinking	Closed basin or restricted sea
Pure limestone	Humid temperate	Low to moderately high or distant	Stable	Warm	Stable	Clear shallow water
Gray shale	Temperate	Low or distant	Moderate uplift	Variable	Stable	Open sea
Black shale	Warm humid	Low or distant	Stable	Warm humid to tropical	Sinking	Restricted zones on shelf or geosyncline
Coal	Warm humid	Low	Relatively stable	Humid tropical	Gently emergent to submergent	Swamp conditions
Red detritals	Humid	Moderate to rugged	Moderate to strong uplift	Humid to arid	Stable	Delta, floodplain, open sea
Conglomerate	Variable	Rugged, swift streams	Rapid uplift	Variable	Sinking	Near shore along unprotected coast
Till or tillite	Glacial	Rugged or moderate	Uplift or stable	Glacial or temperate	Stable or moderately sinking	Continental or glacial marine
Arkose	Arid to temperate	Rugged; granite	Rapid uplift	Variable	Sinking	Shallow water near shore

regions from which rocks were derived, and the places where volcanic activity took place—all deduced from the study of rocks of a given time span. These broad relationships can be deciphered only after the geologist has studied the rocks in great detail, and has applied criteria like the ones presented in Table 15.a.

Such paleogeographic analysis is one of the ultimate goals in historical geology, and although it often involves considerable interpretation on the part of the geologist, a paleographic map is based on factual data derived from geologic observation. A single example will suffice to indicate the lines of reasoning followed in paleogeographic analysis. Although it is an oversimplification, it can be maintained that under normal conditions, sandstone, shale, and limestone are deposited in increasingly deeper water, respectively. Thus, if the geologist encounters a succession of sedimentary rocks, in a right-side up position, in which sandstone is overlain by shale, and the shale by limestone, he

THE KEY TO THE PAST   303

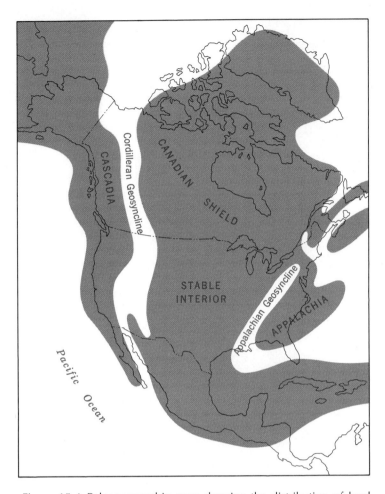

Figure 15.4 Paleogeographic map showing the distribution of land and sea during the middle part of the Cambrian period in North America. This interpretation is generally correct for the interior and the Canadian Shield (land areas) and the Appalachian and Cordilleran geosynclines, but neither Cascadia nor Appalachia were the simple landmasses portrayed here. Modern interpretations of these sedimentary source areas suggest that Cascadia was an archipelago of volcanic islands, and that Appalachia may have been part of the northwest coast of Africa when that continent was nestled against North America before continental drifting caused their separation, as shown in Figure 7.6. (After Schuchert, Charles, 1955, *Atlas of paleogeographic maps of North America,* John Wiley and Sons, Inc., New York, 177 pp.)

deduces that over the period represented by those rocks the water was getting progressively deeper, and probably that the basin of deposition was subsiding at that place. Such a succession is called an *overlap* sequence. The reverse succession, with the sandstone above the shale and the shale above the lime-stone represents a gradual shallowing of the water depth and is called an *offlap* sequence.

As the sediments yield clues not only to the environment but also to the nature of the source area, the geologist may be able to deduce whether the landmass was a rugged mountain range, a chain of volcanic islands, or a low-lying plain. At best, paleogeographic analysis can give only the broadest picture of the earth during past episodes of its history, and it approaches the truth only when the sediments are thoroughly understood.

Figure 15.4 is a classical paleogeographic map portraying the distribution of land and sea on the North American continent during a part of the Cambrian period; Figure 17.1 is a more modern treatment for the same area during a later period and attempts to show details as to source area and various environments of deposition. Figures 18.3, 18.5, and 18.6 are paleogeographic maps portraying more detailed relations during the Mesozoic Era. Figure 13.22 is a paleogeographic map of part of the Great Lakes Region during the waning phases of the Pleistocene Epoch.

The paleogeographic maps in Figures 15.4 and 17.1 reflect the belief that North America was an isolated continent at the times that these maps portray. Although it is by no means proved what the true geographic relations of continents were during such ancient periods of earth history, we have seen in Chapter 7 that the modern evidences for continental drift suggest strikingly different arrangements of continents during the geologic past. Accordingly, it is probable that a more accurate paleogeographic picture for eastern North America is the one suggested in Figure 7.6 showing the juxtaposition of eastern America and northwestern Africa.

Paleoclimatology

Modern paleogeographic analysis includes the study of ancient climates or paleoclimatology, a special field of historical geology

Figure 15.5 Late Precambrian tillite of southern Norway. Large irregularly shaped fragments (clasts) imbedded in a nonstratified matrix is characteristic of glacially deposited rocks. See Figure 13.16 for comparison.

that has received much impetus in the analysis of continental drift (Chapter 7). Paleoclimatology is a complex subject that requires a thorough understanding of climatology and ecology; only a few examples of paleoclimatological inferences can be given in the paragraphs that follow.

The geographic distributions of some marine organisms, such as corals and microscopic animals, are generally limited by water temperature, which in turn is mainly a function of latitude. The detailed study of these organisms preserved in rock layers has provided general temperature distribution patterns as well as some specific oceanic temperature data for various periods in the geologic past.

Plants are other paleoclimatological indicators. Their remains, preserved in sediments, have been used to deduce shifts in climatic zones on continental areas, especially for recent geologic history. Coal beds are generally considered to be indicators of luxuriant plant growth under semitropical or tropical conditions. The extensive coal beds of Antarctica, therefore, must mean that the paleoclimate that prevailed during their formation was very different from the glacial climate that prevails there now.

Rocks with no organic remains have intrinsic properties that can be used in analogy with modern environments in deducing paleoclimates. For example, red sediments have been used as indicators of tropical conditions, and evaporite deposits were probably formed under conditions of high evaporation characteristic of a desert environment. Laterally extensive cross-bedded sandstones of eolian origin record broad expanses of desertlike conditions during the time of their formation. Glacial conditions of the geologic past are indicated by ancient grooved and striated rock surfaces on which indurated tills or *tillites* are found (Fig. 15.5). A comparison between an outwash gravel of recent glacial origin and a lithified sand and gravel some 600 to 700 million years old of glaciofluvial origin is shown in Figures 15.6A and 15.6B.

The recognition of paleoclimates in the geologic record is based on the application of the principle of uniformitarianism to the origin of sedimentary rocks and their organic contents.

A

B

Figure 15.6 *A:* unconsolidated Pleistocene glaciofluvial gravel and sand in southern Norway. *B:* late Precambrian lithified sand and gravel of glaciofluvial origin in southern Norway.

Fossils

Any evidence of past life on earth is a *fossil*. A fossil may be anything from the complete skeleton of an ancient fish entombed in shale (Fig. 19.16) to the tracks of bottom-dwelling aquatic animals preserved in sandstone (Fig. 17.2). It may be the imprint of a leaf in a siltstone or the carcass of a mammoth locked in the frozen ground of Alaska or Siberia. Many fossils represent extinct species (Fig. 15.7), but some modern animals are similar or identical to their fossil ancestors. Extinction of a species is not a requirement for the definition of a fossil, but age is. Generally, an organism must have died in prehistoric time to be classed as a fossil.

Paleontology deals with the study of fossils and is one of the major subdivisions of the geological sciences. Because fossils are integral parts of sedimentary rocks in which they are found, they provide the means whereby the history of life through the ages can be studied. Fossils also provide one of the basic supports to the doctrine of organic evolution, a concept that is discussed later in this chapter.

Fossils were not always regarded as the remains of living things. The ancients considered them "sports of nature," or the works of the devil placed in the ground to mislead mankind. Some writers held that fossils were generated *in situ* within the earth in the same way that minerals were formed. In fact, the word fossil was first applied to all objects dug from the earth's crust, and included metals and minerals as well as organic remains.

Gradually, however, the evidence grew to overwhelming proportions in favor of the view that fossils were once living organisms that were naturally entombed after death and became incorporated in sedimentary rocks. By the sixteenth and seventeenth centuries such men as Leonardo da Vinci (1452–1519) upheld this view, but it remained for Baron Cuvier (1769–1832), the learned French nobleman, to put the science of paleontology on a firmer footing. He recognized that the fossil bones in the sedimentary rocks of France belonged to species of animals that had since become extinct. Cuvier was opposed to the idea of evolution, however, and believed that living things of the past were the result of an initial creative act of God. In Cuvier's view, cataclysmic upheaval of the earth's crust later destroyed parts of the organic world, resulting in a redistribution of the remaining organisms.

Probably no one individual deserves more credit for relating paleontology to geology than William Smith (1769–1839), an astute English surveyor who lacked any formal training in geology. Smith was the first to recognize the fact that fossils could be used as a means of identifying the strata in which they occurred. Smith had plotted the distribution of certain sedimentary rocks in southern England, an accomplishment that produced the first geologic map on record. He was a care-

Figure 15.7 Fossil trilobite from a shale in Yoho Park, British Columbia. These creatures are now extinct, but they flourished in great numbers during the Paleozoic Era. (Courtesy of Geological Survey of Canada. Norford, B.S., 1962, *Illustrations of Canadian fossils. Cambrian, Ordovician, and Silurian of the western Cordillera,* Department of Mines and Technical Surveys, Paper 62-14.)

ful observer, and noticed that the *same group of fossils always occurred in the same rock layers,* and that whenever he could not trace a single stratum from one place to the next, he would examine the isolated rock outcrops for their fossil content in order to know exactly which formation he was working with. Smith was a pioneer stratigrapher who demonstrated that fossils were a valuable tool in the science of geology.

Restoration of Fossils. For an organism to be preserved in the geologic record, two conditions must be met. First, the organism must possess structures that can be preserved in the rocks, and second, the remains of the organism must be quickly buried by sedimentary processes to prevent decay. The hard skeleton of an animal is more likely to be preserved than the softer parts such as skin, tissue, and delicate membranes. Although numerous instances of the preservation of soft-bodied organisms provide very important insights into the myriad of ancient life forms not normally represented in the fossil record, the paleontologist usually must content himself with only skeletal remains. If he wants to relate his fossil discoveries to living forms, he is obliged to restore the appearance of the fossil animal by the addition of flesh and other soft parts. This act of *restoration* is guided by his knowledge of the living relatives of the fossil under study. Restorations of the same organism by different workers may differ, however, especially in the vertebrate group of animals; skeletal parts do not reveal the exact muscular conformation or coloration of the animal, although some exceptional fossil discoveries have revealed the imprints of very delicate features of the animal's skin. Figure 15.8 shows a fossil skeleton and restorations by two different paleontologists.

Organic Evolution. The important discoveries of William Smith led eventually to the establishment of the *law of faunal succession,* which states that the fossil record is an expression of an irreversible time sequence. Fossils occur in the stratigraphic succession in a meaningful order. Although Smith had convinced himself and others of the significance of his discovery, he was not aware of the underlying reason why fossils were so arranged. It remained for the development of the doctrine of organic evolution to provide an understanding of the succession of life forms found in rocks. And, of course, the present-day extensive data from paleontology provide one of the most convincing arguments in support of evolution. We see the development from simple to complex life forms from older to younger strata; among the vertebrates the fishes appear in the record before the amphibians, the amphibians before reptiles, and reptiles before birds and mammals.

Although by no means the originator of all ideas bearing on evolution, Charles Darwin (1809–1882) is generally credited, and justifiably so, with its documentation. His painstaking observations and research over many years led him to propose what was then a shocking affront to the dogma of his day. Darwin noted the great variety of closely related animals, their patterns of dispersal, and their adaptations to a variety of habitats. He reasoned that the reproductive process produced a great number of offspring, among which variations would appear. In the struggle for survival, an individual that possessed variations which had even the slightest advantage would have a better chance to live out its full life span in comparison to individuals who did not possess the advantageous variations. The offspring of the better equipped individuals would pass these advantageous characteristics to their offspring. Darwin described this process as evolutionary change by the action of *natural selection* on the variations produced in a community of individuals of the same species. This remarkable conclusion was contained in his *Origin of Species,* published in 1859, a work believed by some to be the greatest scientific achievement of all time.

Darwin was unaware of the major cause of variations, however. It was not until the work of the Austrian monk, Gregor Mendel (1822–1884), published in 1865 but unappreciated until 1900, that the science of genetics was born. Based on his experiments on peas, Mendel reasoned that heredity

Figure 15.8 Two restorations of an extinct dinosaur, both based on the same skeleton. The animal is known as *Stegosaurus*. The middle drawing is a restoration by Edwin Colbert and the lower drawing is a restoration by Charles Knight.

was controlled by "particles" (later called genes) transmitted by parent to offspring. The genetic makeup of an individual tends to provide for continuation of characteristics (heredity) from generation to generation; at times, however (and seemingly randomly), the reproductive process produces larger scale changes, called *mutations*. Mutations are generally deleterious to survival but, in some fortuitous cases, a certain mutation imbues its possessor with the exact characteristic it requires to survive in a changed environment.

Evolution, then, can be described as the process by which nature will favor individuals with those characteristics that are most favorable for survival in an existing environment, or those that allow an organism to adapt to a slowly changing environment.

Correlation. Correlation is the demonstration of the time equivalence of geographically separated rocks or geologic events. Correlation is an essential requirement in establishing the chronology of earth history. To make correlations from place to place, the geologist must have a firm grasp and understanding of three important aspects of his science. He must make use of the principle of superposition in order to understand the relative age of rocks at any one place; he must recognize the role of organic evolution in the distribution of fossils in space and time. Finally, he must have evidence that the time needed for the dispersal of organisms is "instantaneous" in a geologic sense. If a given species is to be used as an indicator of a given unit of time (an *index fossil*), this species, made up of untold numbers of individuals, must have become dispersed over the area involved in the correlation in a short time relative to the passage of geologic time. Modern studies on organism dispersal, especially during the free-swimming larval stage in marine invertebrates, clearly indicate that, compared to geologic time, the time required for dispersal is very minor.

Rocks containing similar fossil assemblages are of the same geologic age. This principle has allowed the construction of the geologic time scale (Fig. 15.9) which is the basic calendar of earth history the world over.

The Geologic Time Scale

Time scales are of two types, relative and absolute. The relative age of an event fixes its age in relationship to other events and does not imply an age in years. For instance, a certain rock layer is younger than all the layers beneath it, and older than all the layers above it. Relative age can be determined from the geologic relationships of the rocks, but no accurate indication of age in actual years can be inferred from the field relationships alone.

A comparatively recent development in the science of geology, however, now permits the assigning of ages in years to certain rock units. Consequently, an absolute time scale has arisen that not only substantiates the validity of the relative ages of the various strata but also gives a more complete picture of the immense length of time that has elapsed since the oldest rocks were formed. Before investigating this method, however, the geologic time scale as it was originally conceived, and as it has since evolved, will be examined as preparation for a better appreciation of the impact of absolute methods of dating geologic events.

The Geologic Column. The first attempt to subdivide the history of the earth was based on the observations of Abraham Gottlob Werner (1749–1815), a German mineralogist of the late eighteenth century. Werner believed that all rocks were the result of precipitation from seawater and envisioned them as belonging to four chief subdivisions. Werner's ideas on the origin of rocks were later disproved, but the seeds of his idea of subdividing geologic time into major units did not die.

The nineteenth century saw the gradual development and piecing together of a standard geologic column and time scale that now has universal acceptance. The subdivision of geologic time was based on strata exposed in Europe, especially Britain, France, and Germany, and as a result the names given to different geologic periods are chiefly of European origin.

The geologic time scale embodies four major units, each of which is separated into smaller units. The major units do not all encompass the same amount of time, nor do their smaller subdivisions represent similar lengths of time. Except for the Precambrian, each has certain characteristic fossils associated with rocks formed during that particular time. Figure 15.9 shows the geologic time scale now used in the United States.

Basis for Subdividing Geologic Time. The passage of geologic time was accompanied by certain events that are recorded in the rocks. In the piecing together of a workable geologic time scale, it became necessary to pick out recognizable events of such importance and magnitude that they provided natural "punctuation marks" in earth history. The time at which mountains were formed and the shallow epicontinental seas were drained from parts of the continental platforms provided the local "natural" breaks for the initial framework of a chronology for earth history. *Orogeny,* or mountain building, was thus considered the timekeeper of the geologic clock. The growth

ERA	PERIOD	EPOCH	DISTINCTIVE FEATURES	OROGENIES	YEARS BEFORE PRESENT (Millions)
CENOZOIC	QUATERNARY	HOLOCENE	Modern man	Cascadian	— .01
		PLEISTOCENE	Early man; glaciation		— 2.5–3
	TERTIARY	PLIOCENE	Large carnivores		— 7
		MIOCENE	Abundant grazing animals		— 26
		OLIGOCENE	Large running mammals		— 37
		EOCENE	Modern types of mammals		— 54
		PALEOCENE	First placental mammals	Laramide	— 65
MESOZOIC	CRETACEOUS		First flowering plants; climax of dinosaurs and ammonites, followed by extinction	Nevadan	— 135
	JURASSIC		First birds: first true mammals. Many dinosaurs and ammonites		— 180
	TRIASSIC		First dinosaurs; abundant cycads and conifers	Palisades	— 225
PALEOZOIC	PERMIAN		Extinction of many types of marine animals, including trilobites	Alleghanian	— 280
	PENNSYLVANIAN	CARBON- IFEROUS	Great coal swamps; conifers. First reptiles		— 310
	MISSISSIPPIAN		Sharks and amphibians Large scale trees, seed ferns		— 345
	DEVONIAN		First amphibians; fishes very abundant	Acadian	— 400
	SILURIAN		First terrestrial plants		— 435
	ORDOVICIAN		First fishes. Marine invertebrates dominant	Taconic	— 500
	CAMBRIAN		First abundant record of marine life. Trilobites and brachiopods dominant		— 600
PRECAMBRIAN	LATE		Continental glaciation	Grenville Keweenawan igneous activity	
	MIDDLE		Iron formations Continental glaciation	Penokean	
	EARLY		Oldest algae—3.2 billion years Oldest dated rock—3.5 billion years	Algoman Laurentian	

(In the Paleozoic column, a vertical note reads: "CONTINENTAL GLACIATION IN SOUTHERN HEMISPHERE" and a vertical arrow labelled "APPALACHIAN OROGENY". In the Precambrian features column, a vertical note reads "Fossils generally rare".)

Figure 15.9 The geologic time scale as used in North America. Absolute time is given in the column at the right and is based on radiometric dates of strata of known relative age. (Based on the Geological Society Phanerozoic time-scale, 1964, *Quarterly Journal Geological Society London,* 120s, pp. 260-262.)

of mountains was thought to represent an event of worldwide importance. That is to say, the major geologic time intervals were supposedly represented by continuous deposition of sediments in epicontinental seas, and each episode of deposition was brought to a close by alleged worldwide orogeny that caused the retreat of the seas and the subsequent erosion of the sediments deposited in them. The sequence was started anew after the mountains were worn down by erosion and the seas once again invaded the land. Hence, the sediments of the later advance of the sea were laid down on the older rocks that had been subjected to deformation and erosion during the "break" in the depositional history. Such breaks constituted lost records because no sediments were being deposited. The break or gap in the record of the rocks was present in the form of a surface of erosion or nondeposition, an *unconformity* in geologic terminology. Figure 15.10 shows, diagrammatically, two unconformities in a theoretical sequence of layered rocks. It is self-evident that the oldest group was deformed, faulted, and eroded before the middle sequence of sediments was deposited, and that erosion of the middle group and continued erosion of the oldest group took place before the last sequence was laid down.

The general belief in worldwide orogeny

has dwindled until now geologists have come to realize that mountain building is not necessarily a worldwide event but may be restricted to a single continent or even part of one continent during any one interval of geologic time. Definite unconformities in one area may not appear at all in another area but are represented, instead, by a complete sequence of sedimentary rocks. In fact, it is quite possible that every part of geologic time is represented somewhere by a sequence of sedimentary rocks.

Orogeny has been almost entirely abandoned as the basis for partitioning geologic time because it is only of local significance. However, the use of the fossil plant and animal assemblages from the different rock layers, largely those of the European sequence of formations, are employed as the basis for comparison with fossils in strata from other parts of the world. The standard rock column as we know it today is still based on the strata of Western Europe and is universally used as the basis for the geologic time scale.

An Absolute Time Scale

When the European geologists began the long and arduous task of piecing together the strata on which the modern geologic

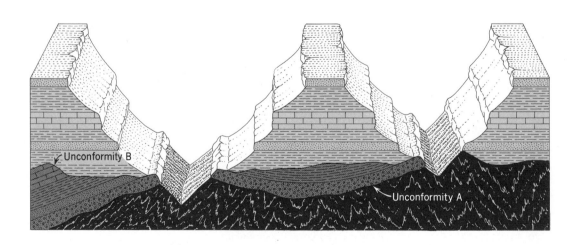

Figure 15.10 Block diagram showing two unconformities. Both represent periods of uplift and erosion, and they are, therefore, gaps in the geologic record.

time scale is based, they had no firm basis for dating an individual rock layer or unit in terms of actual years. The chronology built by the fathers of modern geology was based on the law of superposition and correlation by means of fossils. Thus, it was only a relative chronology.

For many years geologists sought methods whereby the various rocks could be dated in terms of actual years. They tried all sorts of ideas. One involved the rate of sedimentation in the sea. The early geologists thought that the thickness of a layer of rock was directly proportional to the time it took to accumulate. By multiplying the thickness of all rock layers in feet by the time in years necessary to form one foot, some idea of the magnitude of geologic time was arrived at. But this assumption might almost be classed as naive, and it certainly cannot stand the rigorous scrutiny of modern science. The basic assumption that all rock layers were laid down at the same average rates can be disproved simply by observing the widely different rates at which different types of sediments are forming today. A single flood in the lower Mississippi River might deposit 5 feet of mud in one day, whereas many thousands of years would be required for 1 foot of mud to accumulate in the bottom of one of the Great Lakes.

Radioactivity and Geologic Time. An absolute geologic time scale began to be realized when the phenomenon of radioactivity was discovered around the turn of the century. Radioactivity is an inherent characteristic of some chemical elements that causes them spontaneously to emit particles from their nuclei and thereby to become changed into other elements. No outside energy is needed to initiate or keep the radioactive process in operation; in addition, the rate of decay from one element to another is independent of temperature, pressure, or chemical environment. So far as is known, the rate of radioactive decay of an element in the laboratory is the same as in nature.

The principle of absolute age determination is thus quite simple. Element A is radioactive; it decays at a known rate into element B. Therefore to determine the age of a rock containing elements A and B, the total amount of each must be determined. Assuming that all of B was derived by decay from A, the ratio of B to A is an index of time since the decay process began. If it can be proved on geologic grounds that the beginning of the decay process and the "moment" of formation of the rock are the same, we have determined the age of the rock. For example, the radioactive age of a sample from an intrusive granite indicates the time of crystallization of the granitic magma. This date is acceptable as the age of the granite, even though it is probable that not all portions of the magma cooled at the same time and that the period of cooling was a long one.

Uranium (U_{238}) is a radioactive element that decays to lead (Pb_{206}) through a series of radioactive changes. The time it takes for half the original amount of material to change to the end product is defined as the *half-life,* and for U_{238} this constant is 4507 million years. Thus, a gram of U_{238} would be reduced to one-half gram of U_{238} in 4507 million years, and after another 4507 million years only one fourth gram of the original U_{238} would remain. Therefore, to determine the age of a rock containing U_{238}, it is necessary to know the exact amount of U_{238} and Pb_{206}. This analysis requires great skill and precision, partly because the lead formed by radioactive decay of uranium is not the same as ordinary lead. In addition, the total amount of radioactive material may be exceedingly small and easily masked by outside contamination. In spite of these difficulties, however, many dates based on uranium-lead ratios have been determined. All dates determined by the analysis of pairs of radioactively related elements are called *radiometric dates.*

Other radioactive elements that are used for age determinations include thorium decaying to Pb_{208}, and rubidium (Rb) to strontium (Sr). Another method involves the potassium-argon (K-Ar) ratio which has become widely used because of its application over a wide range of geologic time. But since argon is a gas, it can escape from the rock in which it is formed, thereby rendering the age determination unreliable. If some argon has been lost, the "age" will be too young. In fact, all methods are susceptible to error if the system in which the decay and accumulation occurred has not been a closed one. Either leaching of materials from the rock or additions to the rock make the determinations incorrect.

Most radioactive age determinations have been made on igneous rocks, either extrusive or intrusive types, but some have been made on sedimentary rocks. Nevertheless, a sedimentary rock can be dated relative to an igneous rock of known radiometric age, as illustrated in Figure 15.11. It can be seen from the contact relationships that the layered rock unit A is older than intrusives X and Y, and that the layered unit B is younger than intrusive X but older than intrusive Y and the lava flow. If the radiometric ages of X, Y, and the lava flow are known, the ages of the layered rocks can be established within maximum and minimum limits.

Direct radiometric dating of sedimentary rocks is also possible, providing that the minerals on which the date is based can be shown to have "grown" in the sediment at the time of deposition. Such a mineral is an *authigenic* mineral (page 286) in contrast to a *detrital* mineral, one which was worn from some older rock before deposition. A radiometric age of a detrital fragment would provide significant information regarding the provenance of the sedimentary rock, but it would give only a maximum age for the sedimentary layer.

A technique of special importance for dating of materials and events of very recent geologic history and for archaeological materials is the radiocarbon method. This method presupposes that, before the advent of atomic testing, atmospheric abundance of radiocarbon (C_{14}) was constant. Radioactive carbon is produced in the upper atmosphere by the action of cosmic rays on atmospheric nitrogen. All living organisms incorporate both C_{12} (normal carbon) and C_{14} into their tissues through ingestion of carbon dioxide, either directly by plants or indirectly by animals through the food chain.

A constant ratio of C_{14} to C_{12} exists in all live organisms, but when they die, the replenishment of C_{14} ceases and the remaining C_{14} decays radioactively to N_{14} at a known rate. Organic materials contained in sediments are presumed to have died at about the same time that the sediments were laid down. The length of time that an organism has been dead will be reflected in the amount of C_{14} remaining in its tissues. Since the half-life of C_{14} is known, and since it can be detected in very small amounts by laboratory techniques, it can be used to date sediments. Radiocarbon dating is restricted to organic remains that are less than 40,000 years old because the half-life of C_{14} is only 5770 years.

Radiometric dates on rocks of many ages now make it possible to assign absolute ages to the various segments of the earlier established geologic time scale (Fig. 15.9). The radiometric dates confirm the relative positions of the various geologic eras, periods, and epochs, all of which were defined and established prior to the discovery of radiometric dating techniques.

As will be discussed in subsequent chapters, the orogenies indicated in Figure 15.9 represent known episodes of crustal deformation and igneous activity, or more simply, episodes of mountain building. These orogenic episodes are deduced from the presence of dated unconformities, the presence of volcanic rocks associated with dated sequences of sedimentary strata, and the relationships of igneous intrusives of known geologic age to sedimentary strata.

Oldest Rocks and the Age of the Earth. Radioactive age determinations indicate that the oldest rocks of the earth's crust are about 3.5 billion years old. The age of the earth, based on this figure and independent evidence from astronomers and other sources, is estimated to be between 4.5 and 5 billion years. This age is in accordance with the idea that all elements of the earth and solar system originated at the same time. To what extent this generalization will be further substantiated or disproved is a matter to be determined by the future research of geologists, planetary scientists, and astronomers.

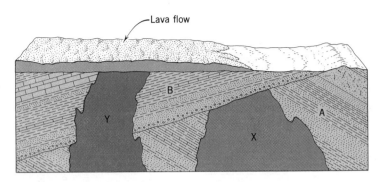

Figure 15.11 Hypothetical geologic cross section demonstrating the principle of dating sedimentary rocks that have been intruded by igneous rocks on which radiometric age determinations have been made. The layered sequence A is older than intrusive rocks X and Y, and the layered sequence B is younger than X but older than Y and the lava flow. If the radiometric ages of X, Y, and the lava flow are known, the ages of A and B can be determined within maximum and minimum limits.

References

Adams, Frank D., 1938, *The birth and development of the geological sciences,* Dover Publications, New York, 506 pp.

Axelrod, Daniel I., and Harry P. Bailey, 1969, *Paleotemperature analysis of Tertiary floras,* Paleogeography, Paleoclimatology, Paleoecology, V. 6, pp. 163–195.

Bell, W. C. *et al.,* 1961, Note 25 — Geochronologic and chronostratigraphic units, *Am. Assoc. Petroleum Geologists Bull.,* V. 45, pp. 666–670.

*Engel, A. E. J., 1969, Time and the earth, *American Scientist,* V. 57, pp. 458–483.

Geological Society of London, 1964, The Phanerozoic time-scale, *Quarterly Journal of the Geological Society of London,* V. 120s, Geol. Soc. London, London, 458 pp.

Ladd, Harry S., 1959, Ecology, paleontology, and stratigraphy, *Science,* V. 129, pp. 69-78.

Libby, W. F., 1961, Radiocarbon dating, *Science,* V. 133, pp. 621–629.

Nairn, A. E. M., 1964, *Problems in paleoclimatology,* Interscience Publishers, Wiley, London, 380 pp.

Rusnak, G. A., Tj. H. Van Andel, J. E. Nafe, B. C. Heezen, and D. B. Erickson, 1960, *Marine sediments,* McGraw-Hill Encyclopedia of Science and Technology, pp. 132–147.

Stebbins, G. Ledyard, 1966, *Processes of organic evolution,* Prentice-Hall, Inc., Englewood Cliffs, New Jersey, 191 pp.

Zeuner, Frederick E., 1950, *Dating the past,* Methuen and Co., London, 474 pp., 24 plates.

* Recommended for further reading.

16 The Precambrian

. . . and some rin up hill and down dale,
knapping the chucky stanes to pieces
  wi' hammers,
like sae mony roadmakers run daft—
they say it is to see how the warld was
  made!
Sir Walter Scott

Approximately 600 million years ago large parts of eastern and western North America were invaded by a shallow sea that left a record of thick deposits of sedimentary materials containing numerous invertebrate fossils. These rocks are called Cambrian because they can be correlated on paleontologic grounds with rocks first studied in Wales ("Cambria"). The rocks *underlying* Cambrian sediments and the rocks from which Cambrian sediments were derived are the subject of this chapter (Fig. 16.1).

Precambrian time encompasses the whole of geologic time from the beginning of decipherable earth history until the earliest fossiliferous Cambrian sediments were deposited. If the earth is 4½ to 5 billion years old, the Precambrian represents 80 to 85 percent of all earth history. Yet this vast

Tehachapi Mountain, California. (Photo by Richard B. Saul. Courtesy of California Division of Mines and Geology.)

317

Figure 16.1 Angular unconformity between Cambrian sandstone and Precambrian rocks north of Kingston, Ontario. (Photograph by A. S. MacLaren, Geological Survey of Canada.)

period of geologic time is among the least-known portion of the geologic record, primarily because of the intense deformation and metamorphism to which these ancient rocks have been subjected (Fig. 16.2). Some Precambrian rocks have suffered several episodes of metamorphic change, and each episode has further obscured the original nature of the rocks.

The geologist working on Precambrian problems is often therefore hampered in his attempt to reconstruct Precambrian history. He must remove the metamorphic marks before the real meaning of the rocks in terms of their original characteristics can be determined.

Precambrian Correlation

Even in those areas where Precambrian rocks are not highly deformed or metamorphosed, they generally lack fossils in comparison with the relatively abundantly fossiliferous post-Precambrian rocks. These two general characteristics of Precambrian rocks, metamorphics and lack of fossils, makes correlation a difficult problem. Historically, Precambrian geologists have relied on similarity of lithology, degree of deformation, intensity of metamorphism, and similarity of sequence in attempts to correlate rocks of different localities.

With the lack of any better methods, such bases for correlation are not wholly unreasonable, but there is no compelling reason why two granites of the same composition in two widely separated areas must necessarily represent the same episode of magmatic activity. Nor is there any reason why quartzites, schists, gneisses, or slates occurring in separate localities must be of similar age simply because they are similar in mineralogic composition. Similarity of rock sequence and similarity of geologic history as inferred from field studies form a more reliable basis for correlation of Precambrian rocks and events, but these too must be used with considerable caution.

Recent advances in radiometric dating of rocks have provided impetus to Precambrian correlation. As a consequence, some of the

Figure 16.2 Strongly metamorphosed rocks of the Canadian Shield in Hastings County, Ontario. (Photograph by M. E. Wilson, Geological Survey of Canada.)

"ages" determined by conventional geologic methods are in serious discordance. Most of the radiometric dates measure episodes of igneous activity and metamorphism accompanying deformation and cannot, therefore, represent directly the time of deposition. Some dates, such as those determined on sedimentary glauconite (an iron silicate mineral) in the late Precambrian Belt Series of Montana provide a direct age for these sedimentary rocks because the glauconite is apparently authigenic. Further radiometric determinations on widely separated Precambrian rocks employed with other more conventional methods may eventually bring about a worldwide basis for correlation of this important segment of earth history. Until this is possible, however, only regional classifications appear to be feasible.

Distribution of Precambrian Rocks

Precambrian rocks occur on all continents including Antarctica. They are exposed to view in three types of occurrence: (1) the large areas of relatively low relief called *shield* areas, (2) the cores of folded mountain areas, and (3) the bottom of deeply carved canyons.

Shield Areas. The shield areas of the world are so named because of their gentle topographic convexity, vaguely similar to the convexity of the shields used by ancient warriors. They comprise the ancient nuclei of all continents, around the flanks of which younger post-Precambrian mountain systems appear to have been developed. Eastern Siberia contains the Angarra Shield, and the Scandinavian peninsula is largely occupied by the Baltic Shield. Most of southern and eastern Africa contains the Ethiopian Shield, and the Amazonian and Guiana shields underlie a great part of eastern and northwestern South America. In central and western Australia the Precambrian Shield is called the Australian Shield. The Canadian Shield (Fig. 16.3) will serve as the example for most of what will be said about Precambrian shields in this chapter.

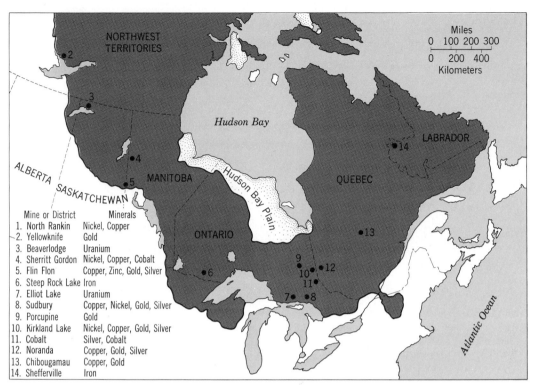

Mine or District	Minerals
1. North Rankin	Nickel, Copper
2. Yellowknife	Gold
3. Beaverlodge	Uranium
4. Sherritt Gordon	Nickel, Copper, Cobalt
5. Flin Flon	Copper, Zinc, Gold, Silver
6. Steep Rock Lake	Iron
7. Elliot Lake	Uranium
8. Sudbury	Copper, Nickel, Gold, Silver
9. Porcupine	Gold
10. Kirkland Lake	Nickel, Copper, Gold, Silver
11. Cobalt	Silver, Cobalt
12. Noranda	Copper, Gold, Silver
13. Chibougamau	Copper, Gold
14. Schefferville	Iron

Figure 16.3 Map of Canadian Shield showing important mining localities. (After Geological Survey of Canada, 1962, *Principal mineral areas of Canada,* twelfth edition, Ottawa, Ontario, scale, 1:7,600,000.)

Within the shield areas complex relationships of sedimentation, volcanic eruptions, intensive deformation, very large-scale intrusive activity, and widespread and intense metamorphism are characteristic. In contrast to the intense tectonism that characterized much of Precambrian time, the shield areas of the world have been remarkably quiescent and tectonically stable since the beginning of Cambrian time, 600 million years ago.

The Canadian Shield

The Canadian Shield covers 2,800,000 square miles (including part of Greenland), of which nearly two-thirds lies in eastern and northeastern Canada. The shield has been under investigation for more than a century largely because of the occurrence of a variety of valuable ore deposits. In spite of this interest, however, and owing to difficulties of exploration, only a small part of the Canadian Shield is covered by detailed geologic maps. Most of these maps are concentrated in mineralized areas. As a consequence of the economic impetus provided by the mineral wealth of the area, most of the detailed knowledge of the shield has grown through the accumulation of geologic data from widely separated localities rather than by a systematic investigation of the area as a whole. This type of approach is to be expected, since without the lure of ore deposits of economic value to justify the geologic mapping programs carried out by governmental surveys and private companies, very little detailed geologic mapping would be accomplished. And, of course, without geologic maps the geologic history of an area would never be deciphered.

Subdivisions of the Precambrian in the Canadian Shield

Ultimately, the geologist expects his investigations to reveal a sequence of geologic events that will allow him to understand the geologic history of a certain area. Such sequences of events are generally divisible into time units of various magnitudes, depending on the details of earth history that can be read from the record in the rocks.

Correlation of Precambrian rocks is difficult, even with the fairly numerous radiometric ages now available. Because of these difficulties, the Precambrian rocks of the Canadian Shield do not permit the establishment of very small subdivisions of time. Agreement has not yet been reached as to how many major divisions of time lie within the Precambrian period. An early view held that the Precambrian should be divided into two major time units, the *Archeozoic* and the *Proterozoic* eras. This division was supported by the widely held view that much of the shield had experienced one truly major episode of mountain building, igneous intrusion, and metamorphism which separated two grand sequences of rocks. The older (Archeozoic) was represented by highly deformed and highly metamorphosed rocks intensely invaded by granites. A great unconformity truncated the older sequence and separated it from the younger, less deformed and metamorphosed rocks of Proterozoic age. It is quite true that many areas in the Canadian Shield reveal two major sequences of rocks. Correlations were made from area to area on the assumption, now known to be false, that the deformational episodes were synchronous over wide areas. Radiometric age determinations now show clearly that this view is no longer tenable.

An example of a more recent classification of the Precambrian is based on the Precambrian rocks of Minnesota. The threefold division of Precambrian time shown in Table 16.a is based on radiometric ages coupled with field relationships. Two things must be stressed about Table 16.a: it is applicable only to Minnesota, and it is probably far too simple in terms of the actual geologic events of the Precambrian. It is, however, an improvement over older classification systems, and undoubtedly will undergo revision as future research reveals more of the details. Inspection of Table 16.a shows three major time units, Early, Middle, and Late Precambrian.

The upper limit of the Early Precambrian is based on radiometric dating of the Algoman granite. The middle Precambrian extends from the end of the Algoman orogeny to the end of the Penokean orogeny about 1800 million years ago. Finally, the Late Precambrian covers the time span from 1800 million years ago to 600 million years ago. The 600-million-year date for the end of the Precambrian and the beginning of the Paleozoic Era is a geologic date that has widespread acceptance the world over. The paragraphs that follow provide an abridged account of the Precambrian history of the Great Lakes region as outlined in Table 16.a.

The Early Precambrian

What is here called the Early Precambrian has traditionally encompassed two major units of rocks. The oldest, the Keewatin Group, including a succession of metasedimentary[1] rocks of uncertain age comprises a considerable thickness of metavolcanics,[2] many of which are "pillow lavas" (Fig. 16.4), lava deposited in a subaqueous environment. The metavolcanic and metasedimentary rocks are deformed and intruded by granites representing the Laurentian orogenic episode, following which the Keewatin Group and the granitic rocks were deeply eroded. They are unconformably overlain by the younger rocks of the Early Precambrian belonging to the Knife Lake Group, a succession of slate, graywacke, conglomerate, and minor volcanics. Following Knife Lake deposition, major igneous activity, the Algoman episode, affected a large area of the shield.

[1] synonymous with metamorphosed sediments.
[2] synonymous with metamorphosed volcanics.

Figure 16.4 Pillow structures in Precambrian lava southeast of Gordon Lake, Northwest Territories. The bottom of the flow is toward the lower right as indicated by the way the pillows are "draped" over each other. (Photograph by J. F. Henderson, Geological Survey of Canada.)

Radiometric dating of the Laurentian granite has not been possible because it has been so completely affected by the later Algoman intrusive event 2600 million years ago that its minerals yield only this younger age. The Algoman episode, which terminates the Early Precambrian, represents one of the greatest invasions of the crust by granitic magma ever recognized.

The 3300- to 3550-million-year dates shown on Table 16.a represent age determinations on a granite gneiss from the Minnesota River Valley, south of the Canadian Shield. The gneiss is not in visible contact with any other Precambrian rocks, so its true relation to the Keewatin or the Laurentian cannot be determined. Its age alone sets it apart as the oldest rock near the southern extremity of the Canadian Shield.

The Middle Precambrian

The rocks of the Animikie Group in Minnesota lie with major unconformity on Algo-man and older rocks. The Animikie Group contains most of the important iron formations of the Lake Superior region. These sedimentary accumulations reflect unique depositional conditions over larger areas and are one of the more puzzling types of sedimentary rocks as to origin (Fig. 16.5). No sedimentary rocks of similar mineralogic composition are known from post-Precambrian rocks.

The Huronian succession of rocks in northern Michigan is a possible equivalent of the Animikie of Minnesota. In Wisconsin and Michigan the Huronian rocks contain significant iron ore deposits as well as quartzite and dolomite, which are also characteristic of the Animikie. In addition, in the Canadian province of Ontario, the Huronian succession contains the Gowganda tillite, a well-documented ancient glacial deposit, one of the oldest yet recognized.

The termination of the Middle Precambrian in the Lake Superior region is marked by mild deformation, a relatively minor episode (Penokean) of granitic intrusion, and

Table 16.a Classification of Precambrian Rocks of Minnesota*

Era	Group†	Formation	Event	Intrusive Rocks
Paleozoic (600 m.y.)				
		——— UNCONFORMITY ———		
		Sandstones		
		———UNCONFORMITY———		
Late Precambrian	Keweenawan volcanics	Flows, tuffs, and sediments	Keweenawan Igneous activity (1000-1200 m.y.)	Duluth Gabbro
		Conglomerate		
		——— UNCONFORMITY ———		
		Quartzite		
		——— UNCONFORMITY ———		
(1800 m.y.)			Penokean orogeny (1600-1900 m.y.)	Granite
Middle Precambrian	Animikie (Huronian) Group	Slates		
		Iron formation		
		Quartzite		
		——— UNCONFORMITY ———		
(2600 m.y.)			Algoman orogeny	Granite
	Knife Lake Group	Slate, graywacke,	(2400-2750 m.y.)	
		Conglomerate, tuffs, lavas		
		——— UNCONFORMITY ———		
Early Precambrian			Laurentian orogeny	Granite
	Keewatin Group	Iron formation Metavolcanics (greenstone) Metasedimentary rocks Older rocks		
			(3300-3550 m.y.)	Gneiss

* Modified from Goldich, S. S., 1968, Geochronology of the Lake Superior region, *Canadian Jour. Earth Sciences*, V. 5, pp. 715-724.

† A group is two or more formations with close geologic affinities.

the development of an unconformity separating these and older rocks from the rocks of Late Precambrian age.

The Late Precambrian

During Late Precambrian time one of the major volcanic outpourings (Keweenawan volcanics) of geologic history took place in the Lake Superior region. More than 30,000 feet of basaltic lavas were extruded, and several thousands of feet of sandstone beds were deposited during nonvolcanic intervals. The Duluth gabbro, a massive sill, was also intruded during Late Precambrian time. This intrusive layer has a maximum thickness of 50,000 feet. All of these rocks are well exposed along the entire north shore of Lake Superior from Duluth, Minnesota

Figure 16.5 A close view of a Huronian (Middle Precambrian) iron formation near Negaunee, Michigan. The dark bands are red jasper and the lighter bands contain mostly iron minerals. (The scale is six inches long.)

Length of the Precambrian

The oldest rocks dated by radioactive methods are about 3500 million years old. The Cambrian began about 600 million years ago, which makes a known total of some 2900 million years for the Precambrian. These radiometric dates are derived from pegmatites and granites that were intruded into older metasediments and metavolcanics. Hence, the 2900 million years represents a minimum time span for the duration of the Precambrian. The estimated age of the earth is between 4500 million and 5000 million years. This age is based on abundances of radioactive isotopes in the Earth's crust and on radiometric dates of meteorites that probably were formed at the same time as the Earth. The time between 3500 million years and 4500 to 5000 mil-

to Fort Williams, Ontario, and they form some of the most spectacular scenery in the Great Lakes region.

lion years ago is thus a gap in the geologic record. The geologic events that transpired during the first 1000 to 1500 million years of earth history can only be guessed at, and their discovery and elucidation remains as one of the great challenges to future students of earth history.

Many Precambrian geologists of the late nineteenth century believed that some of the granites of the Canadian Shield were vestiges of the original crust of the earth. In fact, Laurentian granite from eastern Canada was so interpreted. Subsequent field studies have shown, however, that all the granites of the various shield areas are intrusive into older rocks; no vestige of the original crust of the earth has yet been discovered.

Precambrian Life

The Cambrian seas were inhabited by many of the advanced forms of invertebrate life. Since it is believed that these animals were the products of an evolutionary sequence, one can also assume that the ancestors of Cambrian marine creatures evolved during Precambrian time. But the dearth of fossils in rocks of Precambrian age presents a serious problem. It is quite true that many of the Precambrian strata are so strongly metamorphosed that any organic remains originally contained in them have been destroyed in the process. There are many areas, however, in which the Precambrian sedimentary rocks are largely unmetamorphosed and undeformed. Because these beds are subaqueous in origin, the fact that they should be so devoid of the remains of life is a mystery not easily explained.

Some Precambrian rocks contain fossils and other indirect evidences of life, but occurrences of this kind are generally too rare to be used as a means of correlation. The earliest life forms recognized to date in Precambrian rocks are blue-green algae and algalike "fossils" found in cherts from the Gunflint iron formation (1900 million years) of the Animikie Group from Ontario, and from two series of cherty rocks from South Africa dated at 3100 and 3200 million years old. These occurrences are rare, but clearly indicate the tremendous antiquity of life on earth.

Of greater abundance in the Precambrian rocks are *stromatolites,* which have been discovered in 2700-million-year-old rocks from South Africa. Stromatolites are branching or layered structures of calcium carbonate or silica, both compounds of which are produced by the action of blue-green algae. Stromatolites are particularly abundant in rocks of Late Precambrian age, and eventually may be useful as a correlation tool.

Some Precambrian rocks contain abundant complex carbon compounds of possible organic origin. For example, certain metamorphic rocks of eastern Canada (Grenville Formation), contain large quantities of graphite that may have had an organic origin. Metamorphism has been so intense, however, that all vestiges of the original organisms have been destroyed.

The oldest known fossils of undoubted multicellular organisms are those of the Ediacara fauna[3] of South Australia, discovered in 1947 (Fig. 16.6). This fauna contains the impressions of such forms as jelly fish, segmented worms, sea-pens, and other forms not related to any living organism. All were apparently soft-bodied forms, preserved as a result of rather unique conditions. The Ediacara fauna occurs only about 500 feet below rocks of undoubted Early Cambrian age, and is therefore, latest Precambrian in age.

Here, then, is an unsolved dilemma. The 3200-million-year-old plant fossils are of Early Precambrian age, and the undoubted animals of the Ediacara fauna are no older than latest Precambrian. Practically nothing is known about the life forms that inhabited the earth during the time that separates these two occurrences. Yet, the doctrine of evolution demands a succession of life forms from simple to more complex over the course of time. Thus far, no clear-cut evolutionary progression of plants or animals has been identified from Precambrian rocks, mainly because of the lack of fossils.

Some paleontologists suggest that the paucity of Precambrian fossils is because these ancient animals had no hard parts such as shells, skeletons, or teeth, all of which are more easily fossilized than soft tissues. Although it has to be admitted that soft tissues of jelly fish or the like would, indeed, require special sedimentary conditions for their preservation, it is difficult to imagine that these conditions were so uncommon in the Precambrian that only the Ediacara fauna and its meager correlatives in a few other areas have been discovered. Furthermore, tracks, trails, and burrows of soft-bodied animals are lacking in Precambrian rocks, but are common in younger sedimentary rocks (Fig. 17.2).

One explanation for the lack of Precambrian fossils is that animal life originated and evolved under terrestrial or coastline

---

[3] A fauna is a group of animals living in a particular environment at a particular time.

Figure 16.6 These imprints of Precambrian animals are rare occurrences in Precambrian rocks. These fossils are from the Pound Quartzite (late Precambrian), Ediacara Hills, South Australia. *A: Dickinsonia,* a broad flatworm; *B: Cyclomedusa,* a jellyfish-like form; *C: Spriggina,* a segmented worm; *D: Tribrachidium,* possibly a very primitive echinoderm. (Photographs by Karl Waage with permission of M. F. Glaessner.)

environments, both of which are unlikely to be represented in rocks as old as Precambrian. If these early life forms did not migrate into true marine environments until very late in the Precambrian, then the scarcity of fossils in rocks ranging from 3000 to 600 million years old is accounted for.

Another attempt to account for the rarity of Precambrian fossils is the hypothesis that development of animal life depended on the amount of oxygen in the Precambrian atmosphere. It is generally believed that the earth's early atmosphere was devoid of free oxygen and carbon dioxide, and that the present atmosphere gradually increased in oxygen by the action of photosynthetic algae. It has been suggested that when free oxygen became sufficiently abundant to support animal respiration, the conditions were ripe for the rapid development of the early animals. There remain difficulties, especially the highly advanced state of the Ediacara and Early Cambrian faunas; it is difficult to imagine their development in

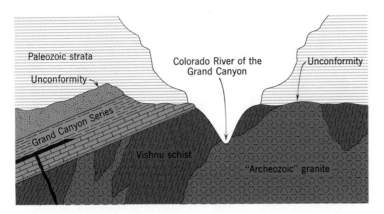

Figure 16.7 Cross-sectional sketch of the lower part of the Grand Canyon of the Colorado River showing the relationships of the various rock units. (From a diagram by L. F. Noble.)

the short span of time that is suggested by some investigators.

Clearly, much more work needs to be done in the search for Precambrian fossils the world over. In addition, geologists interested in this part of the geologic record must concentrate on the organic constituents of Precambrian sediments in order to learn more about the environments in which they were deposited. Once the environmental conditions of deposition are understood, a more intelligent evaluation of life during the Precambrian will be forthcoming.

Some Precambrian Rocks Elsewhere in North America

The Grand Canyon. The mile-deep gorge of the Colorado River cuts through approximately 3000 feet of Paleozoic sedimentary rocks and provides a spectacular lesson in geology to the thousands of tourists who visit this scenic wonder (Figs. 16.7 and 19.6). Near the bottom of the canyon a complex of deformed and metamorphosed rocks lie unconformably below the lowest Cambrian sedimentary rocks.

The oldest rocks in the canyon consist of schist (the Vishnu schist) containing intrusive granite and pegmatite dikes. Both schist and dikes have been called Archeozoic

(Early Precambrian), because of similarity in metamorphism and structural position to the "Archeozoic" rocks of the Canadian Shield. Radiometric dating of the granites indicate an intrusive age of between 1600 and 1800 million years, suggesting only that the schist is pre-Penokean in the terminology of Table 16.a.

Lying unconformably on the Vishnu schist are the strata of the Grand Canyon series, a succession more than 10,000 feet thick of quartzite, limestone, conglomerate, and shale. These sequences have been called Proterozoic by analogy with the rocks of the Lake Superior region. The correlation is probably correct, but the continued usage of Proterozoic as an age term is inappropriate in view of the trend to divide the Precambrian into three rather than two major time units.

Two different periods of erosion are represented by the two Precambrian unconformities shown in Figure 16.7. The erosion surface on the rocks of the Grand Canyon series is remarkably flat with a maximum relief of about 50 feet, but it does contain some monadnocks that rise as much as 800 feet above the general level of the unconformity.

A study of the rocks immediately beneath each erosion surface reveals that feldspars were altered to kaolinite (clay), and biotite

Figure 16.8 The sedimentary rocks exposed in this steep wall in Glacier National Park, Montana, are the Belt Series of Precambrian age and are of marine origin. (Photograph by Tad Nichols.)

was changed to iron oxide. These changes are normally induced by chemical weathering under relatively humid conditions. It is therefore concluded that a humid climate prevailed during both periods of prolonged erosion in the Grand Canyon region.

The Rocky Mountains. The Rocky Mountains from Colorado to British Columbia display large-scale occurrences of Precambrian rocks. In the relatively simple anticlinal arches of the central Rockies in Colorado, Wyoming, Utah, and southern Montana, principally crystalline rocks are exposed. The radiometric ages of these rocks range from approximately 3000 to 1200 million years. Farther to the west, in northwestern Montana, Idaho, and British Columbia, vast thicknesses of geosynclinal sediments of

Late Precambrian age are exposed. In many respects these are similar to the Grand Canyon series.

An exceptionally fine sequence of relatively unmetamorphosed sedimentary rocks is that of the Belt Series exposed in the Glacier National Park area of Montana (Fig. 16.8). In the park area, more than 25,000 feet of essentially flat-lying shale, limestone, and sandstone occur. The limestone units contain abundant stromatolites, described previously. Some of the most magnificent mountain scenery in the northwestern United States is found in the Glacier Park region where rocks of the Belt Series have been carved by valley glaciers into steep-walled valleys and serrated ridges. The rugged peaks forming the Continental Divide or

Garden Wall of Glacier Park expose nearly 2000 feet of Beltian rocks.

The age of the Beltian group of rocks is around 1100 or 1200 million years, and is based on a 1070 million year date of sedimentary glauconite from the upper Belt in Montana, and a 1190 million year date for pegmatite dikes cutting the Beltian rocks in Idaho.

Southern Great Basin. In the Death Valley region of eastern California and nearby areas, rock sequences similar to the Vishnu and Grand Canyon series occur. A succession of unmetamorphosed limestones, quartzites, and shales, the Pahrump Group, lies with erosional unconformity beneath Cambrian rocks, and lies unconformably upon "Archean"-like gneiss and granite. The similarity of sedimentary sequence in the Pahrump sediments, the Grand Canyon series of Arizona, and the Beltian and similar rocks of the Rocky Mountain regions suggests a major geosynclinal trough of Late Precambrian age along the western margin of the continent. This pattern was repeated during the Paleozoic and Mesozoic histories of the western United States, as is discussed in the next two chapters.

Precambrian Glaciation

A feature of Late Precambrian rocks in many parts of the world is the presence of rocks called tillites (Fig. 15.5). The Gowganda tillite of Huronian age has been mentioned previously. Its disbribution in central Canada suggests that an area of several thousands of square miles was affected by continental glaciation during Middle Precambrian time. Even more remarkable is the occurrence of very Late Precambrian glacial deposits on all continents. In North America, these deposits are known from the Pahrump rocks of eastern California and the Beltian rocks of Utah. Glacially derived sedimentary rocks are also known from Scandinavia, Greenland, India, China, Africa, and Australia. This very great distribution of glacial deposits suggests that the Late Precambrian may have been the time of greatest continental glaciation in all of geologic history.

Mineral Wealth from the Canadian Shield

It would be difficult to overestimate the importance of the Canadian Shield to the mineral industries of the United States and Canada. Precambrian rocks of various ages have been veritable storehouses of such important ores as iron, gold, nickel, copper, silver, cobalt, zinc, and uranium. Of these, iron is the chief metal produced from sedimentary rocks; all others occur in relation to igneous rocks or igneous processes. Some of the mines from which these metals have been produced have been in operation for more than one hundred years, and others are the result of more recent discoveries.

Billions of dollars worth of metalliferous ores have been produced from mines in the Canadian Shield. The details of the origin and occurrence of the many different kinds of deposits are far beyond the scope of this introductory textbook. But even in the relatively small space that can be devoted to this intriguing subject, it is possible to gain a brief insight into the discovery and development of an ore deposit.[4]

Iron. Probably no other metal is as important in our modern mechanized age as iron. In the form of steel, iron is literally the backbone of industry. Without it the wheels of industrial progress would most surely collapse. Iron is one of the most abundant metals in the earth's crust. It usually occurs in chemical combination with oxygen as the minerals hematite, magnetite, and limonite.

Iron was discovered in the Northern Peninsula of Michigan near the south shore of Lake Superior in 1844 by a land-surveying party whose magnetic compasses were thrown into wild paroxysms by the iron ore bodies near the surface. By 1865 the discoveries of new ore bodies were extended to Minnesota, a state that is still the greatest producer of iron ore in the United States. Minnesota supplies about 60 percent of all iron produced in the United States.

The iron-bearing rocks in the Lake Su-

---

[4] An ore is a rock that contains a metalliferous mineral or minerals that can be mined at a profit.

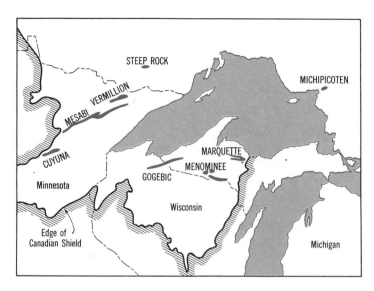

Figure 16.9 Map showing the distribution of iron ranges in the Lake Superior district. All are Middle Precambrian in age except the Vermillion Range of Minnesota which is Early Precambrian.

perior region are all of Precambrian age (Fig. 16.9). The iron ranges in Minnesota and Michigan provide about 75 percent of the total production in the United States. The iron is mined by both open pit and underground methods. The largest iron mine in the world is the Hull-Rust-Mahoning pit on the Mesabi Range (Fig. 16.10). One-fourth of all the iron ore mined in Minnesota has come out of this huge man-made cavity, which covers 1535 acres. It is 3¾ miles long, ½ to 1 mile wide, and has a maximum depth of 490 feet. From the pit the ore is transported to the surface and then goes by rail to the loading docks at Duluth and other ports on Lake Superior. From there the ore is carried by lake vessels to ports on the lower lakes that serve steel centers such as Chicago, Detroit, Buffalo, Cleveland, and Pittsburgh.

Iron ore produced from the Lake Superior district averages about 50 percent iron. The tremendous production of World War II pointed up the fact that the high-grade weathered ores of the Lake Superior district would not last forever. This realization turned attention to the large unused deposits of low-grade magnetite-bearing rocks of the region.

This low-grade ore averages about 20 to 40 percent iron and is known as *taconite*. As it occurs in nature, taconite contains too little iron for profitable exploitation, so a method was devised to increase the iron content by a concentration process whereby the magnetite was separated from such minerals as quartz and other silicate minerals. The taconite industry is rapidly gaining in importance as the high-grade reserves are depleted. By this means the reserves of ore have been greatly increased.

The origin of the iron ores of Minnesota and Michigan has long been the subject of heated debate by Precambrian geologists. The ore represents the results of a leaching process in which the initial iron-bearing rocks lost silica, thereby increasing the percentage of residual iron. The two main questions still unanswered are: (1) What kind of environment favors the deposition of iron minerals and silica (chert, SiO_2) in such large volumes? and (2) was the silica leached by (a) cold surface waters, (b) surface waters that became heated by deep circulation, or (c) water in the form of steam rising upward from underlying magmas? The reader may gain the impression that

330   THE GEOLOGICAL STORY

Figure 16.10 A view of the Hull-Rust-Mahoning open pit iron ore mine near Hibbing, Minnesota. This is the largest open pit mine in the world. Between 1895 and 1957, more than a half billion tons of ore were extracted and nearly half a billion tons of overburden (waste rock) were removed. (Courtesy of Oliver Iron Mining Division, United States Steel Corporation.)

such questions are purely academic, but the discovery of new deposits is vitally dependent on the knowledge of the origin of the known occurrences of ore.

Sedimentary iron ores in Labrador are very similar to those of the Lake Superior district and pose the same problem of origin.

Copper. The Canadian Shield produces about 10 percent of the world's copper. Most of the copper deposits are associated with igneous activity of one kind or another. Some of the copper occurs in association with other metals such as nickel, zinc, silver, and gold.

On the southern fringe of the Precambrian shield in the Upper Peninsula of Michigan, *native* copper occurs in tilted lava flows and conglomerates of Keweenawan age. Most of the high-grade deposits of native copper are now depleted, but some ore containing less than 2 percent copper is still mined. Primitive tribes of Indians used this copper for ornamental and utilitarian purposes about 3000 years ago.

The origin of the Keweenawan copper deposits is attributed by some geologists to ascending hot solutions that deposited the copper in the porous lavas and in the spaces between pebbles of the conglomerates.

Nickel. World nickel production is dominated by the famous deposits of Sudbury, Ontario. There, nickel occurs in chemical combination with iron and sulfur and in association with other metallic minerals that yield copper, platinum, gold, silver, and cobalt. The larger deposits occur near the base of a saucer-shaped basic intrusive about 35 miles long. The origin of the ore deposits is in debate, but most geologists concur that a genetic relationship exists between the basic intrusive and the ore minerals; that is, both may have been derived from the same magmatic source.

Uranium. The atomic age ushered in an intensive search for uranium minerals, and the Canadian Shield received much attention in this respect, especially the Blind River district of Southern Ontario just north of Lake Huron. The chief ore mineral is pitchblende (uranium oxide) and other associated uranium minerals that occur in a Middle Precambrian conglomerate. In spite of the fact that the percentage of uranium is extremely low, the volume of mineralized rock is so enormous that the Blind River district has one of the largest uranium-producing potentials of the world.

Another deposit of Precambrian pitchblende occurs in the veins around Great Bear Lake in the Northwest Territories. There, the uranium minerals occur with quartz as well as other compounds containing native silver, copper, and iron sulfides. The mineralization is probably related to granitic intrusives of uncertain Precambrian age.

Gold. Canada produces most of its gold from the shield, where the gold occurs in quartz veins. The province of Ontario produces about one million ounces of gold annually. The gold-quartz veins of the Porcupine district (Fig. 16.3) are related to the Algoman granites that invaded the Keewatin volcanics and pre-Algoman sediments.

The Kirkland Lake district is another gold-producing region of Canada in which Precambrian rocks were invaded by igneous rocks. Hot solutions given off by the magma were responsible for the implacement of the gold-quartz veins. At Kirkland Lake, gold is now being mined from some of the deepest mines in North America, some of which extend to depths of nearly 8000 feet.

References

Engel, A. E. J. *et al.,* 1968, Alga-like forms in Onverwacht Series, South Africa: Oldest recognized lifelike forms on earth, *Science,* V. 161, pp. 1005-1008.

Gill, J. E., 1952, *Mountain building in the Canadian Pre-Cambrian Shield,* Report of the Eighteenth Session of the International Geological Congress, Great Britain, 1948, part XIII, pp. 97-104.

* Glaessner, Martin G., 1961, Pre-Cambrian animals, *Scientific American,* V. 204, No. 3 (March), pp. 72-78. (Offprint 837, W. H. Freeman and Co., San Francisco.)

Goldich, S. S., 1958, Geochronology in the Lake Superior region, *Canadian Jour. Earth Science,* V. 5, pp. 715-724.

Harland, W. G., 1963, Evidence of Late Precambrian glaciation and its significance, in *Problems in Paleoclimatology,* Narin, A. E. M., Ed., Interscience Publishers, Wiley, London, pp. 119-149.

* McAlester, A. L., 1968, *The history of life,* Prentice-Hall, Englewood Cliffs, New Jersey, 152 pp.

Ross, Clyde P., and Richard Rezak, 1959, *The rocks and fossils of Glacier National Park: The study of their origin and history,* U. S. Geological Survey, Professional Paper 294-K, 439 pp.

Sharp, R. P., 1940, Ep-Archean and Ep-Algonkian erosion surfaces, Grand Canyon, Arizona, *Geological Society of America Bulletin,* V. 50, pp. 1235-1270.

Stockwell, C. H. (ed.), 1963, *Geology and economic minerals of Canada,* Economic Geology Series, No. 1, Geological Survey of Canada, Dept. of Mines and Technical Surveys, Ottawa, 517 pp.

* Recommended for further reading.

17 The Paleozoic Era

Restoration of a Paleozoic sea bottom during the Silurian period as seen in the Museum of Natural History, Washington, D.C. (Courtesy of the Smithsonian Institution.)

The interest in a science such as geology must consist in the ability of making dead deposits represent living scenes.
Hugh Miller

The clarity of the geologic record from the beginning of the Paleozoic to the present stands in great contrast to the vagueness of Precambrian chronology. The explanation for this lies mainly in the presence of abundant fossils in the Cambrian and later sedimentary rocks. Only by making use of the abundant fossil remains in post-Precambrian rocks has the geologist been able to correlate rocks and events in widely separated areas and thus to decipher geologic history.

Much of what follows in subsequent chapters is based on this knowledge of fossils coupled with the study of rocks and their relations to one another. The history deciphered by this means carries the threads of two stories. One is the physical evolution of the earth, the history of deposition, erosion, and orogeny; the other is the organic evolution of the earth's inhabitants, the pageant of life through the ages as revealed by the fossil record.

Although a full appreciation of the fossil story is possible only with a command of the fundamentals of the plant and animal classifications and the principles of biology, the reader can, nevertheless, appreciate the significance of the fossil record with only a rudimentary concept of botany and zoology.

Some aspects of the plant and animal classifications will be supplied in the pages that follow, but because of obvious limitations of space, many details must be left unmentioned. The reader who is intrigued with the paleontologic aspects of historical geology will find further knowledge outside the pages of this book in the selected references at the end of this and subsequent chapters.

The Problem of Paleozoic-Precambrian Definition

Having stated that the record of the Paleozoic and subsequent eras is marked by an abundance of fossils, passing reference must be made to a problem of definition that is common to all geosynclinal areas of the world, and one that has plagued geologists for many years. Each of the geosynclinal areas of the world possesses thick sequences of sedimentary rocks lying conformably beneath undoubted Lower Cambrian rocks and above undoubted Precambrian rocks. These sequences, often several thousands of feet thick, are usually unfossiliferous or contain only sparse fossil remains. Should these rocks be classified as Cambrian or Precambrian? Until some agreement can be reached as to precise biologic definitions for the top of the Precambrian and the bottom of the Cambrian, this problem will remain. In fact, it is likely that such definitions can be reached for some areas where Precambrian successions (with abundant algal remains) are directly overlain by fossiliferous Cambrian rocks; but this will not solve the problem for the geosynclinal areas in which this situation does not occur. Some investigators have proposed the introduction of new names, such as Eocambrian or Infracambrian for these unfossiliferous layers, regarding them as post-Precambrian yet pre-Cambrian divisions of the Paleozoic era.

General Setting in North America During the Paleozoic Era

The North American continent at the beginning of Paleozoic time was vastly different than it is today. Probably the most striking difference was the absence of the Appalachian Mountains of the eastern United States and the great Cordilleran Ranges of western North America. Where these mountain systems now occur, there existed geosynclines in which marine sediments accumulated throughout most of Paleozoic time (Fig. 15.4).

Between the eastern Appalachian geosyncline and the western Cordilleran geosyncline, both of which were characterized by slow but persistent subsidence, lay a broad stable region that at times during the Paleozoic was partially submerged beneath epicontinental seas.

The pioneer geologists of the United States visualized two lofty land masses, *Appalachia* and *Cascadia,* lying marginal to the two geosynclines. Appalachia lay east of the Appalachian trough, and its counterpart on the western edge of the continent, Cascadia, bordered the western side of the Cordilleran geosyncline. Both were thought to have consisted of crystalline Precambrian rocks from which the Paleozoic geosynclinal sediments were derived. Moreover, it was believed that these old landmasses extended across what are now parts of the continental shelves and ocean basins. According to this interpretation, both Appalachia and Cascadia disappeared beneath the seas.

This paleogeographic picture has been modified considerably since the classical interpretation was in vogue. The large masses of Precambrian crystalline rocks that supposedly existed in Appalachia and Cascadia have never been found; most of the supposed Precambrian rocks have now been identified as highly metamorphosed Paleozoic geosynclinal rocks. Furthermore, the seaward extensions of Appalachia and Cascadia have been disproved by geophysical studies showing that the present continental shelves and ocean marginal areas are not floored with

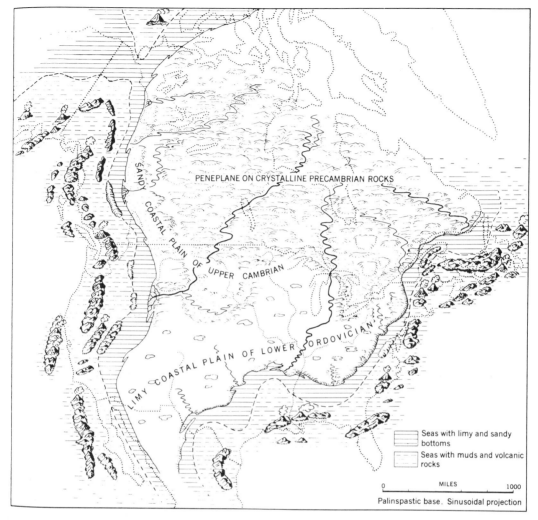

Figure 17.1 Paleogeographic map of North America showing the distribution of land and sea during the Ordovician period. (After Kay, M., 1951, *North American geosynclines,* Geological Society of America, Memoir 48, Boulder, Colorado, Plate I.)

sialic rocks as would be the case if the ancient landmasses of Appalachia and Cascadia had occurred where earlier geologists thought they did.

A more modern view now postulates an arcuate archipelago of volcanic islands in place of the ancient landmasses (Fig. 17.1). This view is supported by the presence of abundant volcanic rocks and volcanic-derived sedimentary rocks in the geosynclinal layers. In fact, a major characteristic of the Appalachian and Cordilleran geosynclines is the presence of large masses of these rocks on the margins adjacent to the volcanic source areas.

The recently developed ideas on sea-floor spreading and plate-tectonics (Chapter 7) afford another interpretation of the nature of landmasses such as "Appalachia." As the reconstructions in Figures 7.6 and 7.16

show, it is possible that "Appalachia" was in reality the northwestern portion of the African continent. The disappearance of "Appalachia" is thus to be explained as a result of the spreading of the continental masses away from the mid-Atlantic ridge.

For most of the Paleozoic it is clear that the bulk of the sediment supplied to the Appalachian geosyncline came from the east, and the source of sediments for the Cordilleran geosyncline was to the west; relatively small amounts were supplied to each from the stable interior. Only during the early history of these geosynclines is there evidence of abundant sediments derived from the stable interior. By the beginning of Ordovician time, the interior of North America ceased to be an important source of sediments for both geosynclines; from then, until the two geosynclines ceased to exist, the sediments came from bordering land areas that no longer exist. The stable interior supplied sedimentary materials in minor amounts at various times during the Paleozoic when it lay above sea level; at other times much of it was invaded and submerged by epicontinental seas.

The complete details of the stratigraphic and tectonic history of the Paleozoic would occupy several volumes. In the discussion that follows, only a few salient features of each of the Paleozoic periods are mentioned. Each period will be treated separately in terms of its general paleogeographic, sedimentary, and biologic history. The orogenic histories will be summarized by geologic eras for reasons of clarity and brevity, except where specific reference to orogenic movements is necessary to explain emergence and submergence of land areas.

Subdivisions of the Paleozoic Era

The framework of Paleozoic history was established by pioneer work in Europe, especially in the British Isles. The time units recognized in Europe have found general use in North America, except that the Carboniferous of the European terminology is replaced by the Mississippian and Pennsyl-

vanian periods. The major time units (Periods) of the Paleozoic Era are as follows:

Permian (youngest)
Pennsylvanian ⎱ Carboniferous
Mississippian ⎰
Devonian
Silurian
Ordovician
Cambrian (oldest)

Cambrian time began about 600 million years ago, and the Permian ended about 225 million years ago. As shown in Figure 15.9, the periods of the Paleozoic are of unequal length. Radioactive dating shows that the Cambrian was about twice as long as the Permian and that the Silurian was a third the length of the Cambrian. The periods are defined on the basis of *natural* geologic episodes recognized in western Europe where the rocks of those ages were first studied. Although some geologists have tried to read a rhythmic concept into the pattern of events, there is no convincing evidence that such was the case. Certainly it is clear that what is recognized as a natural break (unconformity) in the record in one place is represented by a continuous record of sedimentation in another. At best, the subdivisions should be recognized as a somewhat arbitrary method of classifying geologic time according to the rock record. The highlights of that record, period by period, form the main theme for the remainder of this chapter and the rest of the book.

The Cambrian Period

The Cambrian seas began their slow but persistent encroachment on parts of the North American continent along what we now recognize as the two major geosynclinal troughs. With minor lateral fluctuations, represented in the record by alternations of rock type, this geographic pattern persisted through the Early and Middle Cambrian. Most of the Early Cambrian is represented by great thicknesses of detrital rocks, sandstone and shale, with minor limestone and dolomite, a reflection of the shedding of detritus from the stable interior of the continent. Early Cambrian rocks are well displayed in eastern California and western Nevada and in the southern Appalachian region.

Over most of the Appalachian and Cordilleran geosynclines, this pattern of sedimentation was markedly changed in Middle Cambrian time; rocks of this age are dominantly limestone and dolomite, reflecting generally quiet conditions of sedimentation. Middle Cambrian rocks are particularly well displayed in the Canadian Rocky Mountains, but they also occur throughout the geosynclinal areas.

Before the Cambrian period drew to a close, the seaways had spread beyond the confines of the Appalachian and Cordilleran seaways to cover large areas in what is now the upper Mississippi Valley region, and across parts of Texas, Oklahoma, Missouri, New Mexico, and Arizona. Where these seas encroached on emerged land, such as the southern edge of the Canadian Shield and the Ozark region of Missouri, the initial deposits of the Late Cambrian are detrital sediments shed from these lands. By the close of the period, however, the effects of these land areas had diminished, and the latest Cambrian rocks are nearly everywhere calcareous (limestone and dolomite). The Late Cambrian sediments of the stable interior occur in layers that usually are no more than several hundreds of feet thick. The relative thinness of these layers is a reflection of the great stability of the nongeosynclinal areas. Part of the interior, however, began to experience increased subsidence; the region across parts of Texas, Oklahoma, and Arkansas was the site of an incipient trough or geosyncline, called the *Ouachita* trough. It was to be a persistent feature for much of the remainder of the Paleozoic.

Life of the Cambrian. As the curtain of time is drawn back, the Cambrian stage is revealed as a world in which a great variety of invertebrate animals thrived. One of the puzzles of early earth history is the "sudden"

appearance in the Cambrian of representatives of all the invertebrate phyla.[1] Remains of vertebrates are significantly absent and no land plants are known from Cambrian rocks.

In many Cambrian successions, especially in geosynclinal areas, the oldest beds contain remains of primitive shellfish (mollusks) and numerous tracks, trails, and burrows, undoubtedly the result of action by bottom-dwelling organisms who left no other fossil record (Fig. 17.2).

The Cambrian is best known for its abundant *trilobites* (Fig. 15.7; Fig. 17.4M and N), extinct relatives of modern crabs and crayfish, representatives of the *arthropods*. The trilobites and the arthropods are structurally among the most complex forms of invertebrates. They possess segmented bodies and jointed legs as major characteristics. Trilobites were probably the scavengers of the Cambrian seas. Most trilobites were small, a few inches or so in length, but some species attained lengths of more than a foot.

Brachiopods (Fig. 17.4 O and P) were also important inhabitants of the Cambrian seas. They are shelled animals attached to the sea bottom by a flexible stalk or *pedicle,* and they are equipped with an external shell consisting of two valves hinged together. When opened the animal extends two coiled appendages, *brachia,* to entrap food particles. Brachiopods are of minor importance in rocks younger than Paleozoic, but during the Paleozic, including the Cambrian, they comprise one of the most useful tools for correlation available to the paleontologist.

Cambrian rocks also contain rather abundant remains of the extinct *archeocyathids,* a siliceous sponge especially common in early Cambrian strata. Among plant fossils, abundant stromatolites characterize many Cambrian calcareous formations. The remarkable preservation of the impressions of more than 100 species of soft-bodied animals from the middle Cambrian Burgess shale in Alberta

[1] A phylum (plural: phyla) is a primary division of the plant or animal kingdom.

Figure 17.2 These curved grooves on a bedding plane of Lower Cambrian sandstone in the Inyo Mountains of California probably were produced by a bottom-dwelling animal.

reveals that Cambrian life was even more varied and prolific than the abundant trilobites and brachiopods might suggest.

The Ordovician Period

The contact between Cambrian and Ordovician strata in the eastern United States, and in fact over much of North America, gives no indication of widespread land emergence at the end of Cambrian time. Some paleontologists recognize a significant faunal break between the Cambrian and Ordovician, but this does not necessarily require a physical break in sedimentation as well. In many areas the Ordovician strata rest conformably on uppermost Cambrian rocks, indicating that deposition may have been continuous. In northern Wales, however, in the area where the two systems of rocks were defined originally, the Ordovician rocks lie unconformably on folded and eroded Cambrian rocks, indicating a significant break between the two systems (Fig. 17.3).

The Ordovician seaway eventually spread

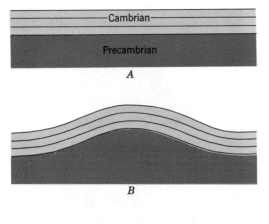

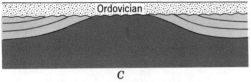

A

B

Ordovician

C

Figure 17.3 Diagram showing the development of the unconformity between the Cambrian and Ordovician strata in North Wales, England. (After Wells, A. K. and J. F. Kirkaldy, 1966, *Outline of historical geology,* 5th ed., Thomas Murby and Company, London, 503 pp.)

over much of the stable interior (Fig. 17.1). The Ordovician sea represents the greatest flooding of the North American continent during all of Paleozoic time and ranks among the largest scale continental seaways for any time throughout the world. At least half of the continent was submerged beneath the warm Ordovician seas before the close of the period, as proved by the widespread distribution of Ordovician rocks. Even part of the southern Canadian Shield was covered by Ordovician marine waters. The Cordilleran geosyncline was occupied from what is now the Arctic coast of Alaska to the Gulf of California, while the Appalachian geosyncline contained a continuous stretch of sea from Newfoundland to the Gulf of Mexico. The Ouachita trough across the southern part of the central United States was again a persistent feature.

The great bulk of Ordovician rocks consists of limestone and dolomite over the North American continent, a reflection of the warm quiescent conditions that prevailed in the seas. Significant departures from this pattern are seen in deltalike accumulations in two parts of the Appalachian region. In New York and southern Ontario, the Queenston delta, and in Tennessee and Virginia, the Blount delta represent westward tapering wedges of detrital materials derived from lands to the east. These deltas reflect orogenic activity to be discussed later in this chapter. This crustal unrest is further reflected in the stratigraphic record of the northern Appalachians by the presence of thick accumulations of late Ordovician volcanic rocks in New England and Newfoundland.

Life of the Ordovician. The panorama of living forms revealed by the fossil remains found in Ordovician rocks is truly outstanding. The organic world had reached a high level of diversification, and a remarkable number of new varieties appeared.

One of the most significant fossil discoveries in Ordovician rocks was made in sedimentary strata in Colorado. There, the fragments of the bony covering of a primitive fish are considered the earliest known vertebrate fossils. The fragmental character of the fossils provides little evidence as to the nature of these early fishes, but they resemble the more complete fossil remains of fish from the rocks of the younger Silurian and Devonian periods. From a study of the sandstone in which the Ordovician fish remains were found, geologists have concluded that these primitive fish were inhabitants of a freshwater environment.

Invertebrates. Although the appearance of the first vertebrates is a significant event in the history of life on earth, it does not overshadow the great diversity of invertebrate marine life of the Ordovician. Trilobites (Fig. 17.4K) were even more diversified and numerous than in the Cambrian, but their importance was lessened by the abundance of brachiopods (Fig. 17.4Q) which dominated the Ordovician seas. In addition, several new kinds of creatures became important during Ordovician times. Among them are the *corals* belonging to a group of animals known as the *coelenterates,* which possess the elements of digestive, nervous, and

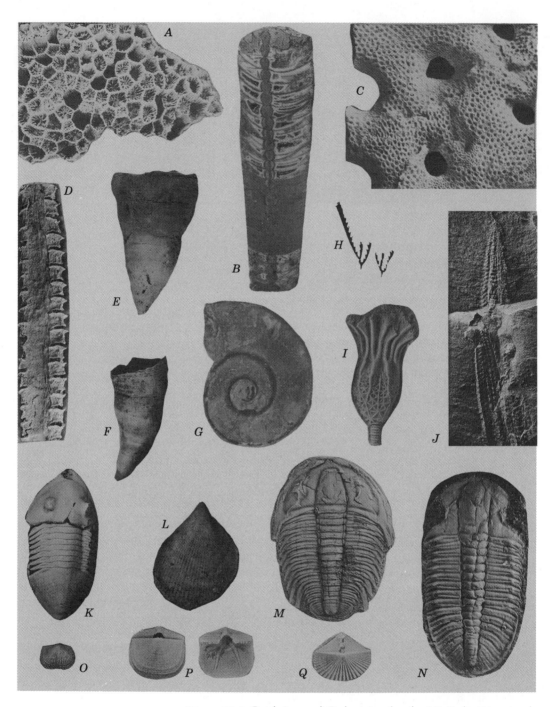

Figure 17.4 Cambrian and Ordovician fossils. (*A*) Ordovician coral, × ¾. (*B*) Ordovician cephalopod, × ¾. (*C*) Ordovician bryozoa, × ¾. (*D*) Ordovician graptolite, × 5. (*E* and *F*) Ordovician horn corals, × ¾. (*G*) Ordovician cephalopod, × ½. (*H*) Ordovician graptolite, × ½. (*I*) Ordovician crinoid, × ½. (*J*) Ordovician graptolite, χ 1½. (*K*) Ordovician trilobite, × ½. (*L*) Ordovician pelecypod, × ¾. (*M* and *N*) Cambrian trilobites, × ¾. (*O*) Cambrian brachiopod × ¾. (*P*) Cambrian brachiopod, complete specimen (left), interior of one shell (right), × 1. (*Q*) Ordovician brachiopod, × 1. (*A, B, C, D, J, M, P,* and *Q,* courtesy of W. C. Bell. *E, F, G, H, K, L, N,* and *O,* University of Michigan Museum of Paleontology. *I* from University of Minnesota Paleontological Collection.)

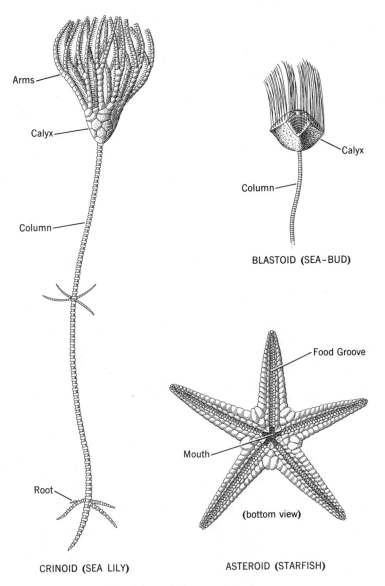

Arms

Calyx

Column

Root

CRINOID (SEA LILY)

Calyx

Column

BLASTOID (SEA-BUD)

Food Groove

Mouth

(bottom view)

ASTEROID (STARFISH)

Figure 17.5 Sketches of three different echinoderms.

muscular systems. Modern examples of co-elenterates are jellyfish and corals.

The coral is well suited to fossilization because it secretes a hard exterior cup (exoskeleton) of calcium carbonate (Fig. 17.4E, and F). All corals are attached to the bottom of the sea. Some of those inhabiting the Ordovician seas were solitary forms and conical in shape; others lived in colonies (Fig. 17.4A). The cup corals are now extinct, but colonial forms are prolific in warm shallow seas of the modern world.

The now-extinct *graptolites*, once considered members of the coelenterate clan, are now thought to be chordates, hence more closely related to vertebrates (Fig. 17.4D, H, and J). Graptolites were widely dispersed in Ordovician seas as free-floating forms, as indicated by their widespread occurrence in Ordovician shales and limestones. Their abundance and wide dispersal have made them one of the best index fossils for the correlation of Ordovician strata.

Another large group of invertebrate animals represented by many fossilized forms is known as the *echinoderms*. This group includes starfish (asteroids), sea lilies (crinoids), sea buds (blastoids), and sea urchins (echinoids) (Fig. 17.5). The echinoderms are generally characterized by a fivefold radial arrangement of appendages such as the five "points" of the starfish.

The sea lilies possess long branched arms attached to a bulbous body which in turn is fastened to the sea floor by a flexible stem or stalk. On death, the segments of the arms and stem (Fig. 17.5) generally separate and become strewn over the sea bottom by currents and scavengers. In rare cases, the nearly complete animal was buried so rapidly by sediment that the arms and stalk were not completely dismembered, as shown in Figure 17.4*l*. The crinoids became diversified in the Ordovician, although well-preserved specimens of nearly complete crinoids are extremely rare. Fossil starfish and blastoids are also very rare.

A final group of animals that became highly diversified in the Ordovician includes such well-known types as clams and snails. Collectively they are known as *mollusks*. Three categories of mollusks are (1) pelecypods (clams and oysters), (2) gastropods (snails), and (3) cephalopods (squid, octopus, and nautilus) (Fig. 17.6).

As a group, the mollusks are more advanced physiologically than the coelenterates. Besides well-developed digestive,

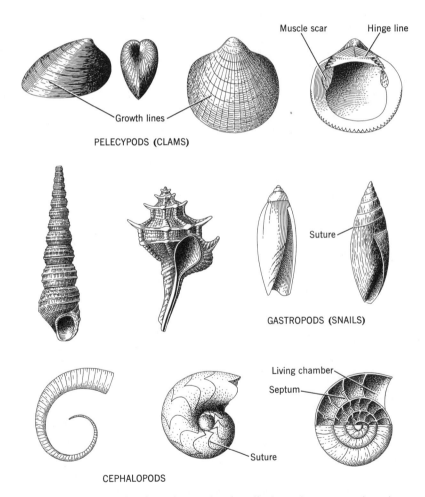

Figure 17.6 Sketches of some fossil mollusks. (After von Zittel, Karl A., 1927, *Textbook of Paleontology,* Macmillan and Company, London, 839 pp.)

344   THE GEOLOGICAL STORY

muscular, nervous, and circulatory systems, the mollusks possess good sensory organs and means of locomotion. Some mollusks have eyes and most are adapted for free movement, either on the sea floor or through the water.

Some cephalopods possess an external shell only, either in the shape of a straight cone or a coiled tube. Although superficially resembling a coiled snail shell, the shell of a cephalopod contains chambers connected by a tube. The soft part of the animal occupies only the outermost chamber, and the animal is able to propel itself by ejecting seawater through a tubular siphon. Other cephalopods possess no outer shell but contain a small internal hard part that serves as a structural element in the animal. All cephalopods possess tentacles around the mouth.

Ordovician mollusks include a large variety of snails, a few clams, and some remarkably large cephalopods, which were probably the largest animals of Ordovician time. Figure 17.4B and G illustrates two types of Ordovician cephalopods; 17.4L, an Ordovician pelecypod.

The Silurian Period

The orogenic movements that characterized the middle and late Ordovician of eastern North America left the northern part of the Appalachian geosyncline emergent long enough for the upended Cambrian and Ordovician rocks to be eroded and planed down. As a consequence, the deposits of the Silurian seas in this area are found lying with structural discordance on older rocks. Elsewhere in North America, the Ordovician-Silurian contact is generally conformable. In many places the boundary between them is difficult to define owing to the common absence of faunas found in the early Silurian strata elsewhere. It is not clear whether this absence is the result of emergence during the early Silurian, or the result of environmental barriers that precluded the development of these faunas.

Over much of the United States, Silurian

deposits bear close resemblance to the underlying Ordovician, especially in the large percentage of limestone formations, many of which represent significant coral-reef development. The Appalachian geosyncline received extensive detrital sediments, derived from mountains raised during the late Ordovician and early Silurian.

Of special interest are the evaporite deposits (salt, gypsum, and anhydrite) of late Silurian age in the northeastern part of the United States. Evaporite deposits accumulated in landlocked arms of the sea that underwent continuous supply of marine waters and excessive evaporation, very probably under conditions of aridity.

Rocks of Silurian age are well exposed in the eastern United States, not only in the Appalachian Mountains but also in the Great Lakes region and southern Manitoba. One of the most famous exposures of Silurian rocks occurs in the gorge of the Niagara River in New York and Ontario. Niagara Falls owes its existence to the resistant layers of the Lockport dolomite of Silurian age (Fig. 17.7). The falls has retreated upstream by the undermining of this resistant layer in the plunge pool of the falls. The Lockport dolomite forms what is called the Niagaran escarpment. It can be traced nearly continuously from New York State across south-

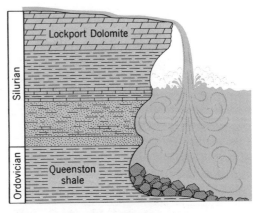

Figure 17.7 Diagrammatic cross section of Niagara Falls showing the stratigraphic units visible in the Niagara Gorge. (Based on the section described by Taylor, F. B., 1933, *Guidebook 4*, 16th International Geological Congress.)

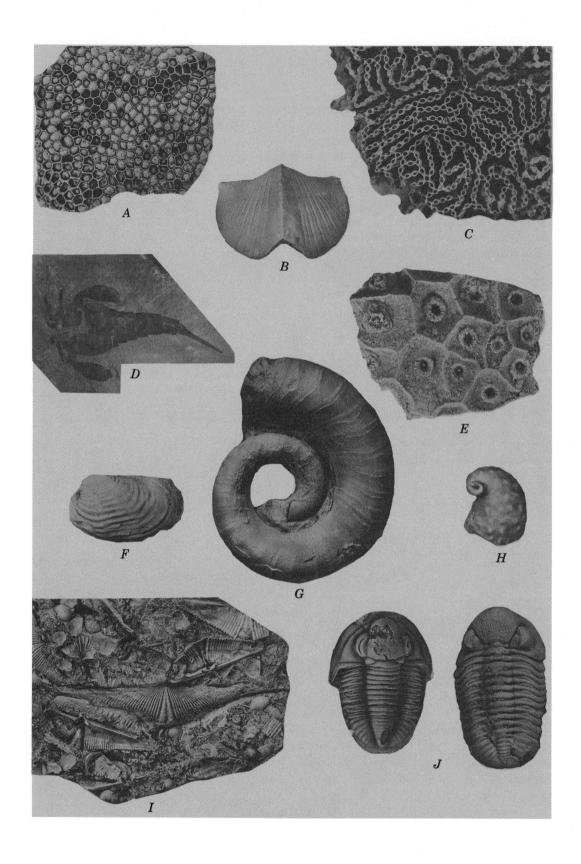

ern Ontario and into the northern peninsula of Michigan, from where it follows the west shore of Lake Michigan into Wisconsin.

The Cordilleran geosyncline received significantly smaller thicknesses of sediment during the Silurian than it had in the Cambrian and Ordovician. These deposits are generally carbonates and appear to be mainly confined to the middle portion of the Silurian.

Silurian rocks are nearly everywhere in North America parallel to overlying Devonian rocks, arguing for a quiet, nonorogenic close to the period. This is in sharp contrast to the situation in western Europe, where a major unconformity separates Silurian from younger rocks.

Life of the Silurian. Outstanding among the Silurian fossils are the *eurypterids*, scorpionlike creatures that inhabited the Silurian sea floor (Fig. 17.8D). Eurypterids are extinct today. In fact, they occupied only a very limited span of geologic time and did not survive the Paleozoic era. These creatures belong to the advanced group of invertebrates known as the *arthropods* and are distant kin to the trilobites. Although most of the Silurian eurypterids were only a few inches long, some attained a length of more than 7 feet.

The Silurian seas also contained corals (Fig. 17.8A, C, and E) of great diversity. The colonial forms grew in such profusion that they formed reefs similar to the great coral reefs fringing many islands of the South Pacific. Other invertebrates common to the Silurian were the brachiopods and crinoids; the crinoids were extremely prolific. Graptolites had already declined by Silurian time, but trilobites were still common. Fossilized fish remains are present but rare in Silurian strata.

A noteworthy new form of life that emerged during the Silurian is represented by scattered remains of very primitive land plants. Although the fossils are rare, they represent the first indication that plants existed on dry land. To what extent the Silurian landscape was covered with vegetation is not known, but probably no large trees or forests existed that early in geologic time. It is interesting to observe, however, that terrestrial plants were established in a land environment long before animals invaded the continents.

The Devonian Period

One of the best-known formations in all geologic literature is the Old Red Sandstone, a name immortalized through the writings of the Englishman, Hugh Miller, who published his account of these strata more than a hundred years ago. The "Old Red" is a conspicuous sequence of red-colored detrital sediments deposited in a number of terrestrial basins in Scotland, Ireland, and western England. These basins occurred as intermountain troughs formed by a major European orogeny, the Caledonian. As a consequence, the Old Red rocks lie with profound unconformity on the upturned edges of Silurian and older rocks.

In North America, Devonian strata are exceptionally well developed in New York State where all the early studies on Devonian rocks of North America were made.

The two major geosynclines received marine sediments throughout much of the period, although the deposits in the Cordilleran area are patchy. In middle Devonian time, however, it is clear that a continuous seaway extended from northern Alaska to southern California, and the stable interior was again inundated by marine waters in which calcareous deposits accumulated.

Beginning with the middle Devonian,

Figure 17.8 Silurian and Devonian fossils. All are about one-third natural size. (A) Silurian honeycomb coral. (B) Devonian brachiopod. (C) Silurian chain coral. (D) Silurian eurypterid. (E) Silurian colonial coral. (F) Devonian pelecypod. (G) Devonian cephalopod. (H) Devonian gastropod. (I) Devonian brachiopods. (J) Devonian trilobites. (A, B, C, E, F, G, H, I, and J, courtesy of G. M. Ehlers and E. Stumm. D, courtesy of R. Sloan.)

Figure 17.9 This restoration of a Devonian sea bottom shows various forms of invertebrate marine animals that lived in relatively shallow water. (Diorama from the University of Michigan Museum of Paleontology.)

the northern part of the Appalachian geosyncline experienced an influx of detrital material of great thickness from an eastern source. These deposits are known as the Catskill red beds of deltaic origin, and are closely analogous to the Old Red Sandstones of Great Britain.

The distribution of Devonian sediments in Europe and North America has been used to reconstruct the distribution of those two continents according to the hypothesis of continental drift (Fig. 7.7). Such reconstructions are of considerable importance in testing the hypothesis of continental drift.

Life of the Devonian. The Devonian period is commonly referred to as "the Age of Fishes." This designation is appropriate because, for the first time in the history of

the earth, a group of vertebrate animals occupied a prominent part of the organic world. To be sure, the invertebrates were present in great numbers, and fossil crinoids, corals, and brachiopods are of great importance in Devonian rocks (Fig. 17.8*B, F-J*; Fig. 17.9). But the student's attention cannot help being focused on this new and quite diversified group of animals, the fishes.

Fishes are cold-blooded aquatic animals that breathe by means of gills. Many of the Devonian fishes were bizarre in comparison with modern types. Some possessed no moveable jaws but instead had a "vacuum-cleaner" type mouth, which suggests that the animal lived near the bottom and gathered his food by a sucking process. These same primitive vertebrates were flat hori-

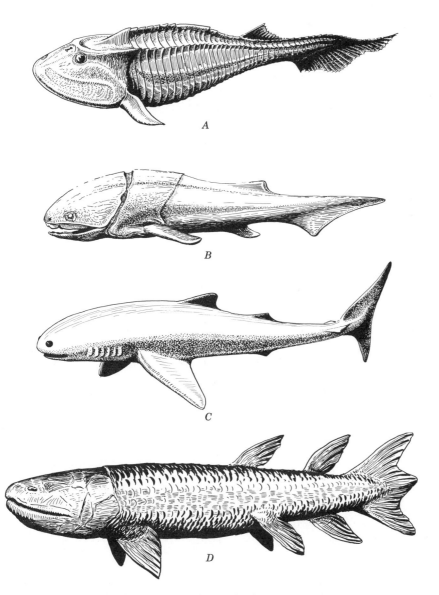

Figure 17.10 Devonian fishes (*A, B,* and *D*) and a shark (*C*). *A,* Ostracoderm; *B,* Placoderm, and *D,* Crossopterygian. (After Colbert, E. H., 1955, *Evolution of the Vertebrates,* John Wiley and Sons, New York, 479 pp.)

zontally rather than vertically, further suggesting bottom-dwelling habitat.

Many of the Devonian jawless fishes had no scales. Their bodies were covered instead with bony plates. The entire early group of jawless and armoured fishes, known collectively as *ostracoderms* (Fig. 17.10A), was indeed diverse and varied. The *placo-*derms (Fig. 17.10B), on the other hand, had primitive jaws with sharp bony projections that functioned as teeth. The placoderms flourished briefly, and most of the Devonian forms did not survive the period, although some attained gigantic size.

Two other groups of fishes appeared on the Devonian scene, the *sharks* and *bony* fishes. The Devonian sharks had stream-lined torpedo-shaped bodies that were supported by skeletons of cartilaginous material (Fig. 17.10C). Then, as now, sharks were confined to a saltwater environment.

The bony fishes possessed heavy scales and a skeleton of true bony material. They had true teeth embedded in articulating jaws. The first fossil bony fishes were discovered in fresh water Devonian sediments, but marine forms are also known.

One group of bony fishes, the *lungfishes,* was apparently able to breathe air when the lakes or rivers in which they lived became stagnant during part of the year. This deduction is based on the similarities between the skeletons of the Devonian lungfish and the modern varieties of these strange creatures that today live in Africa, Australia, and South America. Although the ability to breathe air for extended periods suggests that the lungfish represents an intermediate evolutionary stage between fishes and terrestrial vertebrates, paleontologists consider the living lungfishes and their Devonian relatives (Dipnoi) too highly specialized to be in the direct line of descent from fishes to land animals.

A related group of air-breathing fishes that is the most likely intermediate form between fishes and land-living vertebrates is the *crossopterygians,* or lobe-finned fishes (Fig. 17.10D). The reasons for this statement lie in the many similarities between the crossopterygians and the earliest land dwellers. The Devonian specimens of this group have skulls that closely resemble the skulls of early land vertebrates, and their teeth possess a structure similar in many details to those of their dry-land successors. But above all, the crossopterygians possessed paired lobate fins with a structure that is suggestive of the limbs of early amphibians, the next higher category of vertebrate animals.

Sometime during the Devonian period the transition from water to land was accomplished, an evolutionary change that must be regarded as one of the greatest advances of the organic world. Paleontologists working with Devonian strata in Greenland have discovered the remains of undoubted amphibians bearing many structural features similar to the crossopterygian fishes. These discoveries confirm the important transition of animals from water to land during the Devonian Period.

The statements made occasionally that this or that kind of animal appeared "suddenly" in one geological period or another leaves the unwary reader with the impression that these new forms literally sprang into existence overnight. What is meant is that new fossil forms appear suddenly in the geologic record; since it is known that the geologic record is very fragmentary, the sudden appearance of a new kind of fossil does not necessarily mean that this marks the first occurence of that creature in time. Its previous history might well be lost forever or may yet be uncovered through new discoveries.

Terrestrial Plants. As if the magnificent development of fishes during the Devonian was not sufficient to set this period off from all the rest, still another noteworthy fact is associated with it. Trees from the Old Red Sandstone and other Devonian strata represent the first large land plants. Possibly the Devonian landscape was "forested," although this is speculative. Certainly it is not beyond the realm of possibility that dense stands of primitive trees existed in localities favorable to their growth.

The Mississippian Period

One is likely to become bored with the repeated statement that the Paleozoic seas advanced and then retreated, only to re-advance and withdraw again. Nevertheless this repetition of spreading and shrinking seas further emphasizes the fact that this earth is not a static body and that the features of the continents seen today are transient when viewed in the perspective of geologic time.

As with the previous periods, the Mississippian was a time of widespread encroachment of marine waters across North America. Both the Appalachian and Cordilleran geosynclines were inundated, although the effects of Devonian crustal movements are seen in the fact that the northern Appalachians were not the site of significant marine sedimentation after the Devonian. The Middle Devonian was a time of especially widespread marine conditions with much of the interior submerged; the seaway extended into Alberta, British Columbia, and northern Alaska.

In the Rocky Mountain region, a vast sheet of limestone was deposited. Known as the *Madison Limestone* in the northern Rockies, and the *Redwall Limestone* in the Grand Canyon, this widespread limestone indicates that a remarkably uniform environment prevailed for the central interior of the United States where the Mississippian limestones are more than 2000 feet thick in places.

Eastward from the central region, the limestones give way to greater thicknesses of sandstone and shale, a result of increasing orogenic activity in the Appalachian region. Some of the most prominent ridges of the Ridge and Valley Province of the Appalachian Mountains are developed on the upturned edges of Mississippian sandstone formations.

Mississippian rocks in Arkansas and Oklahoma give evidence that renewed crustal activity in the Ouachita trough region was taking place. Mississippian and younger Paleozoic rocks there are of significantly greater thickness and more detrital in character, testifying to more rapid accumulation in a subsiding basin.

The Mississippian limestones of Kentucky and Missouri contain a remarkable network of underground caverns, of which Mammoth Cave is the most prominent. The Indiana limestone of Mississippian age has been extensively quarried for many years and used as a building stone, especially where intricate carvings and scroll work are required. Other economic products associated with Mississippian strata are discussed briefly at the end of this chapter.

Life of the Mississippian. The warm seas of Mississippian time provided an environment well suited for the development of a great variety of marine forms. The crinoids (Fig. 17.11*E, F,* and *G*) were especially diverse and abundant members of Mississippian organic communities. So prolific was

the crinoid population that their fossil remains contributed significantly to the bulk of some Mississippian limestones. A related form, the blastoids, were also well represented (Fig. 17.11*B*).

Another significant contributor to the making of Mississippian sedimentary rocks were the *bryozoans* (Fig. 17.11*C,* and *J*), minute colonial organisms that first appeared in the Cambrian but did not become important in the fossil record until mid-Paleozoic times.

Cephalopods of the primitive ammonoid variety became significant in the Mississippian (Fig. 17.11*A*), heralding a major diversification in later rocks.

The simplest of all animals, the *protozoans,* were extremely abundant in the Mississippian seas. One category of the protozoans known as *Foraminifera*, tiny one-celled organisms with calcareous external shells, were so abundant in Mississippian time that the accumulation of their remains on the sea floor became discrete masses of limestone. Although the foraminifera date from the Cambrian, they did not become prominent rock formers until the Mississippian.

The Pennsylvanian Period

The Mississippian and Pennsylvanian periods, together, comprise what is known as the *Carboniferous* in Europe, so named because of its content of coal formations. In the United States, the Pennsylvanian was characterized by widespread development of terrestrial swamps, an environment well suited for the growth of luxuriant vegetation. The remains of the forest trees and other types of vegetation went into the making of the thick beds of coal found in central and eastern North America as well as many other parts of the world. Most of the major coal fields of the world had their origin in the Pennsylvanian period.

Although much of the eastern and southern part of the North American continent was emergent during the early Pennsylvanian, a result of the orogenic activity that was to culminate late in the period, marine sediments were accumulating in great thick-

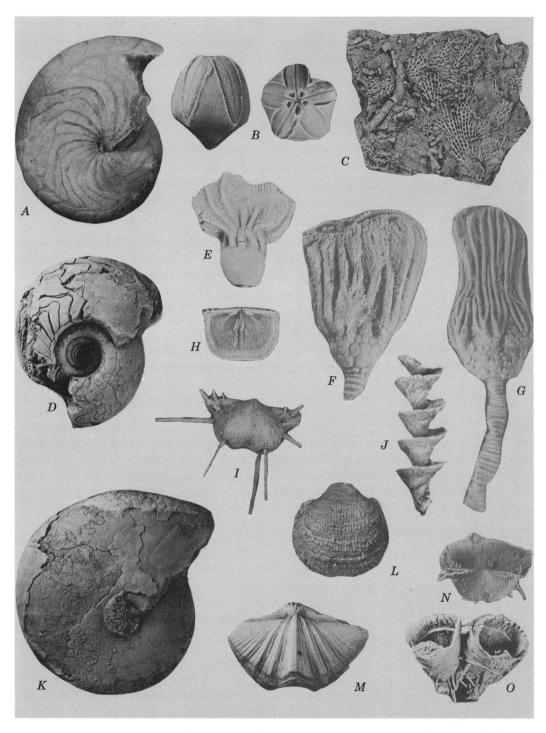

Figure 17.11 Mississippian, Pennsylvanian, and Permian fossils. (*A*) Mississippian cephalopod, × ¾. (*B*) Mississippian blastoid, × ¾. (*C*) Mississippian bryozoa, × ¾. (*D*) Pennsylvanian ammonite, × ¾. (*E, F,* and *G*) Mississippian crinoids, × ¾. (*H*) Pennsylvanian brachiopod (interior), × 1. (*I*) Permian brachiopod, × ¾. (*J*) Mississippian bryozoa, × ¾. (*K*) Permian ammonite, × ¾. (*L* and *M*) Pennsylvanian brachiopods, × ¾. (*N* and *O*) Permian brachiopods, × ¾. (*A, C, G, I, J, L,* and *M,* from University of Michigan Museum of Paleontology. *B, D, E, F, K, N,* and *O,* W. Charles Bell.)

nesses, especially in Arkansas, Oklahoma, and Texas. Similarly, widespread marine inundation took place over much of the Rocky Mountains, the Colorado Plateau, and the southern Great Basin of the western United States.

The coal beds of the central and eastern United States present a special problem as to style of sedimentation. Especially in the coal fields of the Illinois basin, the coal layers occur in repeated successions called *cyclothems* in which alternation of marine and nonmarine shale, sandstone, and limestone occur with monotonous regularity. Approximately 100 of these cyclothems have been reported in Illinois, and many of the individual beds of the cyclothems can be traced for large distances laterally. Cyclothems occur also in Pennsylvania and Kansas. These peculiar deposits reflect alternations of marine and non-marine conditions over a very large area of the United States. These oscillations have been explained by some as exceptionally (in the geologic time sense) rapid tectonic pulses in both the source and depositional areas. Others, convinced that the Permo-Carboniferous glacial episodes (to be discussed later) are largely Pennsylvanian in age, suggest that the fluctuations in sea level required to explain the cyclothems are due to the waxing and waning of the continental ice sheets in the Southern Hemisphere.

Life of the Pennsylvanian. Fossil remains in Pennsylvanian rocks represent several significant evolutionary advances over those of the Mississippian. Of these, the domination of the coal forest types and the appearance of the earliest reptiles deserve special mention.

Marine invertebrates included members of many of the previously mentioned categories, especially brachiopods (Fig. 17.11*H*, *L*, and *M*), cephalopods (Fig. 17.11*D*), and a foraminiferan known as the *fusuline,* an elongate rice-shaped form. These fusulines are exceedingly abundant and diverse in Pennsylvanian and Permian strata and are important index fossils for small subdivisions of both periods.

Land animals were also moderately abundant. The amphibians increased in numbers and varieties, and the insects attained gigantic sizes. Cockroaches, spiders, scorpions, and the many-legged centipedes were common inhabitants of the low swamp lands bordering the Pennsylvanian seas. A winged insect resembling the modern ''dragonfly'' had a wing span of more than 2 feet. The fossiliferous remains of small primitive reptiles make their appearance in late Pennsylvanian rocks, but they are extremely rare.

Plants. The hundreds of species of insects and the sprawling forms of amphibians are impressive, but they do not overshadow the ancient trees and ferns of the Pennsylvanian landscape. The Pennsylvanian period is appropriately known as the Coal Age, and the coal measures yield a fossil flora that is unlike anything before or after. The trees were primitive types, most of which have long been extinct, but a few of them are represented in the modern plant kingdom by small related forms.

The ferns were perhaps the most common inhabitants of the Pennsylvanian coal swamps, having reached their climax of speciation. In general appearance, the ferns of late Paleozoic time resemble the modern ferns that flourish in the shade of the present-day forests. Both the primitive ferns (*Pteridophytes*) (Fig. 17.12*E*) and the extinct ''seed'' ferns (*Pteridosperms*) (Fig. 17.12*G-I*) are represented in fossil plants of Pennsylvanian age. The seed ferns reproduced by bearing seeds, an evolutionary advantage over Pteridophytes, which reproduced by means of spores.

The modern horsetail or scouring rush is all that remains of a tribe of giant coal measure plants known as *Calamites.* Fossilized remains of these plants reveal that they consisted of a single-jointed trunk with vertical ribs and no branches save a ring of short leaves at each joint (Fig. 17.12*B, D*).

The great ''scale'' trees of the Pennsylvanian swamps were probably the most unusual plants of that time. One group, the *Sigillaria,* was characterized by the hexagonal markings on the trunk (Fig. 17.12*C*). These represent cushions or pads on the trunk from which the leaves became detached as it increased in height.

Another one of the scale trees was the *Lepidodendron* (Fig. 17.12*F*). Its leaf scars

THE PALEOZOIC ERA   353

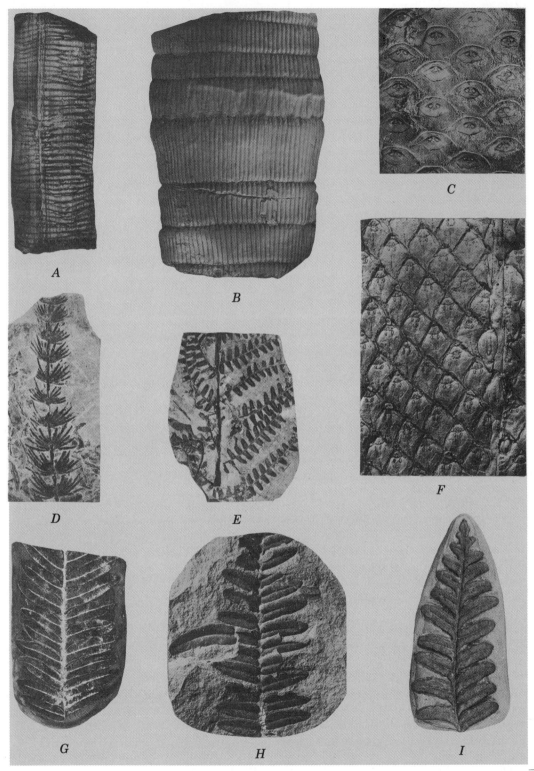

Figure 17.12 Pennsylvanian plant fossils, about one-third natural size. (*A*) Cordaites. (*B*) Calamites. (*C*) Sigillaria. (*D*) Calamites foliage. (*E*) Fern. (*F*) Lepidodendron. (*G, H,* and *I*) Seed ferns. (Reproduction by permission of McGraw-Hill Book Co., from Arnold, C. A., 1947, *An introduction to paleobotany*, McGraw-Hill Book Co., New York, 433 pp.)

were diamond shaped and arranged in diagonal rows that spiraled around the trunk. *Lepidodendron* and *Sigillaria* attained heights of more than 100 feet; *Sigillaria* had no branches, and *Lepidodendron* had a system of paired branches.

The only Pennsylvanian trees that resembled modern trees in general form were the *Cordaites* (Fig. 17.12A). The cordaites were the forerunners of the modern conifers (cone-bearing trees). They averaged 1 to 2 feet in diameter and had a crown of branches with leaves and seeds. Cordaites were about 50 feet high when mature.

There have been many attempts by artists to recapture the arboreal scenes of the Pennsylvanian forests through restorations based on fossil plant remains. Unfortunately, however, because a plant has no counterpart of the skeleton of an animal, it does not lend itself to good preservation. Furthermore, land plants are prone to dismemberment on death, a fact that accounts for the lack of complete plant fossils. But in spite of the rarity of complete specimens, the character of the coal-measure forests has been fairly well established by paleobotanists using fragments of leaves, foliage, roots, and other incomplete specimens.

The Permian Period

The contrast between the Pennsylvanian and Permian periods in terms of climate and physical conditions is very great. In North America, no marine sedimentation of any consequence occurred east of the Mississippi Valley in the Permian. In addition, much of the sediments in the western part of North America are of terrestrial character and contain extensive red beds and evaporites, although important marine deposits occur widely in the far west. Permian strata are exceptionally well exposed in western Texas and New Mexico, and in the Grand Canyon region where they comprise the upper layers of that succession.

Significant climatic change can be seen in the change from the luxuriant foliage of the coal-producing swamps of the Pennsylva-nian to conditions of local aridity as exemplified by the evaporites in the Permian beds of Kansas and Texas. These evaporites are especially well developed in the Guadalupe Basin, which encompassed what is now western Oklahoma, western Texas, and part of New Mexico. The basin was apparently open to the sea, providing repeated influx of marine waters to supply the salts produced by evaporation under arid conditions. Beneath the evaporites occur several thousand feet of shales and limestones, many of which are richly fossiliferous. Some of the best preserved Permian fossils come from the Glass Mountains of western Texas where Permian strata are well exposed.

The Permian *Capitan* limestone of western Texas and southeastern New Mexico is an ancient *reef,* several miles wide, hundreds of miles long, and thousands of feet thick. Modern reefs are accumulated masses of lime-secreting marine algae and corals that thrive in warm shallow seas. The ancient Capitan reef possesses nearly all the features common to such modern reefs as Bikini atoll in the Pacific Ocean. This observation leads to the conclusion that modern and ancient reefs formed under similar environmental conditions.

Permo-Carboniferous Glaciation. A major feature of the Permian and Pennsylvanian Periods is the well-documented continental glaciation of the Southern Hemisphere. (This subject was discussed briefly in Chapter 7). Lithified tills, called *tillites,* have been found all across the Southern Hemisphere in Brazil, Africa, Madagascar, Australia, and Antarctica. The only known Northern Hemisphere occurrence for tillites of this age is in India. Although precise geologic dating of the glacial deposits is difficult, there is little question that they are of late Paleozoic age.

The *Dwyka* tillite is a lithified Permian morainal deposit in South Africa that contains striated boulders. The ancient crystalline rocks over which the Permian glaciers moved still bear the grooves and markings left by the ice sheet. Some of the glacial deposits in Africa are found at a present latitude of 10° south, too close to the equator for the development of lowland continental glaciers. In addition, some of the tillites of South America and Australia indicate derivation by ice from areas now occupied by ocean basins. These anomalies are explained best if a continental arrangement, such as is shown in Figure 7.8, is imagined. This evidence is one of the strong arguments in support of continental drift. It is more plausible to regard the Dwyka and equivalent tillites as the product of a single ice cap centered in southern latitudes, perhaps close to the Antarctic of that time, rather than as the result of separate ice sheets of localized extent.

Life of the Permian. By Permian time, both amphibians and reptiles were well established on land. Amphibians are vertebrates born into an aquatic world that invade the terrestrial environment at maturity. Although amphibians date back to the Devonian, they were not abundant until the Pennsylvanian and Permian periods. The best-known Permian form is *Eryops,* a sprawling amphibian of modest size, perhaps not more than 6 or 8 feet in length (Fig. 17.13).

Figure 17.14 The skeleton of *Edaphosaurus,* a finbacked Permian reptile from Texas. The function of the long spines is not clearly understood by paleontologists. (University of Michigan Museum of Paleontology.)

One odd-looking Permian reptile, *Edaphosaurus* (Fig. 17.14), possessed a row of long spines down its back. Paleontologists are uncertain as to the function of this anatomical structure. One idea is that the spines were connected by a web of skin that may have functioned as a temperature-regulating mechanism for the animal.

Permian invertebrates provide us with the last glimpse of many marine creatures that became extinct by the end of the Paleozoic. The trilobites, horn corals, many of the crinoids, all of the blastoids, and the fusulines did not survive the Permian period. On the other hand, some of the brachiopods lost many of their older characteristic features, especially the pedicle. The Permian brachiopods possessed spiny appendages that served as a means of attachment, either to the sea bottom or to other brachiopods (Fig. 17.11*I, N,* and *O*). These distinct forms grew in such profusion in the Permian seas of western Texas that the rocks in which they occur as fossils consist almost wholly of brachiopod shells cemented together by limestone.

Cephalopods were well represented in the Permian seas by a group known as the *ammonites* (Fig. 17.11*K*). The ammonite shell is separated into chambers, and the fleshy part of the animal occupies the outer

Figure 17.13 *Eryops,* a Permian amphibian. (Restoration after Colbert, E. H., 1955, *Evolution of the vertebrates,* John Wiley and Sons, New York, 479 pp.)

Reptiles are cold-blooded vertebrates like the amphibians but are able to begin life on land without the necessity of a transitional period in an aquatic habitat. Reptiles are known from the Pennsylvanian, but by Permian times they are fairly well represented by fossil remains.

and largest chamber only. Early Paleozoic cephalopods, or *nautiloids,* had simple walls (septa) between chambers. The ammonite septa, on the other hand, were highly irregular in shape, and where the edges of the septa were attached to the shell wall, an irregular line or *suture* was formed (Fig. 17.6). Permian ammonites with highly intricate sutures provide a means of correlating formations from widely separated areas.

Permian Plant Life. Many of the Pennsylvanian forms such as the scale trees and calamites persisted until early Permian time, but eventually they became extinct, possibly as a result of the aridity and cold that prevailed over large expanses of the Permian world.

Probably the best-known and certainly one of the most controversial groups of plants associated with the geologic record is the *Glossopteris* flora. These plants were seed ferns with a very distinctive foliage, and their fossil remains are associated with the Permian beds of the Southern Hemisphere, notably South America, Australia, Antarctica, and Africa, as well as with India in the Northern Hemisphere. The Australian tillites of Permian age yield spores that are the same as the ones that occur elsewhere with the Glossopteris flora living at the time of the Permian glaciation in the Southern Hemisphere.

This fact is used as an argument in favor of the existence of the continent of Gondwana (Fig. 7.16). Those who support the hypothesis of continental drift argue that the widely dispersed Glossopteris flora can be explained better by a physical connection between the continents of the Southern Hemisphere rather than by "land bridges," or chains of islands that provided migratory paths for both plants and animals.

Orogenic History of the Paleozoic

It has been conventional to treat the orogenic history of the Paleozoic in North America as having been largely confined to the Appalachian Mountains and to consider three discrete and separate episodes, Taconic, Acadian, and Appalachian (see Fig. 15.9). It is true that most of the Paleozoic tectonic activity took place in the Appalachians, but it is now known from studies there and elsewhere that the history was far more complex than was previously thought.

Evidence for orogeny is drawn from a variety of geologic relationships. In the regions undergoing active tectonism, the most common evidence is an angular unconformity between two datable sequences of rock. Intrusive granitic rocks or accumulations of volcanics are regarded as so unique to the process of mountain building that their presence can be taken as evidence of orogeny. In sedimentary successions far removed from the site of orogeny, crustal activity is deduced from the presence of deltaic deposits, or *clastic wedges,* which thicken in the direction toward the orogenic region. Armed with such a variety of evidences, geologists have succeeded in unraveling the complex orogenic histories of many regions.

Orogenic activity in the Paleozoic of North America occurred over a time span from at least middle Ordovician to the end of the period. The Blount (middle Ordovician) and Queenston (late Ordovician) deltas have already been mentioned. Dated granitic intrusions and volcanic rocks of Ordovician age occur in several places, and a number of well-documented unconformities, such as the one that truncates Ordovician strata in New York (Fig. 17.15), testify to a major episode of orogenic activity. This is what has been called the *Taconic* orogeny; it produced complex structures, including important overthrust faults in much of the New England region. Although it is generally regarded as having terminated near the end of the Ordovician, the presence of significant volcanic rocks in the early Silurian of New England, as well as the coarse detrital rocks common to the Silurian deposits of much of the Appalachian Mountains, is evidence that the Taconic orogenic episode extended beyond the Ordovician-Silurian boundary.

A minor unconformity in northern Maine and Quebec, and late Silurian detrital rocks in the Appalachians represent the only evidence for tectonic activity at the time of the very widespread European orogeny termed the *Caledonian.* This orogeny produced pro-

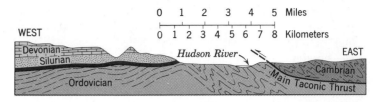

Figure 17.15 Geologic cross section of New York showing the results of the Taconic orogeny. The Ordovician beds were folded as the Cambrian strata were thrust westward. The Silurian beds lie unconformably on the Ordovician. (After King, P. B., 1951, *The tectonics of middle North America,* Princeton University Press, Princeton, N. J., 203 pp.)

found unconformities in the record of the British Isles and Scandinavia, and it was also responsible for the creation of the basins into which the Devonian Old Red sediments were deposited.

The *Acadian* orogeny affected a large area of the New England and Canadian Appalachians. It probably represented the major deformational episode in the Appalachian chain and is characterized by severe metamorphism, large-scale intrusive bodies, abundant volcanics, and the previously described Catskill delta in New York along with similar deposits in New Brunswick. Various lines of evidence from unconformable relationships, radiometric dating, and the age of the detrital rocks suggest that the orogenic episode lasted from early in the

Devonian to the first part of the Mississippian period.

Post-Acadian orogenic movements in the Appalachian Mountains have generally been called the *Appalachian* orogeny (Fig. 15.9), which is based, in part, on an earlier assumption that the movements during late Paleozoic time were mainly responsible for all structures of the Appalachians. It is now known that the Appalachian orogeny which caused the folds and thrusts of the Valley and Ridge Province (Fig. 17.16; 17.17), was not as significant in construction of the Appalachian mountain system as was the Acadian orogeny. The Appalachian orogeny is recorded in clastic wedges of late Paleozoic age, folds and faults of the Valley and Ridge Province, and dated in-

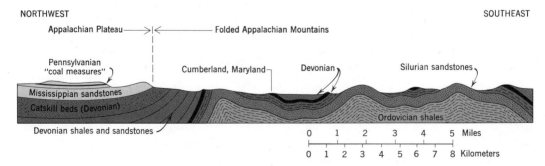

Figure 17.16 Geological cross section of Maryland showing folded Paleozoic beds of the folded Appalachian Mountains. (After Butts, Charles, 1933, *Guidebook 3,* 16th International Geological Congress.)

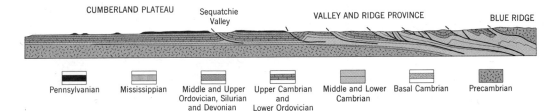

Figure 17.17 Geological cross section of Tennessee showing folded and faulted Paleozoic strata produced by the Alleghanian Orogeny. (After King, P. B., 1959, *The evolution of North America,* Princeton University Press, Princeton, N. J., 190 pp.)

trusive rocks that span a period from mid-Mississippian to earliest Permian time.

The "Appalachian" orogeny thus appears to have been no more than the last phase of a very long, perhaps even continuous, period of mountain building throughout much of Paleozoic time in the region of the Appalachian Mountains. Accordingly, it has been proposed that the complete orogenic cycle from mid-Ordovician through the Permian be termed Appalachian, and that the final phase be referred to as the *Alleghanian* orogeny.

The sporadic nature of the numerous orogenic pulses during the Appalachian period of mountain building is well illustrated by the temporal and geographical distribution of their sedimentary results, the deltas or clastic wedges (Fig. 17.18).

Not all Paleozoic deformation was confined to the Appalachian region. In the area of the Ouachita trough, across parts of Arkansas, Oklahoma, and Texas, large thicknesses of coarse detrital rocks of late Paleozoic age, and fold-and-fault structures equal to those of the Appalachians, indicate crustal unrest over a period from early Mississippian to the end of the Pennsylvanian. This activity has been named the *Ouachita* orogeny in the area described. Similar fold-and-thrust structures that were formed in the Marathon Mountains of west Texas at the close of the Pennsylvanian are ascribed to the *Marathon* orogeny. Because there is a strong possibility that the structures across the southern part of the United States may have had an actual physical connection with the Appalachians, the Ouachita and Marathon orogenies were probably synchronous with the Alleghanian episode. All three represent the final pulse of the Appalachian mountain building sequence.

The late Paleozoic was a time of crustal unrest in the western United States as well. Deep basins with thick detrital accumulations and fold structures in northern New Mexico and Colorado indicate the presence of orogenic pulses dating from late Mississippian to early Permian time. The structures produced by this activity have been called the Ancestral Rockies or the Colorado Mountains.

The Cordilleran geosyncline also experienced orogenic activity during the Paleozoic. In central and western Nevada over large areas, geosynclinal rocks were strongly deformed and thrust eastward over their eastern counterparts, perhaps as much as 50 miles. Although precise dating of this significant orogeny, referred to as the *Antler* orogeny, is not possible everywhere, in some places it can be dated as earliest Mississippian.

Mineral Resources in Paleozoic Rocks

Raw materials are the lifeblood of production in this highly industralized age. Not only does the industrial world need mineral resources for fabrication into useful products but it also needs tremendous quantities of fuel to supply power for all sorts of domestic and industrial needs. Both fuels

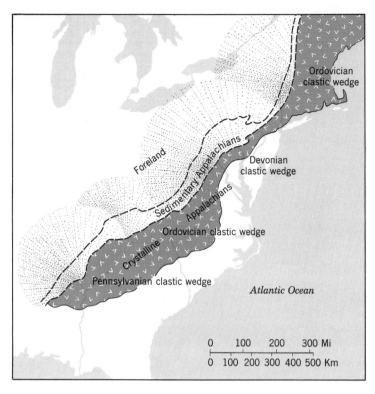

Figure 17.18 Paleogeographic map of eastern North America showing location of clastic wedges formed during various periods of the Paleozoic Era. (After King, P. B., 1959, *Evolution of North America,* Princeton University Press, Princeton, N. J., 190 pp.)

and raw materials are of geologic origin, and they occur in rocks of all geologic ages on all continents.

Fuels. A fuel is a combustible material that produces heat; among the common fuels are wood, coal, gas, and oil. The last three are produced by geologic processes and warrant some discussion here, although full treatment cannot possibly be given in a text of an introductory nature.

Coal. Coal is a rock consisting of plant material that has been chemically and physically altered in varying degrees. The physical and chemical changes account for the different kinds or *ranks* of coal. Three major ranks are recognized: lignite (lowest), bituminous (intermediate), and anthracite (highest). Increase in rank reflects natural processes whereby the carbon content of the coal is increased while the volatiles (chiefly hydrogen and oxygen) are expelled.

From a study of fossil plants associated with coal beds, geologists have concluded that coal represents an aggregation of vast quantities of partially decayed plants that accumulated in the places where they grew. Many different environments suitable for such an accumulation of plant material are known, such as coastal swamps, inland lakes and swamps, high moors, and deltas, but it is generally agreed that the *major* coal fields had their origin in coastal freshwater swamps. This conclusion is based on the fact that extensive coal beds are interbedded with marine strata, indicating an alternation of marine and freshwater conditions common to coastal areas.

The change from lignite to anthracite is a

metamorphic process in which both physical and chemical changes occur. Generally speaking, the older the coal, geologically, the more likely it is to be of higher rank. Pressure on the coal beds, induced by folding, accentuates the metamorphic process and increases rank. Anthracite beds in eastern North America are all associated with strongly folded sediments of Pennsylvanian age, whereas the bituminous coals, also of Pennsylvanian age, occur in gently folded strata. Sedimentary strata containing coal beds are called "coal measures," and they consist of alternating beds of sandstone, shale, and in some cases, limestone.

The Appalachian Coal Field in the eastern United States contains most of the coal produced in North America and includes the states of Pennsylvania, Ohio, West Virginia, and Virginia, as well as Kentucky, Tennessee, and Alabama (Fig. 17.19). Both anthracite and bituminous coals are mined in the Appalachian Field.

The Eastern Interior Field of the United States yields Pennsylvanian age coals that are mined by open pit methods in Illinois, Indiana, and Kentucky.

The coal reserves of the United States are sufficient to last for hundreds of years at the present rate of consumption, and al-

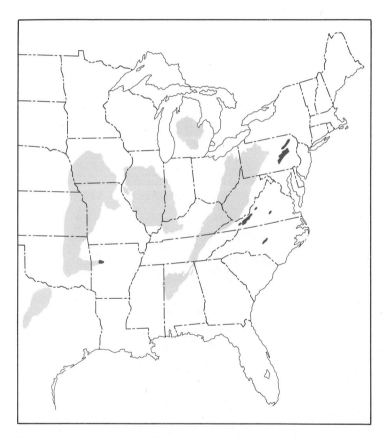

Figure 17.19 Pennsylvanian coal fields of the eastern United States. Bituminous coal occurs in the light-colored areas, and the regions of anthracite coal are shown in dark color. (After Averitt, P., 1967, *Coal resources of the United States, January 1, 1967,* U.S. Geol. Survey Bull. 1275, U.S. Government Printing Office, Washington, D.C., 116 pp.)

though the use of fuel oil and natural gas have made great inroads on the coal industry, coal is still one of the greatest sources of energy in the industrial world.

Oil and Gas. The origin and occurrence of oil and gas is discussed in Chapter 19. In the United States a great deal of crude oil is produced from Paleozoic rocks, but the bulk of the world's oil production comes from rocks of younger geologic age, especially the Tertiary.

Salt and Brine. In ancient times, salt (sodium chloride) was used only as a food preservative and for food seasoning, but in the twentieth century this domestic use has been greatly overshadowed by the use of salt and associated minerals in the chemical industry.

Abundant salt occurs in sedimentary beds that range in age from the Silurian to Recent.

Although the Permian was perhaps the period of greatest salt deposition, most of the production in the United States comes from the Silurian deposits of Michigan, New York, and Ohio.

Silurian salt is mined by underground methods such as those used in the Detroit area (Fig. 17.20), or it is pumped from brine wells drilled into salt-bearing strata. Besides sodium chloride, brine wells yield other salts of the halogen family such as bromine and iodine compounds as well as magnesium and calcium chlorides. These are valuable raw materials in the manufacture of various chemicals used in industry, medicine, and agriculture. Most of the aspirin produced in the United States probably originated in a brine well penetrating a Paleozoic formation.

Limestone and Dolomite. The greatest

Figure 17.20 A salt mine beneath the city of Detroit, Michigan. These deposits are of Silurian age and include salt interbedded with anhydrite and dolomite. (Photo by International Salt Company.)

use for these rocks is in the making of Portland Cement and as a flux in blast furnaces. Many limestones are used as building stones (dimension stones), especially if they are massive and uniform in composition. A good example of this type of deposit is the famous "Indiana Limestone" of Mississippian age extensively quarried near Bedford, Indiana (Fig. 17.21). This deposit is made up of tiny fragments of foraminifera, cemented together to form an easily quarried rock. It is used for intricately carved stones around windows and doorways and as a structural stone in such famous edifices as the Washington Monument.

Gypsum. This evaporite is quarried extensively from Paleozoic rocks of Silurian age in Michigan, New York, Ohio, and Indiana, as well as from younger beds in Texas. Its chief use is in the manufacture of plasters, plasterboard, and quick-setting building cements for interior purposes.

Marble. The famous Vermont marble is an Ordovician limestone metamorphosed during the Taconic and Acadian orogenies. Vermont and other marbles are used chiefly as interior decorative stones in the building industry.

Iron Ore. The major iron-bearing rocks are of Precambrian age, but an important occurrence of Paleozoic sedimentary iron-bearing strata is the Clinton Iron Ore of Silurian age. The deposits occur as sedimentary beds containing 35 to 60 percent iron and crop out in the eastern United States from New York to Alabama. At Birmingham, Alabama, a large steel industry has developed around the Clinton ores, not only because of the ore itself but also because of the availability of coal and limestone (also of Paleozoic age) from nearby localities. Both coal and limestone are needed in the process of steelmaking.

Lead and Zinc. The Tri-State District, comprised of parts of Missouri, Kansas, and Oklahoma, is one of the largest zinc districts of the world. Some lead is mined, since it is generally associated with zinc. Flat-lying cherty limestones of Mississippian age contain many solution cavities and other openings in which the lead-zinc ores accumulated.

Geologic opinion is divided as to the origin of these important strategic metals. Some hold to a cold-water origin whereby groundwater dissolved the metals from older rocks and redeposited them at a later date. The balance of opinion, however, seems to favor an origin that involves the precipitation of the lead and zinc compounds from

Figure 17.21 Quarry in the Indiana Limestone of Mississippian age near Bloomington, Indiana. (Indiana Geological Survey.)

THE PALEOZOIC ERA

hot mineralized waters that originated from an igneous source at depth.

In either case, the deposits are not of Mississippian age because the Mississippian rocks act only as the host in which the ore was trapped. It is probable that the mineralization was accomplished long after the Mississippian beds were deposited, possibly during the Mesozoic Era.

References

* Beerbower, James R., 1968, *Search for the past,* 2nd ed., Prentice-Hall, Englewood Cliffs, New Jersey, 512 pp.

Colbert, Edwin, 1955, *Evolution of the vertebrates,* John Wiley and Sons, New York, 477 pp.

Dunbar, Carl O., and Karl M. Waage, 1969, *Historical Geology,* 3rd ed., John Wiley and Sons, New York, 556 pp.

King, P. B., 1959, *The evolution of North America,* Princeton University Press, Princeton, New Jersey, 190 pp.

Ladd, Harry S., 1961, Reef building, *Science,* V. 134, pp. 703–715.

* McAlester, A. L., 1968, *The history of life,* Prentice-Hall, Englewood Cliffs, New Jersey, 152 pp.

Rodgers, John, 1967, Chronology of tectonic movements in the Appalachian region of eastern North America, *American Jour. Science,* V. 265, pp. 408–427.

Russell, R. J. (ed.), 1955, *Guides to southeastern geology,* Geological Society of America, Guidebook for the 1955 Annual Meeting, 592 pp.

* Woodford, A. O., 1965, *Historical geology,* Appendix A, Classification of animals, pp. 465–466; Appendix B, Recognition of fossil invertebrates and simple plants, pp. 467–492, W. H. Freeman and Company, San Francisco, 512 pp.

* Recommended for further reading.

18 The Mesozoic Era

. . . the organized Fossils
. . . may be understood by all,
even the most illiterate . . .
William Smith

By the end of the Paleozoic, animals had become well established on the lands of the earth. The Mesozoic, which means "middle life," was a time during which the vertebrates became firmly established in a great variety of environments. The Mesozoic Era saw the rise to dominance and the eventual decline of one of the greatest animal dynasties of all time, the reptiles. Not only did the reptiles dominate the lands, but many returned to the aquatic environment, and still others conquered the skies.

In North America, paleogeographic patterns were generally different from those of the Paleozoic. The Appalachian orogenies had eradicated the eastern geosyncline from the region; it was never again the site of marine sedimentation. Although geosynclinal conditions continued in the Cordilleran region, the distribution of land and sea was different from that of the Paleozoic. In addition, a new geosyncline was to develop across the Gulf Coast region.

Cretaceous marine fossils from Shasta County, California.

The subject of this chapter is the geologic history of the three Mesozoic periods, Triassic, Jurassic, and Cretaceous.

The Triassic Period

Triassic rocks were originally defined on the basis of their threefold division in Germany where they were first studied. They occur widely in other parts of Europe as well, and are especially well developed in the Alps. The Mesozoic sedimentary rocks reflect the existence of the Tethyan geosyncline, a large marine seaway that was one of the major marine inundations of the European-Asiatic region.

Eastern North America. The Appalachian orogenies left the eastern part of the North American continent considerably elevated so that stream erosion was the dominant geologic process. Sediments were carried eastward beyond the present Atlantic coast where they lie buried beneath younger strata. Deep drilling on the North Carolina coastal plain has penetrated sedimentary rocks of Triassic age.

The sole record of Triassic sedimentation on the present continental area of eastern North America dates from late Triassic time when a number of structural fault troughs developed. Deposits in these troughs occur discontinuosly along a linear belt extending from Nova Scotia to South Carolina, and include conglomerates, sandstones, and shales, with interbedded lava flows and intrusive rocks (Figs. 18.1, 18.2). The imprints

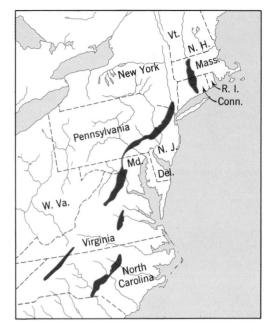

Figure 18.2 Map showing the distribution of the Newark series of Triassic age in eastern United States. (From the *Geologic Map of North America,* 1946, Geological Society of America, Boulder, Colorado.)

of dinosaur tracks and numerous fossil land plants show conclusively that these Triassic beds were of nonmarine origin. The remains of freshwater fishes as well as the red color of much of the sediment point to a continental environment. Good exposures of red Triassic strata and associated lava flows near Newark, New Jersey, account for the name *Newark series* given to these strata.

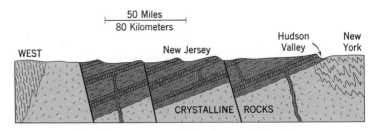

Figure 18.1 Geologic cross section of the Newark beds of Triassic age. The clastic sediments are interbedded with lava flows and intruded by sills. (After Longwell, C. R., 1933, *Guidebook 1,* 16th International Geological Congress.)

The Newark beds were tilted as a consequence of crustal forces, usually referred to as the *Palisades disturbance.* This activity began with faulting that formed the troughs in which the Triassic sediments accumulated. The faulting continued while the sediments were being laid down and was accompanied by the intrusion of basaltic dikes and sills, and the extrusion of lava. These relations suggest tensional forces acting on the crust in this area, and although not strictly orogenic in style, some geologists regard the Palisades activity as the last phase of the Appalachian cycle of mountain building. The Newark beds are typical of "post-orogenic" rocks of other areas and of other orogenies.

The famous and picturesque Palisades of the Hudson River is an eroded edge of a thick sill near the base of the Newark series.

Western North America. Triassic sedimentation west of the Mississippi River was of two types. Marine geosynclinal sediments of great thickness accumulated along with submarine volcanic rocks in Nevada, California, Idaho, and British Columbia. This pattern of geosynclinal deposition was to continue throughout the Mesozoic. This western geosyncline appears to have been separated from areas of Triassic marine and continental sedimentation in the area now occupied by the Rocky Mountains by an intervening land area of uncertain size. The limited marine Triassic in Wyoming is the forerunner of more extensive marine sedimentation in what has been called the Rocky Mountain geosyncline.

Continental beds of Triassic age are common in Arizona, Colorado, New Mexico, and Utah. The brilliant red *Wingate sandstone* may be, in part, a fossil sand dune. The red, pink or yellow *Chinle formation* is famous for its content of petrified logs.

The Jurassic Period

The type locality for Jurassic rocks is in the Jura Mountains, north of the main Alpine chain in Europe. Historically, it is interesting to recall that it was on the basis of collections from the Jurassic rocks of England that

William Smith laid the foundations of stratigraphy and historical geology.

In North America, Jurassic rocks crop out only west of the Great Plains. In the southern United States, Jurassic strata are encountered only at depth in oil wells in the Gulf Coast region. Continental elevation was again the dominant feature of Jurassic time for eastern North America, which accounts for the fact that no sedimentary rocks of that period are known.

Western North America. During the Jurassic, the western geosyncline continued to receive great thicknesses of detrital sediments and submarine volcanic rocks. These are especially well developed in the California Coast Ranges, in parts of the Sierra Nevada, and in British Columbia. Much of the Jurassic sedimentation was going on amidst large-scale orogenic activity, so that the detrital sediments were derived from the mobile belts composed of volcanic islands.

A major marine embayment, the *Sundance Sea,* of late Jurassic time, extended from the arctic regions as far south as New Mexico (Fig. 18.3). Deposits of this embayment are common in Montana, Wyoming, the western Dakotas, and on the Colorado Plateau. Many of the sediments that underlie and interfinger with the Sundance sediments are gypsum-bearing red beds. These indicate shallow lagoon conditions bordered by lowlands across which sluggish streams flowed. The gypsum suggests an arid climate. Before the advance of the Sundance across the region of the Colorado Plateau, this region was the site of extensive eolian sand deposition, represented by the thick "frozen dunes" of the *Navajo sandstone* (Fig. 18.4). This formation forms the white cliffs of the Zion National Park area.

As the Sundance Sea withdrew northward, there was spread out over a vast area, nearly 100 thousand square miles, of the Rocky Mountains and the western Great Plains the well-known *Morrison formation.* It is composed of terrestrial conglomerates, sandstones, and shales, and is well known for the abundant fossils it has yielded. From it have come scores of dinosaur species, numerous primitive mammals, terrestrial mollusks, and land plants.

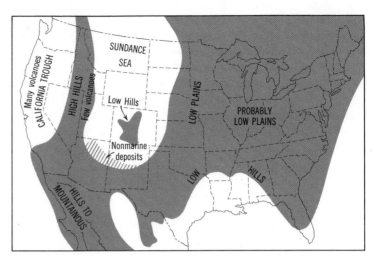

Figure 18.3 Paleogeographic map of the United States showing the distribution of land and sea during the late Jurassic period. (From Imlay, Ralph, 1956, *Paleotectonic Maps, Jurassic System,* Misc. Geol. Investigations, Map 1-175, U.S. Geological Survey, Washington, D.C.)

Figure 18.4 The cross-bedding shown in the Navajo sandstone of Jurassic age in Zion National Park, Utah is believed to be of eolian origin. (Photograph by Tad Nichols.)

Gulf Coast Region. The Jurassic rocks encountered in deep wells in the Gulf Coast region are marine limestones several thousand feet thick. Although their total thickness has not been determined, these sediments appear to herald the beginning of the Gulf Coast geosyncline, a new feature of North American geology. The importance of this new sedimentary basin lies in the thick accumulations of later sediments, largely Cenozoic, which are prolific oil producers.

The Cretaceous Period

The term Cretaceous, from the Latin word for chalk, was first applied to the rocks that form the well-known "White Cliffs of Dover." Cretaceous rocks are among the most widespread of any rock system. During late Cretaceous time, marine inundations occurred on every continent, and were rivaled only by the Ordovician marine invasions of the early Paleozoic. In North America, Cretaceous sediments are widespread in the far west and in the Rocky Mountain and Great Plains areas; they also occur along the eastern and south coastal regions of the United States.

Western North America. In the far western part of the continent, geosynclinal sedimentation continued along the trace of the earlier geosyncline. Some of these deposits, notably the *Franciscan* in California, are of great thickness and give mineralogical evidence of having experienced deep burial during part of their history. Deposits of generally similar type are known in the Coast Ranges from California to British Columbia.

Farther east in western North America the site of the Jurassic Sundance seaway was again inundated, this time by a marine basin of considerably greater proportions. In the early part of the period (Fig. 18.5) the seaway was somewhat restricted and was bordered by areas of nonmarine deposition. At its maximum extent in the Late Cretaceous (Fig. 18.6), this seaway, called the *Rocky*

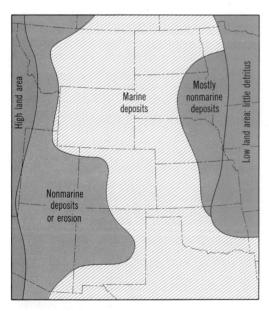

Figure 18.5 Paleogeographic map of the western interior during the early part of the Cretaceous period. (After Geological Society of America, 1957, Memoir 67, *Treatise on marine ecology and paleoecology, v. 2, Paleoecology,* Chapter 18, Paleoecology of the Cretaceous seas of the western interior of the United States, by John Reeside.)

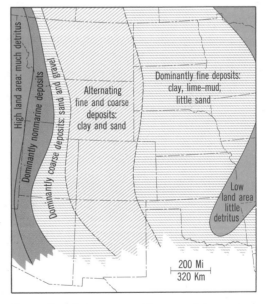

Figure 18.6 Paleogeographic map of the western interior of the United States during the late Cretaceous period. (After Geological Society of America, 1957, Memoir 67, *Treatise on marine ecology and paleoecology, v. 2, Paleoecology,* Chapter 18, Paleoecology of the Cretaceous seas of the western interior of the United States, by John Reeside.)

Mountain seaway, extended from the Arctic to Mexico, and at its widest part it reached from western Wyoming and Montana to Minnesota. The sediments of this marine basin are represented by sandstones, shales, and limestone, with large amounts of coarse detritus and terrestrial materials, including coal, interfingering from the west. This relationship, plus the overall eastward tapering wedge-shape of these Cretaceous deposits (from 20,000 feet in the west to a few hundred feet in the east) indicates a westward source from tectonic lands for most of the materials. On the other hand, the finer-grained shales and clays near the eastern margin indicate that the borderland on that side was low lying.

Eastern United States. The Atlantic and Gulf Coastal Plains contain marine Cretaceous sediments. The maximum inundation is represented by the Mississippi embayment where Cretaceous rocks reach as far north as southern Illinois. The coastal plain Cretaceous sediments extend seaward beneath the cover of younger deposits, and where they have been reached by drilling, they increase in thickness in a seaward direction (Fig. 18.7).

Mesozoic Orogeny of Western North America

In western North America it has been conventional to speak of two major oro-

genic pulses for the Mesozoic Era, both of which were separated geographically and temporally from one another. The earlier, presumed by early workers to be confined to the Jurassic in the far west, was called the *Nevadan orogeny;* the later, said to be a late Cretaceous event in the Rocky Mountains, was named the *Laramide orogeny.* Subsequent work has shown that while the geographic separation may be valid, the temporal separation is not. The terms Nevadan and Laramide will be used in this book but in their more modern guises.

Nevadan Orogeny. Named for the rocks of the Sierra Nevada of California, this orogeny is characterized by large-scale granitic intrusives, set in complexly deformed geosynclinal metasedimentary and metavolcanic rocks. These conditions occur in an elongate belt extending from Baja California northwestward to the Klamath Mountains of California-Oregon, from where the belt swings northeast to include the Boulder batholith of Idaho. It continues to the northwest through the coast ranges of British Columbia and southern Alaska and eventually to the Alaskan Range of south-central Alaska. The large granitic batholiths of this linear belt of deformation are shown in Figure 18.8.

Intrusive relations and unconformities in several areas reveal that much of this intrusive activity occurred during late Jurassic and early Cretaceous time. The range of

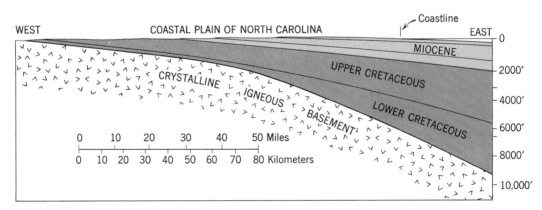

Figure 18.7 Geologic cross section of the Atlantic Coastal Plain through Cape Hatteras, North Carolina. (After Spangler, W. B., 1950, Subsurface geology of Atlantic Coastal Plain of North Carolina, *Bull. Amer. Assoc. Petroleum Geologists,* v. 34, p. 100.)

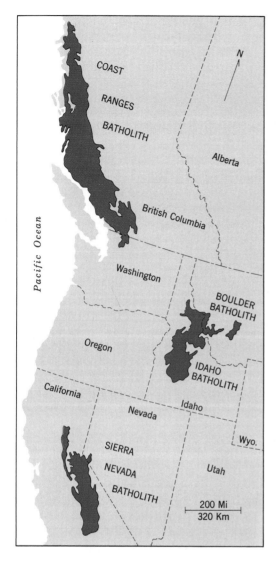

Figure 18.8 Map of the batholiths intruded during the Mesozoic Era. (Based on the *Geologic Map of North America,* Geological Society of America, 1946, Boulder, Colorado.)

radiometric dates obtained from the Sierra Nevada (approximately 200 to 90 million years) indicates that what is Nevadan in style spans a period from late Triassic to mid-late Cretaceous. This is confirmed by the nature of the detrital Triassic and Cretaceous rocks of the Colorado Plateau and western Rocky Mountain regions. These sediments are of the clastic wedge type, and they are attributed to erosion from the orogenic highlands of the Nevadan system.

Laramide Orogeny. The Cretaceous seaway of the Rocky Mountains was eradicated by deformation and uplift that signaled the beginning of the Laramide orogeny. Once thought to be the last event of the Cretaceous, it is now known that the Laramide orogeny was a series of events spread geographically along the trace of the Rocky Mountains from middle Cretaceous time through the early part of the Tertiary period. The Laramide orogenic movements were essentially compressional and resulted in the formation of folded mountains and extensive thrust faults; igneous activity in relation to Laramide deformation is minor.

The Rocky Mountains of Wyoming and Colorado were strongly folded into broad anticlines in which the Precambrian cores were exposed by later erosion. The upturned edges of the late Paleozoic and Mesozoic sediments in Colorado attest to the magnitude of this folding (Fig. 18.9).

The many Laramide thrust faults of the Rocky Mountain region have attracted attention from structural geologists ever since they were discoverd in the last century. Perhaps none has received so much attention as the Lewis overthrust in Montana and Alberta (Fig. 18.10). This fault has a northwest-southeast trend and is gently inclined toward the southwest. During the Laramide orogeny, compressional forces deformed the Precambrian and Paleozoic rocks into broad folds; eventually the folding gave way to faulting, and a large segment of Precambrian rocks was thrust eastward over Cretaceous strata. Chief Mountain in Glacier National Park is an erosional remnant of the overthrust block.

Relations similar to those described for the Lewis thrust occur in a nearly continuous narrow zone extending south from the foothill belt in Alberta into western Montana and eastern Idaho-western Wyoming, thence around the west side of the Colorado Plateau, into southern Nevada. This, along with the Southern Rockies of Colorado and New Mexico, comprises the major belt of Laramide fold and overthrust structures of western North America.

Laramide igneous activity appears to have been limited to the Boulder batholith in western Montana (Fig. 18.8) and a zone of

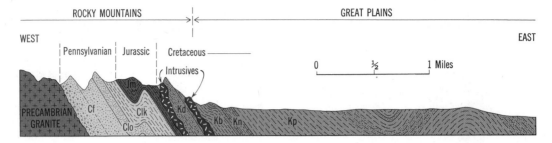

Figure 18.9 Geologic cross section of the Front Range of the Rocky Mountains about 30 miles north of Denver, Colorado. The folding of the Paleozoic and Mesozoic strata was accomplished by the Laramide orogeny. Symbols refer to formational names: Cf, Fountain formation; Clo, Lyons sandstone; Clk, Lykins formation; Jm, Morrison formation; Kd, Dakota formation; Kb, Benton shale; Kn, Niobrara formation; Kp, Pierre shale. (After Sample, R. D. and J. W. Low, 1933, *Guidebook 19,* 16th International Geological Congress.)

small intrusives in the central part of Colorado. The Boulder batholith is clearly related to the great copper deposits of Butte, Montana. The intrusives in Colorado are related to a zone of mineralization containing many of the famous Colorado gold, silver, and lead mines.

Life of the Mesozoic

The panorama of life during the Mesozoic presents an interesting contrast, not only to the fauna and flora of the present day but also to the life of the Paleozoic.

Invertebrates. The Mesozoic seas contained a variety of invertebrate creatures including mollusks, reef corals, crinoids, and others. But two members of the mollusk tribe stand out above all other inhabitants of the Mesozoic seas. These were the cephalopods known as the *belemnites* and the *ammonites* (Fig. 18.11).

The belemnites were squidlike cephalopods with an internal shell that gave rigidity to an elongate body. Like other cephalopods, the belemnites darted backward through the water and, like their relatives the squids, were probably capable of discharging a black inky fluid for the purpose

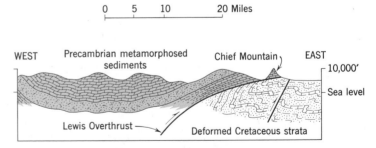

Figure 18.10 Geologic cross section showing the Lewis overthrust in northwestern Montana. Precambrian beds lie on Cretaceous strata. (After Clapp, C. H., 1932, *Geology of a portion of the Rocky Mountains of northwestern Montana,* Montana State Bureau of Mines and Geology, Memoir 4, 30 pp.)

Figure 18.11 Restoration of a late Cretaceous sea bottom showing three kinds of cephalopods common during that time. The large ammonoid in the foreground has a diameter of 19 inches; the intricate sutures are not visible on its shell because of the outer layer of shell material. *Baculites* has a straight shell except for a small coil at the very apex of the cone. *Belemnites* has no external shell but possesses an internal hard part to give the body rigidity. (Diorama from the University of Michigan Museum of Paleontology.)

of camouflage. These unusual creatures make their first appearance in rocks of Triassic age, but by the Jurassic they reached their peak of development and abundance. The belemnites lived on into the Cretaceous period but declined severely in numbers until they finally became extinct at the end of the Mesozoic.

The ammonites are perhaps the most distinctive of all Mesozoic marine invertebrates. These coiled cephalopods were holdovers from the Permian, but the Mesozoic forms far surpassed their Permian ancestors in ornamentation and variation. The ammonite septa were beautifully intricate and amazingly complex, and it is this very feature that permits the paleontologists to classify this important group of animals. After a rise to a high degree of diversification in the Jurassic, the ammonites diminished during the Cretaceous. Today there are no survivors of this ancient race of ornamental shellfish, because they died out *en masse* at the end of the Mesozoic.

Many other invertebrates lived in the Mesozoic seas, but the listing of their names alone would fill a volume. One observation is significant, however, and that pertains to the Cretaceous mollusks. The snails and clams of that final Mesozoic period were remarkably modern in appearance, a fact that clearly demonstrates that many of the present-day creatures have a lineage reaching back nearly 100 million years.

Reptiles. It is no overstatement to say that the reptiles dominated the life scene during the Mesozoic Era. Even though reptiles were already firmly established in the Permian, their climax of development did not come until the Mesozoic.

Reptiles are cold-blooded, egg-laying vertebrates that include crocodiles, snakes, and lizards, as well as the extinct dinosaurs, marine reptiles, and flying reptiles. The word "dinosaur" is over a hundred years old and literally means "terrible lizard." Yet like so many other names in the scientific world, the term dinosaur is not entirely accurate, for many close relatives of these creatures were very small and could hardly be considered "terrible." Furthermore, the dinosaurs were so diversified in their habitats that they scarcely could be classified technically as lizards. Nevertheless, the term dinosaur is useful and is well entrenched in the language.

The first dinosaurs appeared in late Triassic time and were small compared to their successors in the Jurassic and Cretaceous periods. The first Triassic dinosaurs were about 8 feet long and moved about on their hind legs (that is, they were *bipedal*). A long slender tail served as a counterbalance, thereby permitting full use of the front limbs for grasping and tearing food. This anatomical pattern was repeated in many of the flesh eating (carnivorous) dinosaurs of the Jurassic and Cretaceous periods such as *Allosaurus* (Jurassic) and *Tyrannosaurus* (Cretaceous) (Fig. 18.12).

Anatomically, the dinosaurs can be grouped into two categories, the *saurischians* and *ornithischians*. The basic difference between the two kinds is in the structure of the pelvic bones, but other characteristic differences also exist. For example, the ornithischians were herbivorous, as is indicated by their teeth which were highly adapted for cutting or chewing vegetation. *Stegosaurus* (Jurassic) and *Triceratops, Iguanodon* (Fig. 18.13) and *Trachodon* (Cretaceous) are examples of the ornithischians.

The saurischians, such as *Coelophysis* (Triassic), *Allosaurus* (Jurassic), and *Tyrannosaurus* (Cretaceous) were generally carnivorous. An exception is *Brontosaurus* (Jurassic and Cretaceous) who was not only herbivorous but had departed from the general anatomical pattern of other saurischians to the extent that he was quadrupedal.

The dinosaurs reached gigantic sizes during the course of their evolutionary development. The bones of their massive hulks have awed museum-goers the world over, and even the most objective vertebrate paleontologist experiences a thrill when he discovers the fossilized remains of these giants, some of which were 80 feet in length and weighed 50 tons or more.

Marine Reptiles. Although it may seem strange that land animals should return to a marine environment, the paleontologic record indicates that given time, and in the

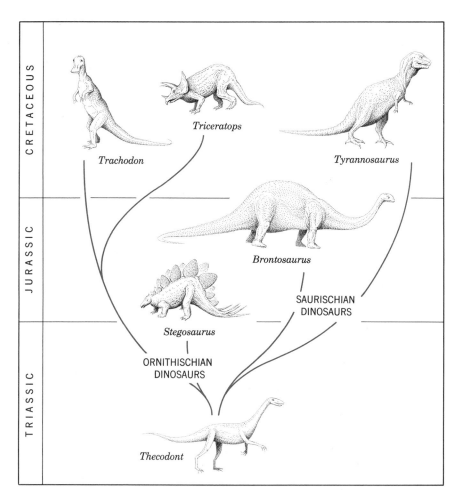

CRETACEOUS

Trachodon

Triceratops

Tyrannosaurus

JURASSIC

Brontosaurus

SAURISCHIAN
DINOSAURS

Stegosaurus

TRIASSIC

ORNITHISCHIAN
DINOSAURS

Thecodont

Figure 18.12 Some Mesozoic dinosaurs that are thought to have had a common ancestor. Not to scale. (After Colbert, E. H., 1955, *Evolution of the vertebrates,* John Wiley and Sons, Inc., New York, 479 pp.)

absence of competition, animals tend to occupy all available ecological niches. The marine reptiles of the Mesozoic evolved from a stock of land reptiles rather than from Paleozoic amphibians, and filled the niche available for highly active marine predators.

Two of the best-known Mesozoic marine reptiles are *Icthyosaurus,* a fishlike reptile that first appeared in the Triassic, and *Plesiosaurus,* a long-necked reptile that vaguely resembled a turtle (Fig. 18.14). *Icthyosaurus* was comparable to a modern porpoise in size and was fishlike in its mode of life, but its skeleton is distinctly reptilian. Both *Icthyosaurus* and *Plesiosaurus* were air-breathing

carnivores, and it is quite probable that their chief source of food was fish.

Flying Reptiles. No less spectacular than the debut of the marine reptiles was the advent of flying reptiles or *pterosaurs* in the Jurassic. To envision the process by which the reptiles acquired their flying ability requires considerable intuitive thought. The process of flying requires a high rate of metabolism because the body is in need of a constant source of energy during flight in order to overcome the force of gravity. By analogy with modern reptiles, it is assumed that the Mesozoic reptiles did not have a high and constant metabolic rate and, hence, were not warm-blooded. This has led to the

Figure 18.13 *Iguanodon,* an herbiverous ornithischian dinosaur from lower Cretaceous rocks near Mons, Belgium. They are also known from the lower Cretaceous of southern England. These animals may have been ancestral to the Trachodonts of the upper Cretaceous. (Courtesy of the Royal Museum of Natural History, Brussels, Belgium.)

Figure 18.14 Icthyosaurs (*A*) and Plesiosaurs (*B*) were marine reptiles that inhabited the Mesozoic seas. (After Knight, Charles, 1935 *Before the dawn of history,* McGraw-Hill Book Co., New York, 119 pp.)

argument that perhaps the flying reptiles were really not good fliers but were capable only of gliding or swooping down from lofty cliffs near the sea. It has also been argued that the flying reptiles may have achieved independently a high metabolic rate.

But whatever their abilities were as heavier-than-air flying animals, the reptiles such as *Rhamphorhynchus* (Jurassic) or *Pteranodon* (Cretaceous) were distinctly reptilian. Their skulls were characteristically like those of other reptiles and their jaws had sharp rows of teeth indicative of a carnivorous diet.

The pterosaur wing consisted of a membrane stretched between an elongated fourth finger and the hind limbs, although the fossil remains are not adequate to establish the exact function of the back legs. The elongated skull of *Rhamphorhynchus* plus the lack of a tail makes one wonder how such an animal ever got aloft. By comparison with modern birds, the flying reptiles must bear the same relationship to birds that the Wright brothers' first crude airplane bears to modern jet aircraft!

Birds. In addition to dinosaurs and flying reptiles, the birds make a startling appearance in the Jurassic. From an anatomical point of view, the bird skeleton is closely akin to the reptilian skeleton. Paleontologists admit that, were it not for the imprints of feathers in the Solenhofen limestone of Germany from which two excellent skeletons were obtained, the fossils would have been classified as reptiles. But the presence of feathers, a singularly distinctive feature of all birds past and present, definitely proves that these Jurassic creatures were true birds (Fig. 18.15).

Archaeopteryx, the earliest known bird was about the size of a raven. Unlike modern birds, it had teeth and a long bony tail with feathers attached on either side. The forelimbs or wings were unlike those of present day birds in that the outer wing bones were separate instead of coalesced, and retained three of the fingers as claw-like appendages.

By Cretaceous time, aquatic birds such as

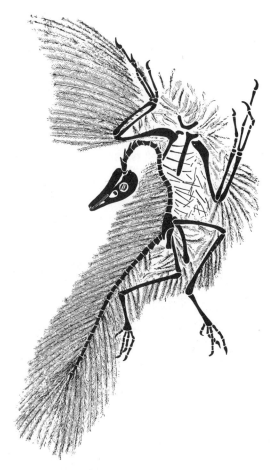

Figure 18.15 Drawing of the skeleton of the first bird, *Archaeopteryx, as it was found in the Jurassic Solenhofen limestone of Germany. (After Colbert, E. H., 1955, Evolution of the vertebrates,* John Wiley and Sons, New York, 479 pp.)

Hesperornis and *Ichthyornis* were present. The former was a diving form whose adaptation to an aquatic life resulted in the loss of the ability to fly, as is indicated by dwarfed wings. *Ichthyornis* had powerful wings and possibly inhabited coastal areas in search of food. The fossil record of birds is woefully inadequate, more so than any other large group of vertebrates. But our knowledge of avian ancestors and their evolutionary history indicates that they did not evolve from flying reptiles, although both birds and flying reptiles may have had a common ancestor.

Mammals. The appearance of birds and flying reptiles in the Jurassic period are important milestones in the evolutionary history of backboned animals. But still another "first" must be added to the Jurassic period, the mammals. Four different kinds of mammals are known from Jurassic sediments, but only two of these groups survived into the Cretaceous.

The distinguishing characteristics of mammals are numerous. Some are anatomical and others are physiological. Among the former are the enlarged braincase as compared to reptiles, and the differentiation of the teeth. Physiologically, the mammals differ from reptiles in that they are warm-blooded with a constant body temperature, and they give birth to their young alive. Young mammals are nourished by milk from the mother and receive parental care, a decided contrast to reptiles who lose interest in their offspring before they are hatched from the eggs.

Even though the Cretaceous was dominated by the dinosaurs, two categories of mammals, the *placentals* and *marsupials,* made their first appearance during that period. Placentals are mammals that nourish the fetus within the body until it is fully developed, whereas the marsupials carry the prematurely born young in a pouch during a period of further growth and development. Most modern mammals are placentals, whereas the marsupials, such as the kangaroos and opossums, are poorly represented except in the isolated areas of Australia, New Zealand, and South America.

Mesozoic Plant Life. The flora of the Mesozoic, until the middle of the Cretaceous period, was predominantly composed of *cycadophytes* or "cycads" which look very much like the modern palms (Fig. 18.16). In spite of this superficial resemblance, however, the Triassic and Jurassic cycads bear no genetic relationship to the palm trees. Ferns were also fairly abundant in the Triassic period, and the conifers of the first two Mesozoic periods were comparable in size to the largest pine trees of today. A spectacular occurrence of fossil Triassic conifers can be seen in the Petrified Forest National Monument of Arizona where coniferous tree trunks up to 7 feet in diameter and 125 feet in length were buried in volcanic ash after they drifted into a shallow lake (Fig. 10.19). Petrifaction of the logs resulted in the brilliant colors displayed in the exhumed giants of a Triassic forest.

The second half of the Cretaceous period is in marked contrast to the previous part of the Mesozoic in terms of the floral communities, because with the beginning of late Cretaceous, the plant world became distinctly modern in appearnace. To be sure, late Cretaceous forests were different from those of today, because some of the older forms were still present, but the great increase in *angiosperms* or the "flowering plants" gave the late Cretaceous plant world a new appearance. Angiosperms bear seeds contained in a closed case and include such diverse forms as the flowers, grasses, legumes, and hardwood trees.

Among the Cretaceous angiosperms were forms such as willows, sassafras, oak, and poplars. Paleobotanists do not know exactly where or when the angiosperms originated because when they first appear in the geologic record they are completely developed. A few angiosperms are known from the Jurassic, but no plant fossil has ever been discoverd that can be regarded as transitional between the angiosperms and earlier types.

Animal Extinction at the End of the Mesozoic

One of the most frustrating problems of earth history is the unsolved question of widespread extinctions of organisms in the geologic past. Why, after millions of years of successful adaptation to various environments did entire segments of the animal world vanish? This has happened several times in the geologic past, but perhaps the most dramatic example is the disappearance of the dinosaurs at the end of Cretaceous time. To be sure, what we describe as an extinction at the end of a period or era did not take place suddenly, but however long it took for the dinosaurs to die out, the evidence is unmistakably clear that they

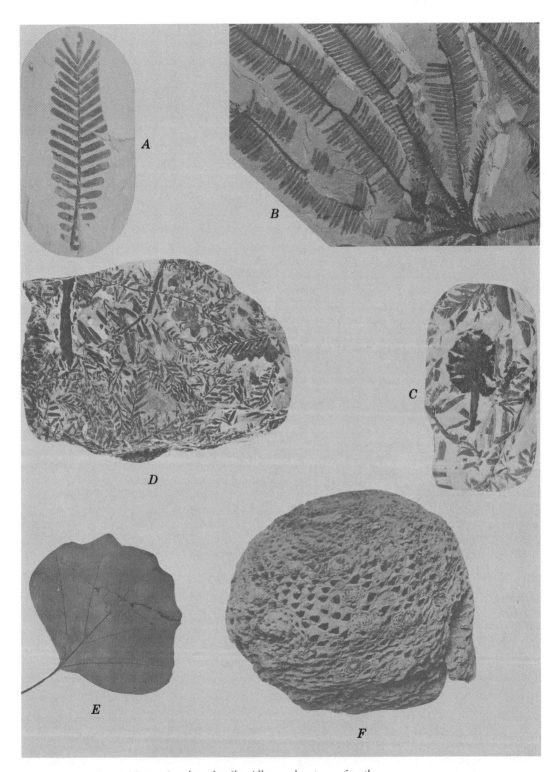

Figure 18.16 Some Mesozoic plant fossils. All are about one-fourth natural size except C, which is about one-half size. (A) Jurassic cycad. (B) Triassic fern. (C) Cone of a Cretaceous conifer. (D) Foliage of Cretaceous conifers. (E) Leaf of a Cretaceous sycamore tree. (F) Trunk of a Cretaceous cycadlike tree. (A, B, and F reproduced by permission of McGraw-Hill Book Co., from Arnold, C. A., 1947, *An introduction to paleobotany,* McGraw-Hill Book Co., 433 pp. C, D, and E by C. A. Arnold, University of Michigan, Museum of Paleontology.)

did not live on into Cenozoic times. The cause for such a dramatic change in the Mesozoic biologic community has not been identified with any reasonable degree of certainty.

Moreover, it is argued that the Mesozoic reptiles were able to adapt themselves to a wide variety of environments, from swamps to the sea, and the air; yet, on the other hand, there must be a logical reason why they were unable to survive the transition from the Mesozoic to the Cenozoic. In addition to the dinosaurs, the marine reptiles and the flying reptiles vanished by the end of the Cretaceous. Among the invertebrates, the same fate overtook the belemnites and the ammonites. There are exceptions, which only compound the difficulty. Lizards, snakes, turtles, and crocodiles among the reptiles lived through the transition. Land plants were not adversely affected, and such marine forms as corals and modern fishes continued to diversify.

A number of explanations for extinctions have been offered: broad fluctuations in sea level producing marked changes in the nature of marginal lowlands; general continental elevation produced by orogeny and its effect on climate and vegetational patterns; changes in the nature of the food chain as a result of large-scale extinctions of pelagic forms; cosmic radiation from a nearby supernova (causing deleterious genetic changes), and so on. None is completely satisfactory in itself.

For the dinosaurs, the most plausible explanation proposes general continental uplift that probably destroyed the swampy lowlands supporting the vegetation favored by such large reptiles. This, together with decreasing equability of the climate during the later Cretaceous, may have produced conditions to which the dinosaurs were unable to adapt. This idea is plausible for terrestrial forms, but it does not apply to marine forms which, presumably, could have migrated to warmer waters with little effort. The true answer may be beyond present understanding, however, and may, in fact, require the discovery of new facts that link biological survival to forces of terrestrial or extraterrestrial origin.

Mineral Resources Associated with Mesozoic Rocks

Cretaceous coal from the western United States and Canada, and petroleum from eastern Texas are among the chief mineral resources produced from Mesozoic rocks. The Mother Lode gold ores associated with the Sierra Nevada batholith, and other metalliferous deposits such as copper, zinc, and silver in the Butte, Montana region are also important economic resources of Mesozoic age.

Since World War II, discoveries of uranium minerals in sedimentary rocks of Mesozoic age in the southwestern United States and Rocky Mountain regions have added a new dimension to the world of economic geology. Important uranium deposits have been discovered in the Colorado Plateau from formations ranging in age from Permian to Tertiary, but the principal sources have been sandstones or conglomerates from Triassic formations known as the Shinarump and Chinle, and the Morrison formation of Jurassic age.

Generally speaking, the uranium of the Colorado Plateau occurs as pitchblende or other dark uranium-bearing minerals associated with fossil organic material such as logs, bones, or plant debris which appears to have acted as a precipitating agent. The ore bodies are most prevalent in sandstones of fluvial origin, especially those which are interbedded with mudstones or shale (Fig. 18.17). The significance of the fluvial host rock is apparently its high permeability, which afforded the uranium-bearing solutions easy lateral passage along the strata. Fluvial deposits are also more likely to contain lenses or "pockets" of organic material washed in during flood stages.

One of the largest single deposits of uranium in the country was discovered in 1955 on the southern fringe of the Colorado Plateau in New Mexico. There at Ambrosia Lake, a black asphaltic substance was encountered during the sinking of a drill hole in search for petroleum. The material was strongly radioactive and turned out to be high in uranium content. The Ambrosia Lake deposits of Jurassic age contain a huge re-

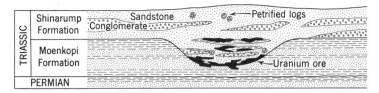

Figure 18.17 Cross-sectional diagram of the uranium-bearing Shinarump conglomerate in the Colorado Plateau. (After Mitchell, T. W., and C. G. Evensen, 1955, Uranium ore guides, Monument Valley District, *Economic Geology*, v. 50, p. 172.)

serve of uranium ore, the mining of which makes New Mexico one of the leading uranium producing states in the nation. In 1970, for example, New Mexico contained about 42 percent of the uranium reserves in the United States. With increasing demands for power from nuclear reactors in the United States and other parts of the world, the need for uranium ores will increase well into the next century.

Age determinations (lead-uranium method) of the uranium from the Colorado Plateau give an age of around 60 million years. This indicates that the uranium deposits are late Mesozoic to Tertiary in age; that is, they are related to the Laramide orogeny.

References

Arkell, W. J., 1956, *Jurassic geology of the world,* Oliver and Boyd Ltd., Edinburgh, 806 pp.

* Axelrod, Daniel I., and Harry P. Bailey, 1968, Cretaceous dinosaur extinction, *Evolution,* V. 22, pp. 595-611.

Bateman, Paul C., and Jerry P. Eaton, 1967, Sierra Nevada batholith, *Science,* V. 158, pp. 1407-1417.

* Bramlette, M. N., 1965, Massive extinctions in biota at the end of Mesozoic time, *Science,* V. 148, pp. 1696-1699.

* Breed, W. J., 1968, *The age of dinosaurs in northern Arizona,* Museum of Northern Arizona, Flagstaff, Arizona, 45 pp.

Colbert, E. H., and M. Kay, 1965, *Stratigraphy and Life History, Chap. 20, Mesozoic life,* pp. 479–499, John Wiley and Sons, New York, 736 pp.

Colbert, Edwin, 1968, *Men and dinosaurs,* E. P. Dutton, New York, 283 pp.

Darrah, W. C., 1960, *Principles of paleobotany,* 2nd ed., The Ronald Press Company, New York, 295 pp.

Dunbar, Carl O., and Karl M. Waage, 1969, *Historical geology,* 3rd ed., John Wiley and Sons, New York, 556 pp.

Eardley, Armand J., 1962, *Structural geology of North America,* 2nd ed., Harper and Row, New York, 743 pp.

King, Philip B., 1959, *The evolution of North America,* Princeton Univ. Press, Princeton, N. J., 190 pp.

* McAlester, A. L., 1968, *The history of life, Foundations of earth science series,* Prentice-Hall, Inc., Englewood Cliffs, N. J., 152 pp.

Nairn, A. E. M. (ed.), 1961, *Descriptive paleoclimatology,* Interscience Publishers, New York, 380 pp.

* Newell, N. D., 1963, Crises in the history of life, *Scientific American,* V. 208, No. 2 (February), pp. 77-92. (Offprint 867, W. H. Freeman and Co., San Francisco.)

* Recommended for further reading.

19 The Cenozoic Era

*Said the little Eohippus,
"I am going to be a horse!"
Charlotte Gilman*

The Cenozoic Era is the shortest of the major segments of geologic time. It has a uniqueness all its own because the events of this era provide us with a link to earlier episodes of geologic history. The clarity of the record contained in Cenozoic sediments is better than any of previous eras, and much of the earth's landscape as it appears today began to take form during the Cenozoic. Furthermore, plants and animals living today ascended from closely related Cenozoic forms whose evolutionary history during the 65 million years of Cenozoic time is better documented than for any previous era. The principle of uniformitarianism is probably more easily invoked in the interpretation of Cenozoic geologic history than for any other division of geologic time.

The Cenozoic Era is divided into two major time units, the *Tertiary* and the *Quaternary* periods. The Tertiary period, which lasted somewhat more than 60 million years, comprises the bulk of Cenozoic time. The Tertiary period contains five subdivisions or *epochs* of unequal length which, from

Wild horses in Wyoming. (Courtesy of the U.S. Department of Interior.)

385

oldest to youngest, are known as the *Paleocene, Eocene, Oligocene, Miocene,* and *Pliocene.* Sir Charles Lyell (1797–1875) proposed the first classification of the Tertiary based on the percentages of modern and extinct fossils in marine strata near London and Paris.

The Quaternary consists of the *Pleistocene,* or "Great Ice Age" as it is more popularly known, and the Recent or *Holocene,* the shortest of all geologic time subdivisions. The Pleistocene began about 2.5 to 3 million years ago, but the exact time of its beginning has not been established. The Holocene embraces the last 10,000 years of earth history. (Continental glaciation during the Pleistocene is discussed in Chapter 13 and will not be repeated here).

Both names, Tertiary and Quaternary, are holdovers from an 18th century classification of geologic time in which they were supposed to represent the last two units of a fourfold time scale in which the Primary and Secondary included all of pre-Cenozoic time (Fig. 15.9). Only the names Tertiary and Quaternary have survived to become incorporated in the modern geologic time scale.

The Cenozoic Era in North America

The most striking difference between the Cenozoic landscape of North America and previous ones was the lack of any extensive inland seas. Only portions of the coastal regions were submerged more extensively than they are today, a fact that can be deduced by the presence of Tertiary marine strata from the Atlantic seaboard to the Gulf of Mexico, all the way to the Yucatan peninsula (Fig. 19.1). The only marine

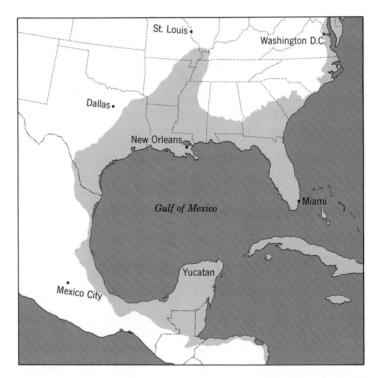

Figure 19.1 Grey area shows maximum submergence of the coastal plain marginal to the Gulf of Mexico in early Tertiary time. (Based on *Geologic Map of North America,* 1946, Geologic Society of America.)

strata of Tertiary age in the interior part of the North American continent are Paleocene in age, and they represent the last time that any part of central North America was submerged beneath sea level.

From Eocene time to the present, therefore, North America closely resembled its modern configuration, although the Tertiary continental shelves were broader than now, and parts of California were flooded by marine waters from time to time. By and large, however, the geography of North America was given its broad outline during the Tertiary. The gross geologic and physiographic characteristics of the North American landscape provide the basis for delineating certain *physiographic provinces*. A brief account of the Cenozoic history of some of these provinces is a convenient means of presenting the Tertiary history of the continent.

The Atlantic and Gulf Coastal Plain Province

The Coastal Plain physiographic province embraces a belt of land between the present shoreline and higher terrain inland (Fig. 19.2). The present coastal plain was actually part of the continental shelf during much of the Cenozoic. From New Jersey to South Carolina the early Tertiary shoreline of the Atlantic was roughly 100 miles farther inland than it is today. Farther south, the Florida peninsula was a broad submarine platform on which fossiliferous limestone accumulated to a depth of several thousand feet. Florida did not become emergent until Miocene time, so in a geologic sense, Florida is the youngest state in the Union.

The presence of older Tertiary marine strata as far north as southern Illinois indicates that the waters of the Gulf of Mexico invaded what is now the lower Mississippi Valley. A great thickness of sediment has been accumulating during the entire Cenozoic Era in the Gulf of Mexico and on the present Gulf Coastal Plain as revealed by fossiliferous rock samples retrieved during the drilling of oil wells in eastern Texas, southern Louisiana, and offshore in the Gulf

of Mexico. Some 40,000 feet of Tertiary sediments lie in the northern Gulf of Mexico, a modern geosyncline that has been subsiding since the end of the Appalachian orogeny.

The Appalachian Mountains and Adjoining Physiographic Provinces

Since no Tertiary sediments occur anywhere in the Appalachian Mountains or adjoining physiographic provinces, the Tertiary history of this region must be inferred from indirect evidence. The "grain" of the Appalachian Mountains is derived from a system of northeast-southwest trending ridges and valleys produced by differential erosion of folded Paleozoic strata. The major rivers draining this area flow generally eastward *across* the regional trend of the upturned edges of the Paleozoic rock layers. To explain this enigma, geologists have proposed several explanations as to why the major rivers flow across the limbs of the folded strata rather than parallel to them.

One of the early ideas that still enjoys considerable popularity among geologists is the following. Sometime during the Cretaceous period, it is postulated, the Appalachian folded mountains were reduced to a peneplain, a surface of low relief sloping gently eastward toward the Atlantic Ocean. In late Cretaceous time this erosion surface, known as the Schooley Peneplain, was transgressed by marine waters that laid down a blanket of sediment over the truncated Paleozoic strata. Epeirogenic uplift in early Tertiary time caused the retreat of the marine waters eastward and left a Cretaceous sedimentary cover exposed to subaerial erosion by streams draining from west to east. As uplift of the Appalachians continued slowly, the network of east-flowing streams completely removed the Cretaceous cover and incised themselves into the reexposed Schooley Peneplain. Thus the modern river courses have been inherited from a previous geologic episode. The rivers were not deflected by the northeast-southwest trending strata because the uplift of the land was slow enough to permit them to retain their

Figure 19.2 Generalized physiographic map of North America. (From Kay, M. and E. H. Colbert, 1965, *Stratigraphy and life history*, John Wiley and Sons, Inc., N. Y., p. 9.)

preexisting courses across the regional structure. The present relief of the Appalachians is ascribed to late Cenozoic weathering and erosion by tributaries of the major river networks, and evidence of the Schooley Peneplain is seen in the form of *accordant* summits of the long ridges that characterize the topography of the Appalachian Mountains. *Water gaps* are notches cut in these ridges by streams that maintained their courses during the slow uplift of the Schooley surface. The Susquehanna River flows through such a water gap at Harrisburg, Pennsylvania (Fig. 19.3).

Some geologists object to this explanation on the grounds that no Cretaceous sediments occur anywhere in the Appalachian Mountains. These critics say that if the Schooley Peneplain was in fact covered by the Cretaceous sea, then remnants of Cretaceous sediments ought to occur on some of the summits of the Appalachian Mountains. Supporters of the theory requiring a Cretaceous sedimentary cover on a peneplained surface believe that the Cretaceous sediments that once covered the Schooley Peneplain have all been stripped by erosion.

Whatever the real explanation is, there is little chance of proving it beyond a reasonable doubt because the evidence is either too subtle to be recognized or simply does not exist in terms that can be unambiguously interpreted.

The Great Plains Province

The Great Plains occupy a broad interior tract east of the Rocky Mountains and stretch from the Arctic Ocean to the Rio Grande. In the Canadian provinces and parts of the northern United States, the Great Plains were covered by Pleistocene glaciers. South of the glacial border, however, this physiographic province possesses a rather simple geologic history. The Great Plains were never the scene of orogeny since Precambrian time

Figure 19.3 Aerial photograph of typical topography of the Valley and Ridge Province near Harrisburg, Pennsylvania. The Susquehanna River flows through "water gaps" in the ridges. (Photograph by John S. Shelton.)

and are underlain by flat or almost flat-lying sediments, ranging in age from Paleozoic to Pleistocene.

During the Tertiary period, streams flowed eastward carrying loads derived from craggy ancestral Rocky Mountains produced by the Laramide orogeny. During the Eocene some of these sediments were trapped in structural basins within the Rocky Mountain Province, but by Oligocene time a blanket of terrestrial sands, silts, and clays was laid down over much of the Great Plains. These terrestrial beds are known as the *White River Series* and are famous for their content of mammalian fossils. In the Badlands National Monument of South Dakota, fossil bones and teeth are easily found right on the surface of the ground where they have weathered out of the soft strata (Fig. 19.4).

In the Pliocene, a network of east-flowing streams originating in the Rocky Mountains spread alluvial sands and gravel over a large part of the Great Plains from Nebraska to west Texas and eastern New Mexico. This deposit, the *Ogallala formation* forms the surface of the so-called High Plains and is an important aquifer for irrigated farming in west Texas. The upper 10 to 30 feet of the Ogallala formation is cemented with hard calcium carbonate which was concentrated there as caliche during normal soil-forming processes that were active over a long time in that semiarid climate.

The Rocky Mountains Province

The Cenozoic history of the Rocky Mountains presents a marked contrast to the earlier geologic history of the same area because the Laramide orogeny brought an end to the Cordilleran geosyncline which had dominated the scene all through Paleozoic and Mesozoic time. A rugged mountain range, the ancestral Rocky Mountains, held center stage from early Tertiary times right up to the present (Fig. 19.5).

Figure 19.4 Geology students hunting for fossils in the Badlands National Monument of South Dakota. The strata in the background are known as the White River Series of Oligocene age.

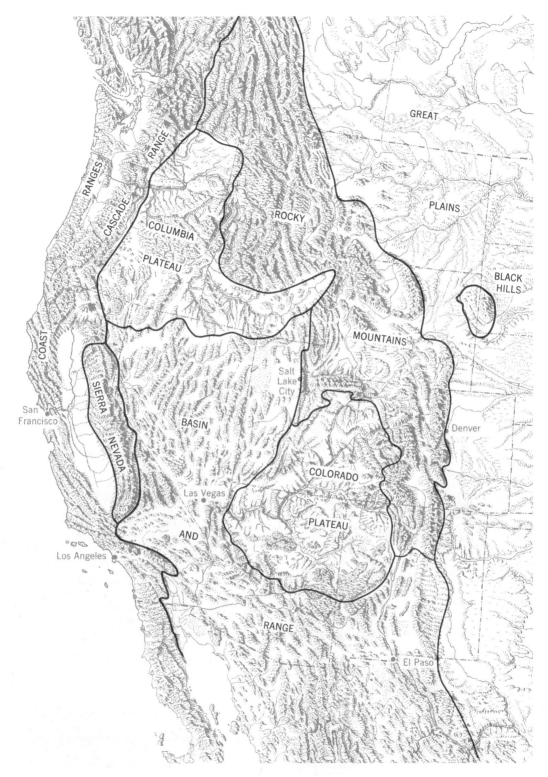

Figure 19.5 Physiographic map of the western United States. (Base map by Erwin Raisz, physiographic boundaries after Fenneman, N. M., 1914, Physiographic boundaries within the United States, Assoc. American Geographers, *Annals,* v. 4, pp. 84-134.)

During the early Tertiary the Rocky Mountains were attacked by the forces of weathering and erosion on the one hand and the forces of internal uplift on the other. At least once during the Tertiary, erosion reduced the Rocky Mountains to a series of widespread pediments (page 232) that were uplifted and dissected later in the Tertiary.

In the late Tertiary, renewed uplift caused the superposition of many rivers across geologic structures that rose athwart the general direction of drainage. The Black Canyon of the Gunnison River and the Royal Gorge of the Arkansas River were incised by those rivers as they maintained preexisting courses during uplift. There seems to be no alternative explanation to account for the courses of the many rivers that flow through canyons across resistant geologic structures instead of following less resistant rock masses lying only a few miles distant. Before the cutting of these canyons, the rivers presumably flowed across pediments and adjoining deep alluvium that had accumulated on the flanks of individual mountain ranges during the long period of early Tertiary downwearing.

During the Pleistocene, many valley glaciers formed on the flanks of the high Rockies. These glaciers advanced and retreated in concert with the expansion and shrinkage of the continental ice sheet in central and eastern North America. Today most of these glaciers have disappeared, but a few shrunken remnants remain in Colorado, Wyoming, Montana, and the Canadian Rockies.

Figure 19.6 The Grand Canyon of the Colorado is a mile-deep gash in the Colorado Plateau. The initial cutting of this erosional feature began sometime during the Tertiary. (Photograph by Tad Nichols.)

The Rocky Mountains Province contains a number of areas that were basins during the Tertiary. Some of these basins were filled with fluvial sediments while others were lake basins into which detrital sediments delivered by rivers were deposited. The *Green River formation* of southwestern Wyoming is a freshwater deposit consisting of thinly bedded shales having an aggregate thickness of about 2000 feet. Their nonmarine origin is inferred from freshwater fossil fish entombed in the shales.

The Colorado Plateau

This physiographic province is something of a geologic wonder. Lying in parts of Arizona, Utah, Colorado, and New Mexico, the Colorado Plateau contains thousands of feet of sedimentary and volcanic rock layers, most of which are nearly horizontal or only slightly deformed. The surface of this province consists of highly angular erosional features representing a long period of Cenozoic denudation that culminated in such magnificent spectacles as the Grand Canyon of the Colorado River (Fig. 19.6).

Tourists by the millions visit various parts of the Colorado Plateau each year and marvel at its scenic beauty and colorful rock formations. The giants of early American geological exploration such as Major John Wesley Powell (1834–1902) and Grove Karl Gilbert (1843–1918) wrote some of their most classic discourses on the geological phenomena that abound in this great outdoor laboratory. So many basic principles of geology are exhibited on the Colorado Plateau that any person who is the least bit interested in his natural surroundings will find little difficulty in learning simple lessons of geology.

In early Cenozoic time the Colorado Plateau was low lying and occupied by a number of basins in which terrestrial sediments accumulated. The Eocene *Wasatch formation* is an example of early Tertiary sedimentation on the Colorado Plateau, and can be observed in Bryce Canyon National. Park in southern Utah (Fig. 19.7). Very few sedimentary strata of late Cenozoic age occur on the Colorado Plateau because erosion rather than deposition has been the dominant process ever since the mid-Tertiary. Extensive uplift is thus implied from the lack of post-Eocene sediments and the widespread degradation that characterizes the Plateau today.

The removal of sedimentary rock strata from the Colorado Plateau by erosion during post-Eocene time has been accomplished on a scale unequaled anywhere in North America. Some estimates of the thickness of rock strata removed from the Colorado Plateau by stream degradation run as high as two miles, but even the most conservative estimates would admit to several thousands of feet. Erosional remnants such as the buttes, mesas, and rock benches of Monument Valley (Fig. 19.8) in northeastern Arizona and southern Utah attest to the magnitude of erosion on that part of the Colorado Plateau.

The details of the erosional history of the Colorado Plateau are very complex and intensely debated by geologists who have worked in the area and hold different views. One major source of considerable difference of opinion centers around the drainage history of the Colorado River and the cutting of the Grand Canyon. The major point of division among geologists arises from the question: How old is the Colorado River? One group argues for the establishment of the Colorado River in its present course in late Tertiary time (Pliocene), whereas another group believes that the Colorado River came into existence much earlier in the Tertiary period after a long and complex series of changes. Such debates are stimulating for the professional geologist and indicate that much work remains to be done by future generations of geologists if opposing points of view are to be reconciled.

Besides the erosional landforms produced by the stripping of sedimentary strata, considerable volcanism during the Tertiary produced many lava fields and several volcanic cones on the Colorado Plateau. These volcanic eruptions occurred from the Miocene to modern times (Fig. 19.9). San Francisco Peak, which lies south of the Grand

Figure 19.7 Bryce Canyon National Park, Utah. This erosional topography is developed in the Wasatch formation of Eocene age. The Colorado Plateau contains many formations that are being stripped slowly away by erosion. (Photograph by R. G. Luedke, U. S. Geological Survey.)

Canyon near Flagstaff, Arizona, is a volcanic cone lying in an extensive area of volcanic rocks. Some of the volcanic activity is very recent in this area as indicated by the date of A.D. 1064 for the age of Sunset Crater, which lies less than 10 miles east of San Francisco Peak in northern Arizona.

Elsewhere on the Colorado Plateau, volcanic features occur in the form of volcanic necks, which are solidified lava throats of old volcanoes whose flanks have been eroded away. Shiprock in northwestern New Mexico is probably the best known example of this phenomenon, but dozens more are known in Monument Valley and elsewhere around the "Four Corners" area of Arizona, New Mexico, Utah, and Colorado.

Pasin and Range Province

The Basin and Range Province lies south of the Columbia Plateau between the Colorado Plateau and the Sierra Nevada (Fig. 19.5). Geologically, the Basin and Range Province is characterized by isolated mountain blocks (the "ranges") and intervening valleys or intermontane basins filled or partially filled with sediments derived from the surrounding mountains during Cenozoic time. The ranges have a subparallel alignment and are uplifted blocks of deformed pre-Cenozoic rocks that were elevated along faults during the Tertiary. Movement along these faults has been recurrent since then, and in some places the movement continues

Figure 19.8 These erosional remnants on the Colorado Plateau in Utah are part of Monument Valley National Monument. (Photograph by Tad Nichols.)

today as evidenced by the numerous Nevada earthquakes.

Early Cenozoic sediments are scarce in the Basin and Range Province, but late Cenozoic deposits in the basins are commonplace. The sedimentary fill in many of these basins is measured in the thousands of feet. These strata are of economic importance because they contain considerable quantities of groundwater that is used extensively for irrigated farming and domestic water supply.

During the Pleistocene many of the basins contained lakes that varied considerably in volume from time to time. The levels of these lakes rose as their volumes expanded in response to greater precipitation during certain phases of the Pleistocene, and lower levels resulted from periods of lesser precipitation. The wet periods or *pluvial periods* were coincident with the glacial maxima, and the dry periods were correlatives of the interglacial stages. Old shorelines provide a means for reconstructing the former extent of these ancient lakes. Some of them have dried up, but others have modern relicts such as Great Salt Lake in Utah, and Pyramid, Carson, and Walker Lakes in Nevada (Fig. 19.10).

Columbia Plateau Province

The Columbia Plateau physiographic province covers much of eastern Washing-

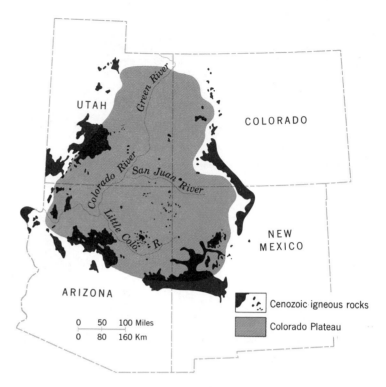

Figure 19.9 Map showing the distribution of Cenozoic igneous rocks on the Colorado Plateau. Volcanic rocks occur mostly around the edges of the plateau, and laccoliths occur in the central part. (After Hunt, C. B., 1956, *Cenozoic geology of the Colorado Plateau,* U. S. Geological Survey, Professional Paper 279, Washington, D. C., 99 pp.)

ton and Oregon plus part of southern Idaho (Fig. 19.5). The Tertiary history of this province was dominated by vast outpourings of basaltic lavas. In some sections, such as the Snake River Plains in southeastern Idaho, volcanism continued into the Quaternary as evidenced by the uneroded character of the lava flows and volcanic cones in the Craters of the Moon National Monument.

Lava tends to flow from the point of emergence toward low areas on the surrounding landscape. Pre-Tertiary river valleys of the Columbia Plateau became choked with lava that oozed from long fracture zones in a series of nonviolent eruptions. The blocking of major drainage ways by highly mobile lava flows, called *flood basalts,* produced large freshwater lakes in which detrital sediments accumulated during periods of non-

volcanic activity. These lake beds and their fossil content occur interbedded with lava flows on the Columbia Plateau, a circumstance that permits the geologic dating of the periods of volcanism.

The Pacific Mountains Provinces and the California Trough

This "province" is a combination of several physiographic units used in more advanced texts. For the purpose of this presentation, however, it will suffice to consider the geographical region between the western boundaries of the Basin and Range and Columbia Plateau Provinces and the Pacific Ocean as a single unit. This region consists of the mountains of western Washing-

THE GEOLOGICAL STORY

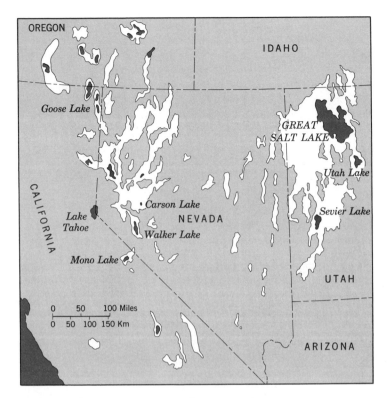

Figure 19.10 The white areas on this map are basins that were occupied by lakes during the Pleistocene. The existence of these former lakes is evidence of a period of higher rainfall in what is now a semiarid environment. The periods of higher rainfall are known as pluvial periods, and they are correlative with the times of glacial advance in the Great Lakes region. (After Meinzer, O. E., 1922, Map of Pleistocene lakes of the Basin-and-Range Province and its significance, *Geol. Soc. America Bull.*, v. 33, pp. 541-552.)

ton, Oregon, and northern California; the Sierra Nevada of eastern California; the Coast Ranges and other mountains of southern California; and the California Trough lying between the Sierra Nevada and the Coast Ranges (Fig. 19.5).

The Tertiary history of this area in the simplest terms involved the uplifting of two parallel mountainous belts that were in juxtaposition in Washington and Oregon, but were separated by a sinking trough in California that contains about 50,000 feet of Tertiary marine and nonmarine sediments.

The Sierra Nevada is a rugged mountain range consisting of a granitic fault block that was tilted westward along a 400-mile fault forming its eastern border. The western flank of this massif was the source of the thick sediments that accumulated in the California Trough lying to the west.

The granitic rock that forms the main mass of the Sierra Nevada was intruded into Jurassic and older rocks in late Jurassic and Cretaceous time. Through most of the Tertiary the Sierra Nevada was a source of sediments for the California Trough. By Pliocene time the Sierra Nevada had become a distinct unit, and continued uplift more than kept pace with erosion. The present topographic configuration of the range was accomplished by erosion during the Pleistocene. Valley glaciers then were much more extensive than now.

The Cascade Range is composed of uplifted pre-Tertiary granitic and metamorphic rocks in the north and Tertiary lavas in the south. Volcanic cones in the Cascades are among the most spectacular volcanic peaks in North America. Many of them are snow-covered, and the higher ones contain systems of glaciers flowing down their flanks. These volcanic cones were built during the Pliocene but some erupted during the Pleistocene. Mt. Shasta, whose snow-clad peak rises more than 14,000 feet above sea level, contains a hot spring near its summit. Lassen Peak, in northern California, erupted in 1914 to 1915 and spewed hot volcanic ash on its slopes, thereby melting hugh quantities of snow which caused extensive mud flows at lower elevations (Chap. 5).

Crater Lake, Oregon (Fig. 5.10) lies in the center of what was once a towering, glaciated, volcanic peak, Mt. Mazama, that existed in the Pleistocene. An eruption during the late Pleistocene blew part of the crater away and caused the remainder to collapse (Fig. 5.11). The depression resulting from the eruption is a *caldera,* and it is now occupied by the waters of Crater Lake. The lake is 4000 feet deep and some five miles across. Wizard Island, a small volcanic cone near one edge of the caldera, represents the last stage of volcanism in the sequence of events.

Cenozoic Orogeny

A wide range of orogenic activities in North America can be identified with the Cenozoic. The normal faulting which has produced the characteristic structures of the Basin and Range Province can be dated from the middle Cenozoic to historic time; extensive outpourings of early Tertiary to recent basaltic lavas occupy extensive segments of the Columbia Plateau, the Basin and Range Province, and the Colorado Pla-

teau. Those of the Columbia Plateau are surmounted by large volcanoes such as Mt. Rainier and Mt. Shasta. The California Coast Ranges are well known for the intensity of their deformational structures affecting strata as young as Pleistocene; and the San Andreas and other major faults are active fractures testifying to the recentness of this period of deformation. It is most common to refer all this orogenic activity to the *Cascadian* orogeny, the episode of mountain building extending from middle Tertiary time to the present.

Although not truly orogenic in nature, the regional uplift that areas such as the Colorado Plateau, the Rocky Mountains, and the Appalachian Mountains have experienced, as exemplified by the superposition of rivers previously described, has been an important related phenomenon.

Such important mountain areas as the Alpine Chain and the Himalayan Range also experienced their major deformation and uplift in the Cenozoic era.

Life of the Cenozoic

Animals. The animal kingdom of Cenozoic times was unquestionably dominated by land mammals as shown by the great variety of their remains excavated from terrestrial deposits of Tertiary and Quaternary age. Many of the early Tertiary forms became extinct by the end of the Eocene, and not until the late Tertiary did animals that closely resembled modern forms appear. Another wave of extinction came in the Pleistocene.

Generally, the history of Cenozoic animals is one of evolution and migration. Evolution progressed rapidly as the mammals replaced the Mesozoic dinosaurs in all ecological niches on land. With the end of the dinosaur dynasty at the close of the Mesozoic, the primitive mammalian stock that survived the Mesozoic apparently underwent a kind of "explosive" evolution that led ultimately to the great variety of species inhabiting the globe today.

Many of the modern animals such as the

horse, camel, rhinoceros, and elephants began their climb up the evolutionary ladder in the early Cenozoic. The first recognizable ancestors of these modern creatures were generally smaller and less specialized than their living counterparts.

One of the best documented accounts of evolutionary change in a single group of animals is the development of the horse in North America. A brief description of some of the specific changes encountered in this evolutionary sequence will not only illustrate the anatomical transformations of each step from the Eocene to the Pleistocene but also will demonstrate the degree to which the sequence is documented in the fossil record.

Figure 19.11 summarizes the evolutionary history of the horse during the Cenozoic. The most obvious difference between *Hyracotherium* (sometimes called *Eohippus*) and *Equus,* the modern horse, is the change in size. *Hyracotherium* (Eocene) was about the size of a small fox and hardly would be recognizable as a member of the horse family if seen in a zoo today. His muzzle was short, his teeth were stubby and low-crowned, and he had toes instead of hoofs on his short legs. Each of these changed as conditions for survival changed during the course of Cenozoic time. The muzzle increased in length and the jaws deepened to accommodate an increased length in the teeth. The legs became proportionally longer and the four functional toes on the forefeet and three on the hind legs evolved to a single hoof on each leg.

The factors that favored the selection of slight changes brought about by mutations during the millions of years separating *Hyracotherium* from *Equus* probably will never be known, but reasonable speculations have been made. *Hyracotherium* very likely browsed on shrubs whereas *Equus* grazes on grass. This difference in feeding habits is suggested by the difference in dentition of the two species. *Equus* also is probably more fleet footed than his three-toed ancestor. This conjecture is based on the assumption that longer, hoofed legs make for faster locomotion than short ones with three-toed feet. *Equus* apparently needed to run faster than his precursors because of a greater danger from swift-moving predators.

This series of adaptive changes in the horse is a remarkable example of the results of evolution. No less remarkable is the fact that, for unknown reasons, the horse became extinct in North America before the end of the Pleistocene. However, through migrations to the Old World across the Siberian land bridge (Fig. 19.12), *Equus* survived in Europe and Asia to the present. The wild horses of the western plains in the United States were progeny of horses introduced into Mexico during the invasion of that country by the Spanish conquistadors in the sixteenth century.

Other animals that lived in North America during the Cenozoic include the rhinoceroses and camels, both of which evolved from smaller forms that appeared in early Cenozoic time, and both of which moved to the Old World later in the era (Fig. 19.12). Pleistocene camels ranged from Florida to Alaska and were represented by several species. The modern Asiatic camels and the smaller South American llamas are both descendants of Pleistocene forms that no longer exist.

Some of the Cenozoic animals terminated their evolutionary careers without leaving any modern successors. An example are the *titanotheres,* a tribe of Cenozoic creatures that resembled the rhinos in gross form but were only remotely related. Just before the titanotheres became extinct in mid-Oligocene time, they were the largest land animals in North America. A reconstruction of a titanothere known as *Brontops* is shown in Figure 19.13.

The evolutionary development of the elephants began in Africa, but during the Miocene some of the stock migrated to North America where they flourished well into the Pleistocene. Before the final disappearance of the continental glaciers, however, these animals succumbed to some unknown force that caused their extinction in the New World. Two elephantlike animals roamed the North American continent during the Pleistocene, the mastodon and

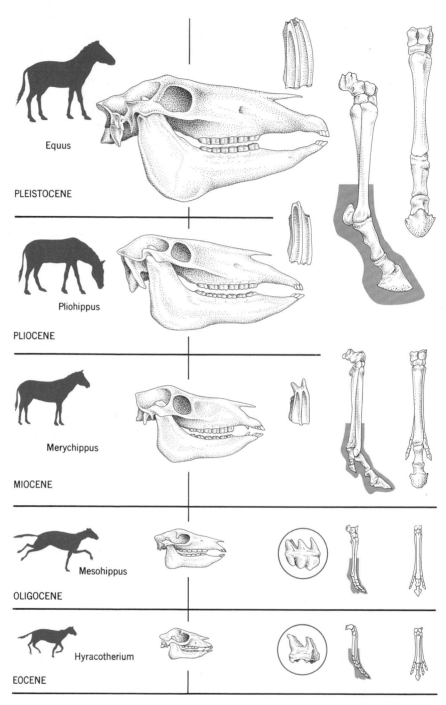

PLEISTOCENE

Equus

PLIOCENE

Pliohippus

MIOCENE

Merychippus

OLIGOCENE

Mesohippus

EOCENE

Hyracotherium

Figure 19.11 Evolution of the horse during Tertiary time in North America. Drawings in each vertical column are to scale except the teeth of *Mesohippus* and *Hyracotherium* (circled) which are enlarged relative to the other teeth in the drawing. (Bodies and limbs after Simpson, G. G., 1953, *Life of the past,* Yale University Press, New Haven, 198 pp. Teeth after Stirton, R. A., 1940, Phylogeny of North American Equidae, *Bull. Dept. Geol. Univ. Calif.,* v. 25. Skulls after Romer, A. S., 1966, *Vertebrate paleontology,* 3rd ed., University of Chicago Press, Chicago, 468 pp.; Colbert, E. H., 1969, *Evolution of the vertebrates,* 2nd ed., John Wiley and Sons, New York, 535 pp.; and specimens in the University of Michigan Museum of Paleontology furnished by C. W. Hibbard.)

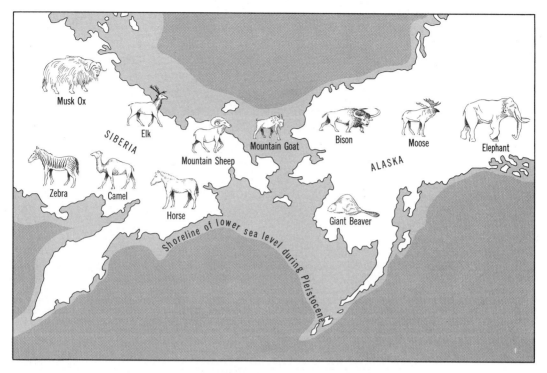

Figure 19.12 The Alaskan-Siberian land bridge functioned as a migration route along which an interchange of animals took place between Asia and North America during the Pleistocene. (After Hibbard, C. W., 1951, Animal life in Michigan during the Ice Age, *Michigan Alumnus Quarterly Review,* v. 57, pp. 200-208.)

Figure 19.13 *Brontops,* a titanothere from the Tertiary of western North America. (After Osborne, H. P., 1929, *The Titanotheres of ancient Wyoming, Dakota, and Nebraska,* United States Geological Survey, Monograph 55, v. 2.)

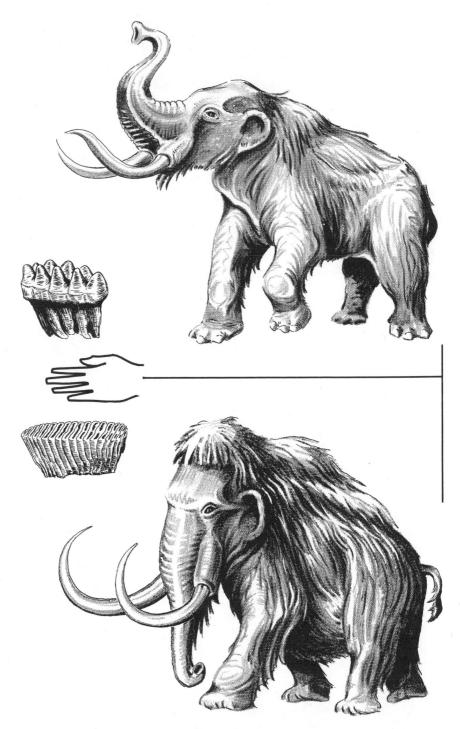

Figure 19.14 Restorations of a Pleistocene mastodon (above) and a Pleistocene Wooly Mammoth. A tooth of each is shown with a human hand for scale. The crowned tooth of the mastodon was suitable for a diet of twigs, branches, and cones, whereas the infolded structure of the mammoth tooth was more adapted to a grass diet. (After Knight, C. R., 1935, *Before the dawn of history*, McGraw-Hill Book Co., New York, 119 pp.)

the mammoth (Fig. 19.14). Both have a superficial resemblance to the modern elephants of Africa and India, but only the mammoth was a true elephant; the mastodon was a remote relative.

The Woolly Mammoth is probably the most famous of all fossil mammals because of the discovery of nearly complete carcasses in the permafrost of the Siberian and Alaskan tundra. The Berezovka mammoth is one of some three dozen such discoveries and, because of its nearly complete carcass, is mounted and on display in the Zoological Museum of Leningrad. This specimen was discovered near the Arctic Circle in eastern Siberia where it was thawing out of the permafrost in which it had been entombed since the late Pleistocene (Fig. 19.15). Except for some skin missing on the face and part of the trunk the animal was found intact, and undigested food was removed from the mouth and stomach.

Popular accounts of the Berezovka and other finds have led some writers to the conclusion that the woolly mammoths were victims of a climatic catastrophe that caused them to be "frozen alive." Such speculation is unwarranted by the facts. There is no direct evidence that any mammoth died by freezing. A more likely explanation for the demise of these cumbersome beasts is that they perished by asphyxiation from live burial in a mudflow or riverbank slide. The broken bones of the Berezovka mammoth give credence to a death caused by a fall from a riverbank or other escarpment made unstable by the melting of frozen ground during the summer months. Rendered helpless and unable to move, the unfortunate creature may have been engulfed by further mudflows and bank cave-ins that buried him alive.

The Pleistocene mammoths were not restricted to an Arctic environment. The Jefferson Mammoth inhabited meadowed areas of the eastern woodlands and the river valleys of Kansas and Nebraska. The Columbian mammoth ranged from Florida

Figure 19.15 The Berezovka mammoth, displayed in the Zoological Museum of Leningrad in the position in which it was discovered, is a nearly complete cadaver. Only the front of the face and part of the trunk was missing. Broken bones in the animal's legs testify to an accidental fall. Undigested food in the mouth and stomach demand sudden death, but not by freezing. (Courtesy of the Zoological Museum, Leningrad, USSR.)

to Mexico during glacial times and as far north as Nebraska in interglacial times. The great Imperial Mammoth stood 13 feet at shoulder height and ranged the Great Plains in Texas and Mexico, and some remains have been discovered in the famous Rancho La Brea tar pits in Los Angeles.

The mastodon, in contrast to the mammoths, was an inhabitant of eastern forests and timbered valleys of the west. His main diet as a herbivore consisted of twigs and foliage of coniferous trees. The mastodon was thus a browser instead of a grazer, a difference that accounts for the different tooth structure in the mammoth and mastodon (Fig. 19.14).

In the sea, Eocene whales represent the first adaptation of mammals to a marine environment. In the air, birds must have flourished in great abundance, but their fossil remains are extremely rare. Lakes and streams contained many fishes that closely resemble modern forms, and they are among the best preserved fossils of the Tertiary; those of the Eocene Green River formation of Wyoming are examples (Fig. 19.16).

Pleistocene Extinction. Both the mammoth and mastodon became extinct before the end of the Pleistocene. Other animals that failed to survive to the present include the large bison, ground sloth, giant beaver, sabre-tooth cat, large jaguars and wolves, woodland musk-oxen, and giant moose. The skeletons of many of these animals have been recovered from the famous Rancho La Brea tar pits in Los Angeles and are on display in the Los Angeles County Museum. These tar pits contain a natural asphaltic material in which thousands of animals become entrapped, either in attempting to cross it or in preying on other animals already caught in the death trap (Fig. 19.17).

The extinction of a race or species of animals always poses a problem for the paleontologist. Since the true explanation can never be proved, the best one can do is to speculate. A number of causes, separate or in concert, have been proposed to explain extinction during various times in earth history (page 380). Some of these are reasonable while others are bizarre. Reasonable explanations involve factors such as gradual climatic change, increase in the number of predators, disease, and the like. Paleontologists who lean toward this kind of answer to the question of extinction tend to invoke the doctrine of uniformatarianism. They believe that the answers will be provided by extrapolation of the principles of ecology as revealed by modern environmental studies. Others conjecture that animal extinction in past geologic time was caused by one or

Figure 19.16 An Eocene fish skeleton from Two Creeks, Wyoming. (University of Minnesota Paleontological Collection.)

Figure 19.17 Animals from the Rancho La Brea tar pits, Los Angeles County, California. The photo inset shows a modern squirrel trapped in the present asphaltic material, just as the larger animals like the bison and elephants were caught in the death trap during the Pleistocene. (Photograph from the Los Angeles County Museum. Animal silhouettes from Stock, Chester, 1953, *Rancho La Brea*, Los Angeles County Museum.)

more catastrophic events, but in most cases the evidence for such happenings is not at all convincing.

One aspect of animal extinction in the late Pleistocene is the question of man's role as a direct or indirect cause. Modern man has brought about the demise of individual species during the twentieth century, and he is capable of even greater carnage in future years. Some respected scientists believe that prehistoric man was the culprit in the elimination of a number of animal species in the

very late Pleistocene. Man, so the argument goes, was a predator par excellence because he had learned how to use projectiles as weapons and was thus able to slaughter thousands of animals in his relentless search for food. There is no doubt that North American Paleo-Indians were able to prey successfully on large animals because archaeological sites in the United States contain man-made projectile points in close association with the rib cage of mammoth skeletons. Whether or not this clear-cut evidence of early man's ability to kill mammoths with primitive weapons warrants the conclusion that he wiped them out entirely is open to question. As one critic of this view pointed out, even though prehistoric man hunted bison and killed them in vast numbers, that particular species of bison did *not* become extinct. They survive today if only in small numbers.

The subject is an intriguing one, and no doubt will be the basis for continued study and many discourses for years to come. The question of the role of early man in animal extinction is more than academic. Man's influence on his environment has increased as his technology has expanded, and there is little doubt that he now holds the key to the survival of the plant and animal world, including himself. If we can gain some insight into the behavior of man when his technology was primitive, it might help us to understand his behavior in the modern world of much greater technological complexity.

Cenozoic Plants. The ancestors of modern land plants were already established by the end of the Cretaceous, and their Cenozoic history is merely a continuation of the evolutionary trend that began in the late Mesozoic. The Tertiary trees closely resemble modern forms if one can judge by fossil leaves (Fig. 19.18). Flowering plants (angiosperms) increased in kinds and numbers and are found fossilized in fine-grained sediments, for example, those that accumulated in the intermontane lakes of the Basin and Range Province in late Tertiary time.

By Pleistocene time practically all of the modern plants were in existence, but their distribution was different than today because of the waxing and waning of the ice sheets. Plant species that live today in the northern United States and southern Canada had a more southerly range during the glacial advances, and those that today have a southerly range migrated to more northerly latitudes during the interglacials. The shifting of vegetation due to the climatic changes of the Pleistocene is documented from fossil plant remains in sediments that accumulated in a single basin while the climatic changes were taking place.

In the Great Lakes Region, for example, it is possible to recognize the general climatic warming that produced the final retreat of the Wisconsin glacial stage by examining the pollen grains entrapped in lake sediments. Pollen grains are extremely resistant to natural forces of decay, and they remain intact and identifiable when buried in lake or bog sediments. The pollen grains in successive layers of such sediments are isolated by laboratory techniques and identified under the microscope. The percentage of each of the pollen types for a given stratum is determined, and a pollen diagram is constructed such as the one shown in Figure 19.19.

Theoretically, the pollen content of any one layer should reflect the forest composition in the immediate vicinity of the basin in which the pollen accumulated. In turn, the forest cover of an area is closely related to the prevailing climate. In the pollen diagram of Figure 19.19, it can be deduced that 11,000 years ago the climate of southern Michigan was colder than today because of the preponderance of fir and spruce pollen in the lower part of the stratigraphic section. These trees are not native to southern Michigan today. This pollen profile also reveals that the spruce-fir forest gave way to a pine forest about 5000 years ago, and that by 4000 years ago oak became the dominant species. From knowledge of the modern geographical ranges of these trees, it is deduced that the change from a spruce-fir forest to an oak forest between 11,000 and 4000 years ago was caused by a warming trend in the general climate of the area.

Mineral Resources of the Cenozoic

Petroleum. The world's leading source of energy is petroleum, a fossil fuel whose occurrence by no means is restricted to the rocks of Cenozoic age. Sedimentary strata from Cambrian to Pliocene are the main reservoirs of liquid petroleum. Petroleum is a fossil fuel, and it occurs in three physical states, liquid, gaseous, and solid. The liquid form of natural petroleum is *crude oil,* a mixture of hydrocarbons that vary greatly in their physical properties and chemical com-

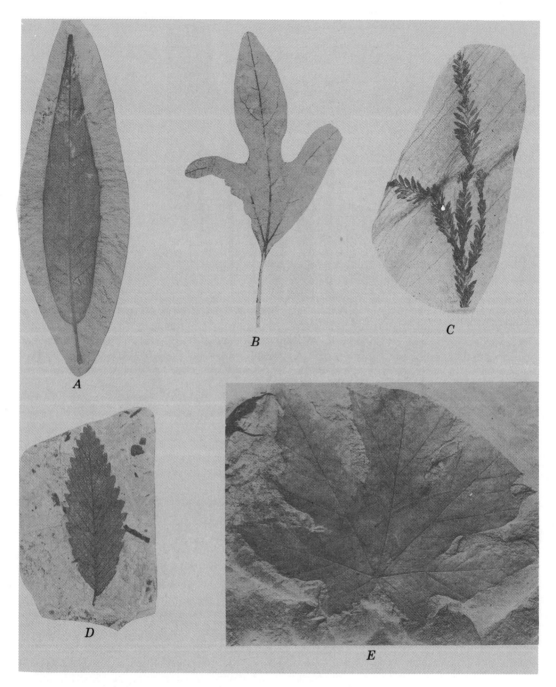

Figure 19.18 Fossil leaves from some Tertiary trees. (A) Miocene willow-oak. (B) Miocene sassafrass. (C) Oligocene sequoia. (D) Oligocene birch. (E) Miocene maple. (A, C, D, and E by C. A. Arnold, University of Michigan Museum of Paleontology; B reproduced by permission of McGraw-Hill Book Co., from Arnold, C. A., 1947, *An introduction to paleobotany*, New York, 433 pp.)

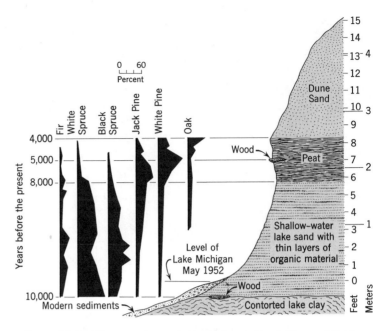

Figure 19.19 Diagram of a geologic section at South Haven, Michigan showing the change in the percentage of tree pollen recovered from various layers. The peat bed was formed when the water level of the Lake Michigan Basin was considerably below the present level. This pollen profile is truncated at 4000 radiocarbon years ago, but other pollen profiles from nearby indicate that the oak peak was reached about 3500 years ago, after which the percentage of oak has decreased and the percentage of white pine has increased. Glacial geologists attribute this to a revertance to a colder climate during the last 3500 years. (After Zumberge, J. H. and J. E. Potzger, 1956, Late Pleistocene chronology of the Lake Michigan basin correlated with pollen profiles, *Bulletin Geol. Soc. America,* v. 67, pp. 271-288.)

position. Natural gas is chiefly methane (CH_4), but other gaseous hydrocarbon compounds are generally present in greater or lesser amounts. *Tar* is a naturally occurring liquid hydrocarbon of such high viscosity that it cannot be extracted from oil wells, and *kerogen* is a true solid that occurs as a component of some shales.

Crude oil and natural gas commonly occur together with saltwater in underground reservoirs. Although oil is said to occur in "pools," the use of this term is a misnomer because oil, like groundwater, occupies the voids between mineral grains or other pores in sedimentary rocks. The term *porosity* is an index of void space in a rock, and the term *permeability* is an index of the size of the inter-

stices and the degree to which they are interconnected so as to permit the flow of oil, gas, or water through them.

The origin of oil is not known with any degree of certainty, but it is believed that liquid petroleum substances are produced through the accumulation of organic remains in a marine environment. The exact way in which the organic compounds are reduced to petroleum compounds is not fully understood, but it is quite likely that burial beneath a sedimentary cover is a basic requirement. Regardless of the chemistry involved, it is a noteworthy fact that oil and gas do not remain in the sedimentary rocks (source rocks) in which they originate. Instead, they migrate into reservoir rocks where they may become entrapped in a geologic structure called an *oil trap*. Oil migrates through rock pores because it occurs with water. The oil or gas, being of lower specific gravity than water, and for all practical purposes insoluble in water, tends to separate itself from the associated water by moving in the direction of least resistance, usually in an upward direction. If no impermeable barrier stops the migrating oil, it eventually reaches the earth's surface in the form of an *oil seep*. However, if geologic conditions are favorable, the oil accumulates in an oil trap where it remains until it is discovered and extracted from wells.

The search for oil is really a search for oil traps. The trained geologist does the searching and the petroleum engineer does the extracting. There is no way whereby the actual presence of oil in a trap can be determined by surface methods alone. Surface oil seeps are the most positive indicators of oil beneath the ground, and many producing oil fields have been discovered from them. However, modern methods of oil exploration do not depend on the telltale clues of oil seeps in the search for more oil. Instead, the geologists and geophysicists use highly sophisticated means of "looking" deep beneath the surface of the earth for the purpose of identifying geologic structures conducive to the accumulation of oil and gas. Figure 19.20 illustrates some of the more common oil traps, not all of which can be deduced from surface observations alone. In the final analysis only the drilling of one or more holes will determine the presence of oil or gas.

"Tar sands" such as occur in northeastern Alberta, Canada are another potential source of petroleum products. These are not to be confused with the "oil shales" of Eocene age in Colorado, Utah, and Wyoming which contain kerogen. The tar sands can be mined and then processed by existing oil-refining techniques because they are actually true liquid hydrocarbons but of high viscosity. Kerogen, on the other hand, is a solid hydrocarbon from which liquid petrolem products can be generated by special refining processes.

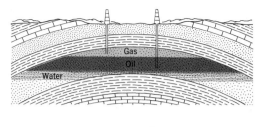

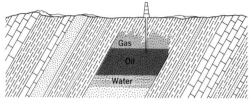

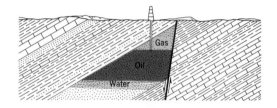

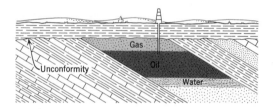

Figure 19.20 Geologic cross sections of some common types of oil and gas traps.

Production and Reserves. A highly developed nation of the modern world needs large amounts of petroleum to sustain its industrial might and high standard of living. Even though coal, hydroelectric power, and nuclear reactors provide additional sources of energy in many countries, petroleum products are likely to be a major source of power well into the twenty-first century.

In estimating future production of petroleum in the world, one has to deal with a number of unknown factors such as the rate of discovery of new oil fields. This is obviously a difficult thing to do, but by using past discovery rates and a knowledge of world geology, estimates of the future production of

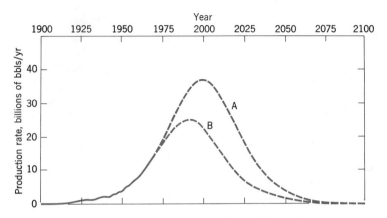

Figure 19.21 Graph showing the production of crude oil with time according to two different estimates. Curve *A* is based on a total ultimate world production of 2100 billion barrels during the two-century period 1900 to 2100, with the peak production of about 37 billion barrels per year occurring in the year 2000. Curve *B* is based on an ultimate world production of 1350 billion barrels, with the peak production of 25 billion barrels per year occurring in about the year 1990. In either case, world reserves will be depleted before the end of the twenty-first century. (Adapted from *Resources and man: a study and recommendations,* by the Committee on Resources and Man of the Division of Earth Sciences, National Academy of Sciences-National Research Council, with the cooperation of the Division of Biology and Agriculture, W. H. Freeman and Company, Copyright © 1969.)

petroleum have been made by a number of experts.[1] Two estimates have emerged: one says that the ultimate total crude oil production for the world since the beginning of the twentieth century will be 1350 billion barrels and the other sets this figure at 2100 billion barrels. Without going into the details of how these figures were arrived at, both provide a basis for predicting the production of petroleum on an annual basis until the supply is depleted (Fig. 19.21).

Total world production of crude oil for 1970 was on the order of 10 billion barrels. According to the smaller estimate, curve B in Figure 19.21, this rate will increase to a maxi-

mum of 25 billion barrels in 1990; the larger estimate, curve A, predicts a production rate of 37 billion barrels in the year 2019. Both curves predict that by the year 2100, or before, the world supply of crude oil will have been exhausted.

The foregoing analysis is extremely significant because it demonstrates that all the world's mineral resources, be they petroleum, iron, or copper, exist in finite quantities. Their extraction cannot go on indefinitely at ever increasing rates or even at fixed rates. In the case of petroleum, man must be prepared with a substitute source of energy well before the end of the twenty-first century.

The oil reserves of the world are shown diagrammatically in Figure 19.22. The overwhelming position of the Middle East in terms of future production is a factor of great significance in world politics.

[1] For an excellent nontechnical summary, see *Resources and Man*, 1969, W. H. Freeman and Company, San Francisco, Chapter 8 by M. King Hubbert, "Energy Resources," pp. 169–200.

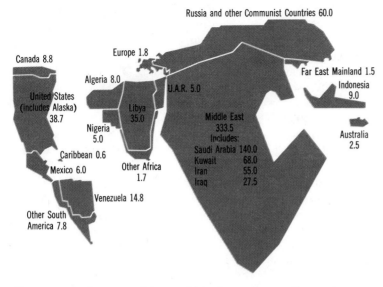

Figure 19.22 This map of the world is distorted according to the reserves of crude oil contained in the various oil-producing countries of the world. The estimated reserves are given in billions of barrels under the name of each country. (From Esso Middle East, 1970, *Oil and the Middle East*, Standard Oil Co., N. Y. Based on data from the *Oil and Gas Journal*, December 29, 1969.)

Some Other Cenozoic Deposits of Economic Importance

Salt. Along the Gulf Coast of Louisiana and Texas, and on the continental shelf of the Gulf of Mexico, salt occurs in the form of vertical pipelike masses that were forced upward through overlying Cenozoic rocks from great depth. The salt was deposited as a sedimentary evaporite in pre-Tertiary times and was intruded upward into the Cenozoic sediments in a plastic state. The strata intruded by the salt domes were arched upward or bent into dome-shaped structures that provide excellent oil traps. The salt itself is mined from deep shafts.

Metals. Many metallic ores in the Rocky Mountain states are mined from rocks older than Tertiary, but evidence points to a Tertiary age for the ore-bearing solutions themselves. Of historical interest are *placer* gold deposits of California, which are accumulations of stream gravels derived from gold-bearing rocks in the Sierra Nevada during the Tertiary. The California gold rush of 1848 involved the panning of stream gravels for their content of native gold particles.

References

Atwood, W. W., 1940, *The physiographic provinces of North America,* Ginn and Co., New York, 536 pp.

Colbert, Edwin H., 1955, *Evolution of the vertebrates,* John Wiley and Sons, Inc., 479 pp.

* Committee on Resources and Man, 1969, *Resources and Man,* National Academy of Sciences-National Research Council (pub. by W. H. Freeman and Co., San Francisco), 259 pp.

Hinds, Norman E. A., 1952, *Evolution of the California landscape,* California Division of Mines, Bull. 158, San Francisco, 240 pp.

* Hunt, Charles B., 1967, *Physiography of the United States,* W. H. Freeman and Co., San Francisco, 480 pp.

Hunt, Charles B., 1956, *Cenozoic geology of the Colorado Plateau,* U.S. Geological Survey, Professional Paper 279, Washington, D.C., 99 pp.

Kay, Marshall, and Edwin Colbert, 1965, *Stratigraphy and life history,* Chap. 24, Cenozoic Life, John Wiley and Sons, New York, pp. 606–630.

* Krantz, Grover S., 1970, Human activities and megafaunal extinctions, *American Scientist,* V. 58, pp. 164–170.

McKee, Edwin D. and others, 1967, *Evolution of the Colorado River in Arizona,* Museum of Northern Arizona, Flagstaff, Ariz., 67 pp.

Thornbury, William D., 1965, *Regional geomorphology of the United States,* John Wiley and Sons, Inc., New York, 609 pp.

Van Houten, F. B., 1961, Maps of Cenozoic depositional provinces, western United States, *American Jour. Science,* V. 259, pp. 612–621.

* Recommended for further reading.

Index